Mathematics for Sustainable Developments

Aligned with the Springer Nature's SDGs program, this series aims to publish books on diverse fields exploring the intersection of mathematics, statistics, and sustainable societal developments. It will not only cover emerging mathematical trends but also address essential topics for human and societal well-being. The goal is to advance sustainability science globally by transcending traditional research boundaries and addressing complex challenges. Scholars from across disciplines will be harnessed to tackle obstacles across all human endeavors, fostering a holistic understanding of sustainability and promoting innovative solutions for a resilient future.

Ashish Ghosh

Data Science and Cases in Sustainability

Pattern Recognition and Machine Learning

 Springer

Ashish Ghosh
Machine Intelligence Unit
Indian Statistical Institute
Kolkata, West Bengal, India

ISSN 3004-9016   ISSN 3004-9024   (electronic)
Mathematics for Sustainable Developments
ISBN 978-981-96-8361-1   ISBN 978-981-96-8362-8   (eBook)
https://doi.org/10.1007/978-981-96-8362-8

Mathematics Subject Classification: 62R07, 68T10, 68T09, 94A16

This Springer imprint is published by the registered company Springer Nature Singapore Pte Ltd.
The registered company address is: 152 Beach Road, #21-01/04 Gateway East, Singapore 189721,
Singapore

If disposing of this product, please recycle the paper.

Dedicated to my beloved daughter,
SHINJINI, and wife, SUSMITA

Prologue

In today's world, everything is smart and interconnected. With the help of technology, humans now have the capability to sense, measure, implement, and connect with anything that we wish to, including physical objects like cars, household electronics, roadways, and traffic along with complete ecosystems like healthcare networks, business organizations, and even entire cities. Unprecedented connectivity among objects result in the generation of massive amounts of data. Human beings are already failing to keep up with the data that is being generated each moment. In this scenario, tools and techniques from **Data Science** could be used to manage the huge volumes of data.

It chiefly comprises tools and techniques that are used to analyze these huge amounts of data. As a whole, data science is a consolidation of multiple bits from miscellaneous areas of research. For tools used in this discipline, the main impact is drawn from computer science, where the problems related to algorithmic efficiency and storage flexibility form the focal point. However, for the analysis part, the influences are far more diverse. The so-called hard sciences (statistics, mathematics, physics) and the social sciences (sociology, economics, political sciences, etc.) have lent their modern methods in the formation of data science. Naturally interdisciplinary class of techniques like machine learning are also very popular in today's scenario.

"Big Data" is a center of interest for various organizations (public and private) like governments, establishments dealing with security and health, research institutions, companies handling commercial recommendations, future trends, detection of anomalies, etc. It can be seen as an umbrella term that circumscribes all things from digital data to healthcare organizations information to the data collected for decades of paperwork issued and filed by the government. Recent trends indicate that data is incoming at a faster pace, from diverse sources in assorted forms and it exceeds the abilities of any existing system to intake, store, analyze, and process. Databases with petabytes of data can be found in enterprises and research organizations, where invaluable information and knowledge is "hidden." The prime objective is to correctly identify and extract meaningful information (like structure, underlying relationships, etc.) from the large quantities of raw heterogeneous data available to us. This complicated task requires novel, significant storage and processing infrastructure due to the unprecedented volume of data. A big wave of revolution has been started for the advancements of Artificial Intelligence for such tasks.

Artificial Intelligence is a very extensive term for anything that tries to mimic human reasoning process for perception, recognition and execution of several tasks. These tasks may range from simple activities like planning, predicting, cooperating with other entities, to complex activities like learning to operate and coordinate hands in order to pick objects. To cope up with this, new improved algorithms are being devised in fields like Pattern Recognition, Machine Learning, and Data Mining. Well-known techniques are also being revisited and tailored to overcome these new challenges. Some may argue that the concepts of these three fields are quite overlapping. All these fields have evolved and borrowed ideas from each other so much that in the current scenario it is actually impossible to clearly demarcate the line between them. However, there are few pattern recognition tasks where learning is not required and thus set it apart as a distinct field from machine learning. Data mining, on the other hand, combines machine learning and pattern recognition techniques and wrap them with databases, non-numerical data, and data visualization techniques. This book is an attempt to explain all these data handling techniques and the differences between them. The surge that is new to all these fields is "Big Data" that is being generated and the issues of handling them.

Data Science, broadly speaking, is the modern science that will drive the decisions of future. This science is all about finding valuable knowledge from the overwhelming big data deluge by making a machine learn like a human, find patterns in the data, mine information from the data, and then analyze the information to gain knowledge. This book, thus, aims to present a link between the major fields of Data Science like Machine Learning, Pattern Recognition, Data Mining, and Big Data Analysis.

Acknowledgments It gives me immense pleasure in acknowledging all my students who helped me in preparing this manuscript and learn many tools and techniques and topics embodied here. Special thanks are due to Debasrita Chakraborty, Anwesha Law, Rahul Roy, Kunal Srivastava, and Subhadip Boral who contributed a lot in writing this text. Thanks are due to other PhD students who have worked with me for the last ten years updating me on a number of new topics related to this manuscript. Space does not allow me to mention the names of innumerable number of interns and master's students who worked with me during the course of preparation of this text. I gratefully acknowledge the contributions of my teachers and colleagues who inspired me to write this text.

My sincere thanks are due to my family members, my wife, Susmita, and daughter, Shinjini, for bearing with me for a long period during the preparation of this manuscript.

I also thank the readers of this book who will probably never notice that this page existed unless someone tells them.

Competing Interests The author has no competing interests to declare that are relevant to the content of this manuscript.

Kolkata, India   Ashish Ghosh
January 2025

Contents

Part IV   Machine Learning

List of Symbols

N : Total number of data points.

$\mathbf{X} = \{\mathbf{x_1}, \mathbf{x_2}, \cdots, \mathbf{x_N}\}$: Dataset.

$\mathbf{x_i} = \{x_{i1}, x_{i2}, \cdots, x_{id}\}$: ith pattern in the dataset.

x_{if} : fth feature in ith pattern.

$\bar{\mathbf{x}}$: Mean.

x_f : The fth feature as a random variable.

$\bar{x}_f$: Mean of the fth feature.

σ_f^2 : Variance of the fth feature.

F : Set of features $\{f_1, f_2, \cdots, f_D\}$

D : Number of dimensions before feature selection.

d : Number of dimensions after feature selection.

Ω : Number of classes.

$\{\omega_1, \omega_2, \cdots, \omega_\Omega\}$: Set of all classes.

k : Number of clusters.

C : Number of clusters in Fuzzy C-means algorithm.

$\{c_1, c_2, \cdots, c_k\}$: Set of all clusters.

w_{ij} : Weights in neural network, between ith neuron of one layer and jth neuron in the next layer.

$w_{ij}^{(r)}$: Weights in neural network, between ith neuron of $(r-1)$th layer and jth neuron in the rth layer.

μ : Fuzzy membership function.

μ_{ij} : Fuzzy membership for ith pattern in jth cluster.

$dist_{ij}$: Distance of ith data point from jth cluster.

$dist_{Rij}$: Radial distance of ith data point from jth cluster.

$dist_{Pij}$: Perpendicular distance of ith data point from jth cluster.

$dist_{GKij}$: Distance of ith data point from jth cluster according to GKC algorithm.

$\mathbf{v_j}$: Point representative for the jth cluster.

E_j : Ellipse representative for the jth cluster.

$\mathbf{ev_j}$: Center of the ellipse representative (E_j) for the jth cluster.

t : Number of iterations.

$\mathbf{\Sigma}$: Covariance matrix.

$\mathbf{\Lambda}$: Positive symmetric matrix, determining major and minor axes lengths and orientation of the ellipse (E_j).

J : Objective function or cost function.

h : Radius of a kernel.

HS : Hopkins statistic.

$dist(.,.)$: Distance norm.

D_k : Dunn index for clustering tendency.

DB_k : DB index for clustering tendency.

s_i : Scatter/spread of ith cluster.

s_i' : Silhouette of ith data point.

S_i' : Silhouette of ith cluster.

R_{ij} : Similarity between ith and jth cluster.

V_{PC} : Partition coefficient.

A,B : Elements of set.

H : Entropy.

$\mathbf{v_j}$: Cluster centers.

m : Fuzzifier.

I : Measure of fuzziness.

$\mathbf{U}$: Matrix denoting fuzzy membership function for all data points and all classes/clusters.

H: Hopkins Test.

R_{ij} : Similarity index between cluster c_i and c_j.

ϕ : Kernel function.

N_s : Number of support vectors.

C' : Parameter for nonlinear SVM.

ξ : Slack variable for nonlinear SVM.

$\mathbf{w}$: Weight vector for SVM.

w_0 : Bias.

w_{AB} : Weight of the edge connecting node A and B.

η : Learning rate.

$\mathbf{c}$: Center in RBF.

$\mathcal{H}(.,.)$: Neighborhood function for SOM.

p : Degree of parallelism.

List of Figures

List of Tables

Chapter 1
Evolution of Data Science

The goal is to extract knowledge from data and use knowledge for humankind.

The word **Data** comes from the Latin word "*Datum*," dated back to the mid-eighteenth century, which literally means "something given" [34]. Though the storage and analysis of data have been done since 20,000 B.C., the first known use of the word "Data" dates back to 1646 [34]. It is the collection of facts such as numbers, words, measurements, observations, or descriptions of a thing. It can be considered as a form of raw "information," which needs to be refined for further analysis and use. Data can be represented at different levels of abstraction. At its basic level, it merely consists of numbers, figures, characters, or symbols. At higher levels (through abstraction), the collection represents images, text, music, video, graphs, etc.

Methods of storing and accessing data changed largely with time. Initially, data were stored and accessed manually in the form of ledgers, registers, or files (entirely paper-based). It was not until 1940s when people started storing and processing it for wide utility with the advent and advancement of electronic equipments. The arrival of digital electronic computers provided the mechanism by which computations of different kinds (processing of data) could be automated with increasing efficiency. These devices began to input, store, process, and output data fast and efficiently. The amount of data also grew larger in quantity gradually. As more and more data were gathered from all possible sources, different new types of data (like hypertext, graphs, and networks) emerged to cope with the demand of the society. The main question that stands now is—"*What to do with all these data?*" How will we extract knowledge/information from them? In other words, how to manage these huge data for the betterment of human being [1].

American astronomer Clifford Stoll said, "Data is not information, information is not knowledge, knowledge is not understanding, understanding is not wisdom" [31]. This forms the basis of a hierarchical model of data (Fig. 1.1). Data, being raw materials, make up the foundation at the bottom of the pile, and information, knowledge, understanding, and wisdom represent the consecutive higher levels—

© The Author(s), under exclusive license to Springer Nature Singapore Pte Ltd. 2025
A. Ghosh, *Data Science and Cases in Sustainability*, Mathematics for Sustainable Developments, https://doi.org/10.1007/978-981-96-8362-8_1

Fig. 1.1 Hierarchical model
of data analysis

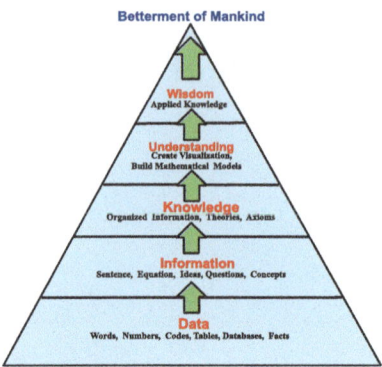

some processed versions of raw data. The final question is "*how to use this wisdom for improving the quality of life*"?

Data in their raw form represent the facts. Hence, it needs to be processed to discover the hidden structures and information inherent to it. Then, we may gain information and draw some inferences. This may increase our knowledge to be used for further improvement of mankind. This sort of inductive inference from the raw data has been prominent since the time of Aristotle. However, the mathematical frameworks for inductive inferencing (from specific cases to generalization) were devised in the early eighteenth century. This is currently known by the name "Statistics" [23, 32]. It is the science of collecting and analyzing data in large quantities, especially for the purpose of inferring some outcome on the whole from the representative samples. Statistics can be described under two of its taxonomies—**Descriptive Statistics** and **Inferential Statistics** [23].

Descriptive Statistics consists of procedures to find out statistical regularities to summarize and describe important characteristics of a set of measurements, whereas **Inferential Statistics** mainly aims at making inferences about the characteristics of an unknown set of data from information obtained from a small portion of it.

As statistics is mainly concerned with large volumes of numerical data, analyzing it manually was both tedious and erroneous. Also, data collection was not easy due to the lack of communication medium. Data analysis was therefore localized and limited. Early days statistics was mainly used for summarization of data. Many inferential statistical studies were carried out in the nineteenth century. These were done manually with small datasets. Advancement of science and technology in every scientific field like medical, agriculture, physical chemical, civil, mechanical, space, and communication generated more data in their respective areas, and the data thus became more domain-specific. Manual manipulation of data became monotonous and thus came the urge to develop machine-based analysis tools. With the advent of computing machines, the inferential statistics gained a lot of prosperity. The conventional mechanisms were modified for processing large amount of data with these computing machines. The quest to extract and to have a compact representation of knowledge from data automatically in reasonable time emulating the principle of human reasoning gave rise to the field of **Artificial Intelligence** [15].

Artificial Intelligence (AI) [19, 30] evolved as a noble field whose objective was to recreate the human intelligence artificially in machines. In addition to doing number crunching, AI tries to mimic human reasoning process. It basically studies and designs intelligent agents or machines to mimic the way a human responds or behaves intelligently. Evolution of different types of non numerical data like image and video made the field of AI broaden up. It started using statistical approaches to analyze data where the goal was not only to represent the data but also to find some interesting patterns or rules from them. In due course, **Pattern Recognition (PR)** [8, 11, 16, 17, 35, 36] and **Machine Learning (ML)** [2, 3, 24, 27] emerged as major fields to automatically look for similarity in patterns and categorize them accordingly in groups. Machine learning is a branch of AI that aims to provide computers the ability to learn without being explicitly programmed. It focuses on the development of computer programs that can train themselves to grow and adapt when exposed to new data. Although PR is considered to be nearly synonymous with ML, it avoids the complex mathematics involved in machine learning. It aims to classify data (patterns) based on either a priori knowledge or on statistical information extracted from the patterns. During 1960–1980, a lot of researchers were interested in ML and PR and designed many efficient algorithms to find some interesting patterns in data [6]. Though the subject PR has attained maturity during the past four to five decades, ML still remains an area of interest for researchers who try to crossbreed the disciplines like computer science, physics, statistics, engineering, psychology, biology, economics, mathematics, and cognitive science.

In recent years, data have grown in size (Kilobytes to Petabytes) and complexity (from numerical to categorical, text, audio, video, graphs, etc.). Also large sections of the world population have easy access to computers. Thus, "manual" data analysis has increasingly been replaced with automated data processing. Nowadays, almost all scientific, government, commercial, or business organizations use computers for computational purposes as well as generation and storage of huge amount of data in their massive databases. Data collected are stored in a structured manner in databases, and if we want to access the data, we have a database management system [9, 12]. Common people also started generating data with the evolution of the internet in 1995. So data collection, manipulation, or analysis not only remained industrial companies' business goals. It became a store house of information for common people also. Internet paved a new path of exchanging information. Thus, there was a need to develop a science that can deal with the delivery and exchange of data. This was named as **Information and Communication Technology** (ICT) [5, 29]. People started connecting themselves globally via Internet. A new way of business called Electronic Commerce [28, 33] thus evolved. Electronic Commerce, commonly known as E-commerce or eCommerce, started the trade of products or services via computer networks, such as the Internet. E-commerce diverged out to technologies such as mobile commerce, electronic fund transfer, supply chain management, internet marketing, online transaction processing, electronic data interchange, inventory management systems, and automated data collection systems. People started connecting to others on distant parts of the world via emails (electronic mail). Consumers in a way started consuming data and also started to generate data. Huge amount of data thus started to be stored in databases. Companies started open database centers to keep those data. Google came up with a strong machine learning algorithm whose mission statement from the outset was "to organize the world's information and make it universally accessible and useful." Orkut, Facebook, Twitter, and other social networking sites enabled people to connect via internet and generate huge amounts of data. In early twenty-first century, internet became an important source of data generation. Every data generating device (sensors, camera, video camera, mobile phone, etc.) was also connected through internet. Large-scale computing and networking ensured that data become accessible to more and more people.

In order to bring out a mechanism, by which people could access these data and extract knowledge from it only through a few clicks of their keyboards, there was a need for an efficient management system. These quests brought out disciplines like Database Management Systems [9, 12] and Knowledge Discovery in Databases [7, 10, 14, 20, 21, 37]. The massive databases generally store much more than numeric data. Data can be in forms of audio, video, text, and symbols. They may also contain errors or redundancy. In 1989, Gregory Piatetsky Shapiro used the term **Knowledge Discovery in Databases (KDD)** to describe such processes by which efficient analysis could be done over these massive and heterogeneous data. Pattern recognition principles play a significant role in KDD to discover natural structures within such massive datasets. In 1990, a company used the term **Data Mining** [14, 20, 21] as a trademark to represent their work. Today, data mining and KDD are used interchangeably. Data mining is the core part of knowledge discovery, which deals with the process of identifying valid, novel, useful, and understandable patterns in data. It excludes the knowledge interpretation part of KDD. The overall goal of the data mining process is to extract information from a dataset and transform it into an understandable structure for further use. The terms data mining and knowledge discovery in databases (KDD), pattern recognition, machine learning are hard to separate as they largely overlap in their scope. However, the term data mining became more popular in the business and press communities. Later, it was given the name **"predictive analysis"** , and nowadays, people know it as **"Data Science"** [4, 13]. Pattern Recognition and Machine Learning are still popular in academic community that is trying to discover fast, efficient, robust and scalable algorithms for data mining and data science.

Present world is wealthy in terms of smart sensor–based instruments, which generate data of large volumes at a much higher rate. Advancement of communication technologies has made people interconnect with each other around the globe anytime. Also, due to inexpensive small integrated circuits, we are able to develop smart portable machines like mobile phones, laptops. These are all equally responsible for an exponential growth in the volume of data (in order of Exabyte), variations in data types (like traditional relational data, raw data, semistructured and unstructured data from web pages, log files, search indexes, social media forums, emails, documents, and sensor data) and velocity (streams of data flowing constantly) of data in the past decade. Internet of Things (IoT) enhanced the data generation process manyfold.

Data come from various sources like scientific experiments, space science, finance, banking and insurance, public services, manufacturing, aerospace and defense manufacturing, oil and gas production and refining, chemical manufacturing, transportation and logistics organizations like airlines, trucking and rail, hospitality, gaming and entertainment, travel, marketing, advertisement, public relations and market research, publishing and broadcasting, telecommunications service providers, internet and managed-network service providers, energy production and distribution utilities, end consumer retail and wholesale distributions, medical devices and supply, and pharmaceutical productions. Social network like Twitter generates 7 terabytes (TB) of data every day, Facebook generates 10 TB, Flicker requires around 4 TB space every day to store all its images, and Google has more than one billion queries every day. Some enterprizes generate terabytes of data every hour throughout the year. The rate is so high that this number will reach 5 Zettabytes (ZB) by 2022 (Fig. 1.2).

In short, we are entering the era of **Big Data** [4, 18, 38, 39]—a term that refers to the explosion of available information [22]. The term big data (BD) applies to the data that can be "processed or analyzed using traditional data mining methodologies or tools." BD usually deals with the datasets of huge volume of instances and dimensions beyond the ability of commonly used machines and software tools to capture, organize, manage, and process them within a tolerable elapsed time. Computers and the Internet certainly aid BD by lowering the cost of collecting, storing, processing, and sharing information; but, at its heart, BD is only the latest step in our quest to understand and quantify the world in digital form. Ultimately, BD marks the moment when the information society finally fulfills the promise implied by its name. All those digital bits that have been gathered can now be

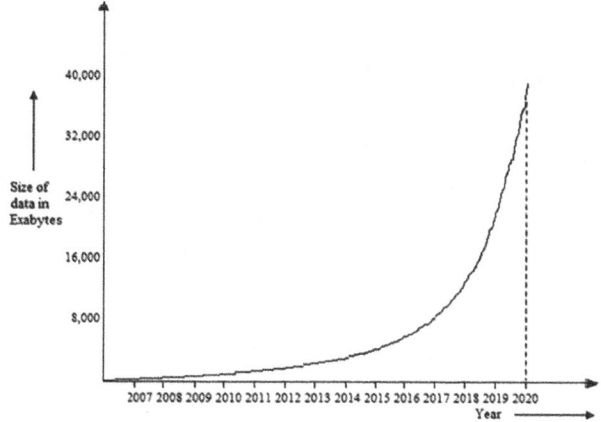

Fig. 1.2 Exponential growth of data with time

harnessed in novel ways to serve new purposes of the society and unlock new forms of value in reasonable time—the birth of **"Data Science (DS)"** [25, 26].

References

1. Ali, A. B. M. S., & Wasimi, S. A. (2007). *Data mining: Methods and techniques.* Cengage Learning Australia.
2. Alpaydin, E. (2014). *Introduction to machine learning.* Adaptive computation and machine learning series. MIT Press.
3. Anzai, Y. (2012). *Pattern recognition and machine learning.* Elsevier Science.
4. Baesens, B. (2014). *Analytics in a big data world: The essential guide to data science and its applications.* Wiley.
5. Borko, H. (1968). Information science: What is it? *American Documentation, 19*(1) , 3–5.
6. Chen, D., & Cheng, X. (2013). *Pattern recognition and string matching* (Combinatorial optimization). Springer US.
7. Cios, K. J., Pedrycz, W., Swiniarski, R. W., & Kurgan, L. A. (2007). *Data mining: A knowledge discovery approach.* Springer.
8. Duda, R. O., Hart, P. E., & Stork, D. G. (2012). *Pattern classification.* Wiley.
9. Erickson, J. (2009). *Database technologies: Concepts, methodologies, tools, and applications: Concepts, methodologies, tools, and applications* (vol. 4). IGI Publishing.
10. Freitas, A. A. (2003). A survey of evolutionary algorithms for data mining and knowledge discovery. *Advances in Evolutionary Computing, 1,* 819–845.
11. Fukunaga, K. (2013). *Introduction to statistical pattern recognition.* Academic.
12. Garcia-Molina, H. (2008). *Database systems: The complete book.* Pearson Education India.
13. Grus, J. (2015). *Data science from scratch: First principles with python* (1st ed.). O'Reilly Media, Inc.
14. Han, J., Kamber, M., & Pei, J. (2006). *Data mining, Southeast Asia edition: Concepts and techniques* (The Morgan Kaufmann series in data management systems). Elsevier Science.
15. Haugeland, J. (1989). *Artificial intelligence: The very idea* (A Bradford book). MIT Press.
16. Jordan, M., Kleinberg, J., & Schólkopf, B. (2005). *Information science and statistics.* Springer.
17. Kittler, J. (1986). Feature selection and extraction. In T. Y. Young & K. S. Fu (Eds.), *Handbook of pattern recognition and image processing* (pp. 59–83). Academic.
18. Kotu, V., & Deshpande, B. (2018). *Data science: Concepts and practice.* Elsevier Science.
19. Kurzweil, R. (2012). *How to create a mind: The secret of human thought revealed.* Penguin Publishing Group.
20. Liu, H., & Motoda, H. (2012). *Feature selection for knowledge discovery and data mining* (The Springer International series in engineering and computer science). Springer US.
21. Maimon, O., & Rokach, L. (2010). *Data mining and knowledge discovery handbook* (Series in solid-state sciences). Springer.
22. McLean, M. (1985). *The information explosion: The new electronic media in Japan and Europe* (Bibliographies and indexes in economics and economic history). Greenwood Press.
23. Mendenhall, W., Beaver, R., & Beaver, B. (2008). *Introduction to probability and statistics* (Available 2010 titles enhanced web assign series). Cengage Learning.
24. Murphy, K. P. (2012). *Machine learning: A probabilistic perspective* (Adaptive computation and machine learning series). MIT Press.
25. Ozdemir, S. (2016). *Principles of data science.* Packt Publishing.
26. Pearl, J., & Mackenzie, D. (2018). *The book of why: The new science of cause and effect.* Basic books.

27. Reimann, P., Spada, H., & Foundation, E. S. (1996). *Learning in humans and machines: Towards an interdisciplinary learning science* (Collection of Jamie and Michael Kassler). Emerald Group Publishing Limited.
28. Reynolds, J., & Mofazali, R. (2000). *The complete E-commerce book: Design, build, and maintain a successful Web-based business.* CMP Books.
29. Rubin, R. E. (1998). *Foundations of library and information science.* ERIC.
30. Russell, S., & Norvig, P. (2010). *Artificial intelligence: A modern approach* (Prentice Hall series in artificial intelligence). Prentice Hall.
31. Saltz, J., & Stanton, J. (2017). *An introduction to data science.* SAGE.
32. Savage, L. (1972). *The foundations of statistics* (Dover books on mathematics series). Dover Publications.
33. Shaw, M., Blanning, R., Strader, T., & Whinston, A. (2012). *Handbook on electronic commerce.* Springer.
34. Stevenson, A. (2010). *Oxford dictionary of English* (Oxford dictionary of English). OUP.
35. Theodoridis, S., Pikrakis, A., Koutroumbas, K., & Cavouras, D. (2010). *Introduction to pattern recognition: A Matlab approach.* Academic.
36. Tou, J. T. L., & Gonzalez, R. C. (1979). *Pattern recognition principles.* Addison-Wesley.
37. Tseng, V. S., Ho, T. B., Zhou, Z. H., Chen, A. L. P., & Kao, H. Y. (2014). *Advances in knowledge discovery and data mining* (Lecture notes in computer science, No. pt. 2). Springer.
38. Warden, P. (2011). *Big data glossary.* O'Reilly Media, Inc.
39. Zikopoulos, P., Eaton, C., et al. (2011). *Understanding big data: Analytics for enterprise class Hadoop and streaming data.* McGraw-Hill Osborne Media.

Part I
Learning from Data

Chapter 2
Learning

Our primitive idea about learning is that we "learn" by interacting with our environment. Whenever a child does some basic activities like playing, waving its arms, or looking around, it needs no explicit training; but it uses its direct connection to its environment as a learning mechanism. Such connections produce a wealth of information regarding the causal effect relationships of events, consequences of actions, and specific actions needed to achieve goals. Whether we are learning to hold a cup or a conversation, we are acutely monitoring how our environment responds to our actions, and we modify ourselves with this feedback. "Learning from interaction" is, therefore, the foundational idea underlying nearly all the advanced theories of learning [31] and intelligence. In an easier language, **learning is the process using which a system modifies its parameters like speed of execution, performance, and efficiency such that its future performance is improved.**

The brains of living organisms are genetically designed for solving certain types of learning problems. For example, if we consider the visual system in human brain, there are numerous optical neurons present in the visual system that makes it possible for the humans to recognize objects. The process of learning is not only limited to humans, it is diversified to plants, animals, etc. A child learns to speak, a bird learns to fly, plants learn to adapt to the environmental conditions, and so on. Our very survival actually depends on our ability to learn and adapt the environment. Similarly, machines can be made to learn and update itself for improving its performance by imitating the natural process of learning, which is termed as "machine learning." We acquire knowledge either from supervised learning, i.e., learning from our teachers, supervisors, trainers; or by self-learning, i.e., learning on our own (unsupervised/reinforcement). When a child learns to speak or to write, she learns it through examples from parents and others a kind of supervised learning. But when a child is asked to sort a set of balls of different colors, he simply sorts them according to his visual perception without knowing which color is called red or which one is green a kind of unsupervised learning.

© The Author(s), under exclusive license to Springer Nature Singapore Pte Ltd. 2025
A. Ghosh, *Data Science and Cases in Sustainability*, Mathematics for Sustainable Developments, https://doi.org/10.1007/978-981-96-8362-8_2

Fig. 2.1 Methods of learning

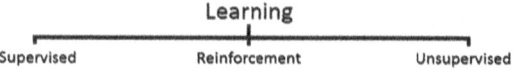

There is another kind of learning in which a child learns his behavior. The parents do not give any examples for learning, but tell him whether the activities he is doing are right or wrong, "reinforcing" the child to modify his behavior called reinforcement learning. Thus, learning (including machine learning) can be mainly classified into three ways: supervised, reinforcement, and unsupervised (Fig. 2.1).

Artificial Intelligence (AI) defines several methodologies or techniques for solving a range of problems, one of which is **Machine Learning** (ML). It might be the goal of an AI system to search for consciousness, (which seems to be so exciting), but it still relies on machine learning heavily. Machine learning is all about developing algorithms that can give a computer the capability to learn in a self-evolving fashion and take decisions on its own.

In this sense, machine learning may be considered as finding statistical regularities or inherent patterns of data. Thus, many researchers consider ML algorithms to be an application of statistical data processing. Following the statement, this is why the machine learning algorithms may not always resemble the human approach of learning a task. Therefore, machine learning becomes most effective when it gets intertwined with human intelligence, rather than just completely replacing it. The sole idea of machine learning gets highlighted with the fact that computers and humans have different capacities: computers are much faster in doing arithmetic and counting operations, while humans excel in logical and reasoning functionalities. These contrasting forms of intelligence are compatible, not diametrically opposite. Thus, machine learning strives to find the learning process in humans.

2.1   Supervised Learning

The simplest form of learning is supervised learning. It is the task to find out a mapping between the input and output data pairs. The task for any supervised learning algorithm is to find out a function to correlate input–output data using labeled training data samples called training data . Each training example is a pair consisting of an input sample (that has several attributes) and a desired output (which denotes the specific category to which the sample belongs) [2]. For example, a teacher wants her students to know two types of fruits: apple and orange. She is interested to teach them the difference between these fruits.

She initially shows a basket full of these two types of fruits and tells the student which of these are apples and which are oranges. This is the "training set of labeled samples" for the student (Fig. 2.2). Following this, the students somehow manage to recognize and remember them and learn/discover a technique for distinguishing them. Now, whenever the students see an apple or an orange somewhere, they can tell if it is an apple or an orange. If any student makes a mistake, the teacher gives

Fig. 2.2 Examples of apple and orange given by the teacher

Apples   Oranges

Table 2.1 Input–output data table

Input	Output
1	1
2	4
3	9
4	16
⋮	⋮
10	100

a feedback of the correct answer to the problem, so that the learner identifies it in future and solves it as it was taught to.

Let us take another example of supervised learning. Consider the following input–output data table (training set) shown by the teacher to the students (Table 2.1). Now if any student is asked to tell the output when the input is 11, the obvious answer is 121. The student learns that the output is the square of the input.

Supervised learning splits into two broad categories: **Classification** for responses that can have just a few known labels (also called class labels) or values, such as "true" or "false," male and female, odd and even. On the other hand, **regression** outputs responses that are real numbers [1].

Classification problems have a primary goal to get the computer classify samples into a fixed set of categories. As a common example, handwritten digit recognition is a classification problem where images of handwritten digits are classified into ten categories (0–9). In general terms, classification is opted for a problem where deducing a category label is needed. To do so, the classification algorithm is often built to optimize an objective function with respect to the given input–output pairs. These input–output pairs are often known through the training set. The agent tries to learn from these examples. It is generally assumed that the distribution of the training set is identical to that of the original data. If such assumptions do not hold, then the classification may be faulty even if it optimizes the objective function. For instance, if we want a person to learn how does a banana look like, we should not show bananas of only one size. Then the learning will not be "general" and will be "overfit". Similarly, with machine learning algorithms, overfitting the training data is considered a disaster, and it essentially means that the system is just memorizing the training set perfectly. Such a system usually fails to classify new data correctly. The detailed description for classification can be found in Chap. 7.

Regression tasks on the other hand are used more for forecasting and predicting scenarios. For example, predicting the age of a person based on certain factors is considered a regression task rather than a classification task as the age can take any real-valued data. However, predicting the gender of a person is a classification task as it has discrete class labels. The detailed description for regression can be found in Chap. 7 of this book.

2.2 Unsupervised Learning

The supervised learning approach does not seem plausible always, since we need a good training set of labeled samples. In unsupervised learning, there is no concept of corrections by external agents. Unlike supervised learning methods, unsupervised means there is no teacher in such learning scenario. Such scenarios arise for events where we have no idea about the solution we are looking for. Unsupervised learning is a kind of learning on its own. Some actions of living beings need no prior teaching. It just gets into our system in an unsupervised manner. For example, we do not need someone to teach us how to smile at a beautiful picture and frown to the bad. That is something we have acquired by grouping pictures as good or bad only based on whether they are pleasant or not and, accordingly, take actions to show our emotions.

Let us consider an example of unsupervised learning. The following set of 10 numbers are given to a class of students:

$$\{2, 3, 9, 16, 45, 75, 99, 100, 278, 516\}.$$

They are asked to separate these numbers into groups using any criterion they want. Since the type of grouping or the number of groups is not given beforehand, the students are free to choose the conditions. There are numerous possibilities. Some students may group using the number of digits present in a number; and the outcome is:

$$\{2, 3, 9\}\{16, 45, 75, 99\}\{100, 278, 516\}.$$

Another set of students can group even and odd, producing the output as:

$$\{2, 16, 100, 278, 516\}\{3, 9, 45, 75, 99\}.$$

Other students may group prime and non-prime and so on . . .

Unsupervised learning systems learn to represent the presented input patterns in a way that captures the statistical structure of input data. Quite contrary to supervised or reinforcement learning, the unsupervised system is not provided with any explicit target outputs associated with each input [42]. Supervised methods rely on training examples with analyst-established categories, but unsupervised methods avoid using pre-established categories. The power of unsupervised machine learning, as it is

Fig. 2.3 Separating two
types of shapes

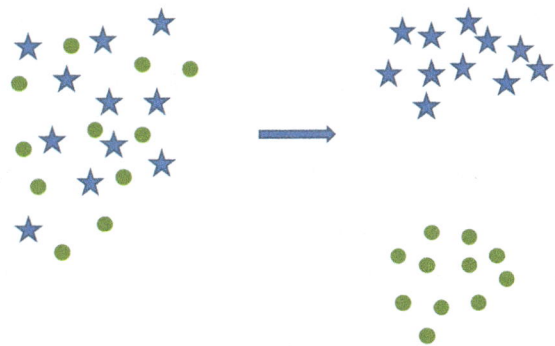

properly known, is that it can find out similarities and connections between data
points and group them accordingly without having any prior knowledge of the data
distribution. A child is able to separate the two types of shapes without the help of
any teacher or explicitly knowing the names of the shapes or colors (Fig. 2.3).

Unsupervised learning is considered to be more elegant as it does not assume
any pre-discovered classification of examples. Unsupervised methods include clustering, density estimation, principal component analysis, data summarization, etc.
The simplest of these is clustering, in which the goal is to partition the given dataset
into a number of groups or clusters depending on the similarities in the data. The
assumption is that there always exist a classification setting for which the clusters
discovered will match with a classification labels. For example, clustering people
based on demographic feature could result into a cluster of wealthy people in one
group and the poor people in another and will match the classification labels of
wealthy and poor.

Although these clustering algorithms do not assign a name to the clusters, it can
identify and assign new examples into one of the clusters. For instance, Amazon
recommends items to its customers based on the buying habits of similar groups
of people. It basically clusters its customers into groups. However, this is a purely
data-driven process, which will work well only when there is sufficient amount of
data. It is quite unfortunate to say that unsupervised learning, being such an elegant
technique, also suffers from the problem of overfitting. The detailed description for
clustering can be found in Chap. 10 of this book.

2.3 Semi-supervised Learning

Semi-supervised learning [29] is a continuum between supervised and unsupervised
learning models. It is a special form of classifier modeling technique. In traditional
supervised learning methods, the classifier is trained only with labeled data. If
the number of labeled data is too few to design a classifier, then semi-supervised
learning comes into picture. Semi-supervised techniques use large amount of

unlabeled data together with only a few labeled data to build better classifiers. A simple example can be given to understand the semi-supervised learning method. An infant listens to a word "apple" and actually sees an apple. He sees many pictures of apples and oranges and hears the names too. If the word apple is heard by him many times before he saw it, then the association of the word apple with the object apple will be much stronger in his memory. If the word is not heard before, the association is weaker. It is similar to cluster and then assign a label to each cluster. Most researchers argue that the earliest idea about the use of unlabeled data samples to enhance classification results is self-learning [34] (also known as self-labeling, self-training, or decision-directed learning). This is a wrapper algorithm that repeatedly uses the decisions made by the supervised learning algorithm method. It begins by training with the few labeled data. At each step, a small portion of the unlabeled data are classified and labeled according to the current decision function. The newly classified data are then augmented with the originally labeled samples, and the supervised method is retrained till the error is minimized.

2.4   Reinforcement Learning

As AI seems to be synonymous with self motivated robots, driverless cars, smart object, etc., there must be a method where the system can learn the causal effect relationship of actions and evolve its parameters. Reinforcement learning is that exciting learning mechanism, which allows any system to automatically learn from its environment by receiving rewards and punishments and determining what to do next from these feedbacks.

Consider a tiny reinforcement learning problem. There are six options A, B, C, D, E, and F for answers to a question that has more than one correct option, and a student can choose more than one option. He does not know the marking scheme or how rewards are earned.

Suppose the grading works as follows:

- Opting only A: +5
- Opting A and C: +7
- Opting A, C and E: +10
- Opting A and any one wrong choice (other than C and E): +2
- Opting A, C and any one wrong choice (other than E): +3
- Opting A, E and any one wrong choice (other than C): +3
- Opting C as one of the options (provided other options chosen are neither A nor E): +1
- Opting E as one of the options (provided other options chosen are neither A nor C): +1
- Opting all wrong: 0

The student opts for a wrong choice and gets zero. He tries another choice in the next examination. Maybe this time he opts C and gets +1. Accordingly, he tries other

Fig. 2.4 Characteristics of reinforcement learning

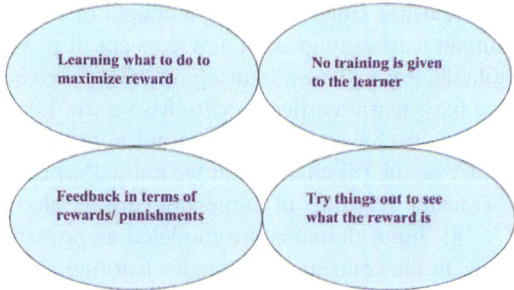

choices to get rewards. The student answers the same question again and again till he gets a reward of +10. Thus, the student learns the actual answer to the question. The basic characteristics of reinforcement learning are shown in Fig. 2.4.

In reinforcement learning mechanism, there lies a critic who gives a binary assessment of performance. For example, a cook is graded by a food critic. If he cooks well, he is given good remarks; otherwise, there is just a bad review about the food. The critic never tells how much or what the cook needs to correct his recipe. It should be noted that the rewards or punishments may not be given after every step but may come after a number of steps also.

2.5 Transfer Learning

The insight behind transfer learning (TL) is that generalization may occur not only within tasks but also across tasks [15]. Most of the machine learning methods give good performance only if the training and test data are drawn from the same distribution. When the distribution of the test data differs from that of the training data, the trained model is expected to fail. In many practical applications, data labeling is expensive or almost impossible. So, re-collecting the needed training data or rebuilding the model is not an option. In such cases, knowledge from the trained model can be transfered to similar domain tasks. This is known as transfer learning between task domains.

The need for transfer learning arises due to many reasons. There may be a case when the data become easily outdated, or in another case, the output is required at a faster rate. In such cases, the data at a time period may evolve and may not follow the same distribution as the training data at a later stage. Transfer learning is just a process of reusing the knowledge from the past related tasks to do a new task. The objective behind transfer learning is using the past learning and experiences to adapt to new, however, related, ideas. The usefulness of transfer learning is typically measured by the decreased number of training examples required to achieve a desired performance level on a set of closely related learning scenarios, compared to the number required for unrelated scenarios: i.e., reduced sample complexity. In many real-life situations, only a couple of training examples is fairly adequate for a human learner to grasp a new concept, if prior knowledge of related problems is available. For instance, figuring out how to drive a truck turns out to be a lot simpler if we have learnt earlier how to drive a car. Learning French is a bit simpler in the event that we have effectively learnt English (vs Chinese), and learning Spanish gets simpler on the off chance that we know Portuguese (vs German).

Transfer learning in some domains is also referred to as domain adaptation [27, 28]. Input domains are modeled as probability distributions over an instance space. In the transductive transfer learning setting, the source and target tasks are equivalent, while the source and target domains are different. In this circumstance, no labeled samples are available for the target domain; but many labeled data are available for the source domain. One builds the models from some fixed source domain; however, wishes to convey them crosswise over at least one diverse target domains. For instance, large-scale speech recognition systems need to function admirably crosswise over arbitrary speech, regardless of background noise or accents. Text processing systems trained on news are frequently needed to be used for web journals or discussion forums. Gene finders are trained on a particular organism (say a chimpanzee), but many a times our aim is to identify the genes of another organism (say a human) or even a group of organisms (such as the set of all apes). Face recognition systems might get trained under certain posture, lighting (day, night, shadow, etc.), and occlusion settings, but is applied to any arbitrary pose, occlusion, and lighting conditions.

Note that, in domain adaptation, the input distribution changes but the labels remain the same; and in transfer learning, the input distributions remain the same, but the labels change. The two problems are quite similar, and indeed, we see similar (if not identical) techniques being applied to both the problems. (Incidentally, we also see articles that are really solving one of the two problems but claiming to be solving the other.)

2.6 Deep Learning

Until 2006, the word "deep" was not discussed much in the machine learning literature. Deep architectures of neural networks were abandoned because of their poor training and generalization. To learn about thousands of objects from millions of images, we need a model with a fast and huge learning capacity [16, 33].

"Deep learning" [4] is a larger subsection of machine learning methods that aim to learn features via different levels of abstractions. This refers to learning feature hierarchies, i.e., low-level features compose higher-level features. This kind of feature learning allows a system to automatically learn complex functions that maps the input to the output directly using the available raw data. Hence, the use of human-crafted features is not necessary. This is conceptually similar to what humans do. The automatic learning of powerful features holds great importance as the amount of data grows and as the range of applications to machine learning methods become more diverse.

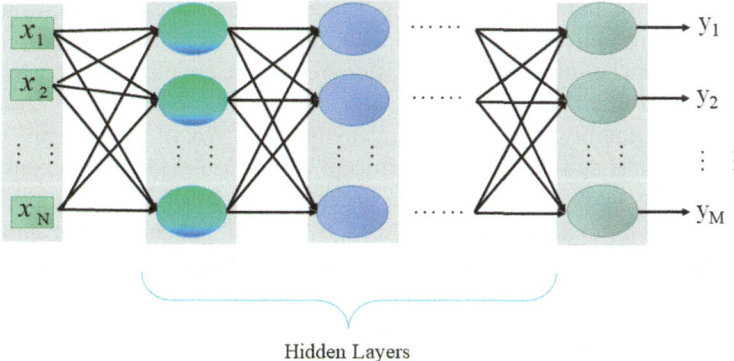

Fig. 2.5 An example of deep network

What sets a deep model apart from everything else is its concept of depth. There are multiple (more than two) layers of feature transformations. This is a major step away from previous "shallow learning" algorithms. As we can see from Fig. 2.5, the input vector $\mathbf{x} = \{x_1, x_2, \ldots x_d\}$ undergoes a nonlinear transformation through each hidden layer. It is argued [41] that deep networks extract features at each hidden layer. The initial hidden layers extract some low-level information like lines, edges, etc., which the latter layers combine and transform to build up complex features like textures and shapes.

Deep learning emphasizes the kind of model (e.g., a deep stacked auto-encoder or a deep convolutional neural network [18]) we might want to use. As the deep learning starts with a practical data from the real world (high dimensional data), it needs a lot of data ("big data") and a lot of parallel processing or high-performance computing facilities (GPUs). Chapter 21 gives a detailed description of these concepts.

2.7 Inductive Learning

Inductive instructions make use of students' "generalizing capability." Rather than explaining a given concept and following clarification with examples, the teacher presents to his/her students many examples indicating how the idea is utilized. The purpose is by using examples, make the students "notice," how the idea actually works.

Utilizing the grammar example, the teacher might present the students various examples of punctuation without giving any fixed rules about how to put a punctuation at a particular place. As the students perceive how the idea of punctuation is utilized, it is expected that they will grasp how the idea should be utilized, and then they would decide the guidelines. At the end of the activity, the teacher can ask the students to explain the punctuation rules and watch that they comprehended the idea.

One example of inductive learning is a learning model called decision trees. Decision trees are built to find rules hidden in the given data. For example, loan prediction task can be solved using a decision tree. There are plenty of data available to the bank regarding many attributes (e.g., "Name," "Age," "Gender," "Address," "Occupation," "Monthly Income," "Company Name," "No. of Dependents") of the past loan applicants and the decision ("Accept"/"Reject") made by the bank about them. These data are used to construct a decision tree, which finally outputs some generalized rules for making decision ("Accept"/"Reject") about the loan applicants. A set of rules like the following ones may come as output.

If monthly income < 8000 and number of dependents > 3, then decision = reject.

If 25 <= age <= 45 and monthly income > 8000 and number of dependents < 3, then decision = accept; and so on.

2.8 Deductive Learning

In contrast to the inductive learning technique, a deductive approach to instruction is a more teacher-oriented approach. In this approach, the teacher introduces the students with a new concept, explains it to them, and after that asks them to practice it. The teacher tells or shows directly what he/she wants to teach. This is also referred to as direct instruction. For instance, when teaching a new mathematical concept, the teacher will introduce that concept, provide the rules related to its use, and finally, the students would practice the concept in various ways. In case of teaching mathematical concept of matrix theory, the teacher will state the rules of matrix addition, scalar multiplication, and matrix multiplication. Afterward, the students will try to solve the problems of matrix operation maintaining those rules.

Medical diagnosis using expert systems is an example of deductive learning. An expert system has three basic parts, e.g., knowledge base, inference engine, and user interface. The knowledge base contains many rules accepted or designed by human experts. User interface accepts queries from the users. The inference engine applies rules on query data received from the user to find the decision/answer, replies back to the user, and also updates the knowledge base if necessary. A doctor may ask a query regarding the diagnosis on the basis of some symptoms of patient to the expert system. The expert system uses it's knowledge base and inference engine to reply to the doctor's query with its diagnosis result.

2.9 Active Learning

The main concept behind active learning is that a machine learning algorithm could perform better using less training if it is permitted to pick the information from where it learns. An active learner might present "inquiries" or "queries," for the most part as unlabeled samples, which is to be marked by a human annotator who has prior idea about the problem. This kind of methodology is well-motivated in various advanced data mining and machine learning applications, where unlabeled data are present in abundance but labelling those data become an expensive, troublesome, or a tedious process.

When the size of data with unknown label is big enough to incur high cost of manual labelling, then data are divided into parts; data with known labels, data with unknown labels, and data chosen to be labeled. So, the learner algorithm is free to choose a portion of data to label. Email classification systems use active learning algorithms for spam detection, where the system learns by interacting with the users asking them to label the mails from time to time.

2.10 Manifold Learning

A manifold, in easier words, can be thought of as an arbitrary surface of any shape. It need not necessarily be a plane. It may be shaped like a folded or rolled sheet with all the curves. This can be generalized to "n" dimensions and is formalized as "manifold" in mathematics. Imagine that a collection of seeds are fastened on a glass plate, which is resting horizontally on a table. Now, because of the way the concept of space is generally described, more or less it would be safe to say that the bunch of seeds lie in a two-dimensional space, since each seed can be identified using two values, which give that seed's actual position on the surface of the glass.

Now if the plate is tilted diagonally upward, such that the surface of the glass no longer stays horizontal with respect to the ground, so to locate one of the seeds, there lies now a couple of options (Fig. 2.6). If it is decided that the glass will be ignored, then each seed would appear to be floating in the three-dimensional space above the table, and so each seed's location needs to be specified using three coordinates, one for each spatial direction. However, just by tilting the glass, the fact that the seeds still live on a two-dimensional surface does not get altered. So, it can be explained how the surface of the glass lies in three-dimensional space, and then the locations of the seeds on the glass could be described using the original two dimensions.

Fig. 2.6 Manifold learning

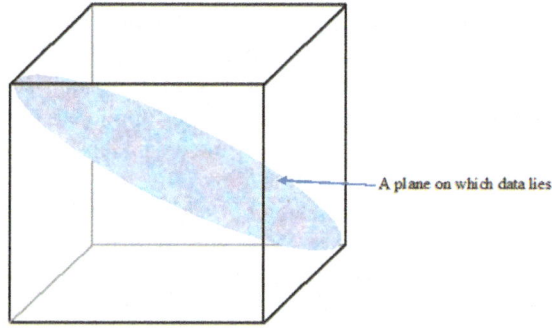

A plane on which data lies

In this experiment, the glass surface is akin to a low-dimensional manifold that exists in a higher-dimensional space, it does not matter how the plate is rotated in three dimensions, the seeds will still live along the surface of a two-dimensional plane.

Raw or natural data are usually high dimensional. However, in many cases, the properties of the data can be explained as an outcome of some process, which involves only a few degrees of freedom. The manifold assumption in machine learning is that, instead of assuming that data in the world could come from every part of the possible space (e.g., the space of all possible one mega pixel images, including white noise), it makes more sense to assume that training data come from relatively low-dimensional manifolds (like the glass plate with the seeds). Then learning the structure of the manifold becomes an important task. Additionally, this learning task seems to be possible without the use of labeled training data.

2.11 Subspace Learning

In linear algebra and related fields of mathematics, a linear subspace, otherwise called a vector subspace, or, in the more established literature, a linear manifold, is a vector space that is a subset of some other (higher-dimension) vector space. A linear subspace is generally called a subspace when the setting serves to recognize it from different sorts of subspaces. Multilinear subspace learning is a way to deal with dimensionality reduction problems.

With the progresses in data collection and data storage techniques, big data (or huge data sets) are being produced daily in various fields of emerging applications. A large portion of these big data are typically high-dimensional, largely redundant, and provide only a small information of the entire input space. Accordingly, dimensionality reduction is frequently employed where high-dimensional data get mapped to a low-dimensional space while information is retained as much as possible.

Linear subspace learning algorithms are conventional dimensionality reduction methods that depict the input samples as vectors and solve for an optimal linear mapping to a lower-dimensional space. Inspite of that, they generally turn out to be failure in dealing with massive multi-dimensional data. They result in high-dimensional vectors, prompting the estimation of countless parameters.

Multilinear subspace learning utilizes various sorts of data tensor analysis tools for dimensionality reduction. It can be applied to observations whose measurements were vectorized and organized into a data tensor or whose measurements are treated as a matrix and linked into a tensor.

In visual object tracking, a target is generally connected with a few fundamental subspaces, each of which is spread over a set of basis templates. Therefore, subspace learning–based generative appearance models focus on obtaining these fundamental subspaces and their respective basis templates effectively using various techniques for subspace analysis. For instance, eigenvalue decomposition [37] or linear regression for subspace [6] and multiple subspaces [24] are utilized to model the distribution characteristics of the object appearance.

2.12   Distance Metric Learning

A distance metric is a function that defines the distance between elements of a set as non-negative real numbers. Various machine learning algorithms (like kNN) are dependent on the distance metrics for the input dataset. For example, let us consider that k-Nearest Neighbors (kNN) algorithm is used for image classification. Here the distance metric that finds significantly low distance value between a pair of similar images and considerably high distance value between a pair of dissimilar images is a desirable distance metric.

Thus, we can say that distance metric learning [47] is a procedure to learn an appropriate distance metric corresponding to the input data (considering pairs of similar/dissimilar points). The suitable distance measure must preserve the distance relation among any pair of data points. Metric learning methods can be formulated in a supervised, semi-supervised, or unsupervised way. In a supervised setting, the metric learning methods mostly try to learn a suitable distance metric from the extra information. This extra information is often available as pairwise constraints, i.e., pairs of similar data points and pairs of dissimilar data points. The similarity or dissimilarity between a pair of examples can be easily calculated from the labeled samples. With labeled data, our aim is to derive a metric that fits our prediction task in a better way. Distance metric learning can be implemented in a supervised manner, where this problem is mathematically formulated as a convex optimization problem. It learns a global distance metric that minimizes the distance between the data pairs with the equivalence constraints and maximizes the distance between data pairs with the in-equivalence constraints. Choosing the proper distance metric is very important for the successful implementation of many real-world problems like content-based image retrieval (CBIR), image classification.

2.13   Codebook Learning

A codebook primarily encodes information on spatial distribution and appearance of the object parts. It has the capability to adaptively catch the information about the dynamic appearance from both foreground and background. Recently, codebooks are considered to build robust discriminative appearance models for tracking objects. Yang et al. [44] developed a tracking system by constructing two codebooks of image patches using two different features such as RGB and local binary pattern (LBP). This approach can handle some problems like occlusion, variation in scale, and rotation during tracking. To capture more discriminative information, an adaptive class specific codebook is built for object tracking [13].

Some developed algorithms in this direction are found in [32, 45, 46, 52]. However, it is very difficult to construct a universal codebook for different scenes or objects. As a result, it is necessary to collect different training samples for different scenes or objects, leading to inflexibility in practice. In addition, determining the size of the codebook is also a difficult task in practice.

2.14   Randomized Learning

Recently, techniques based on randomized learning (e.g., random forest [8, 22, 26, 35]) have been successfully applied in the area of computer vision. Due to selection of random input and random feature, randomized learning techniques are able to build a diverse classifier ensemble. In contrast to boosting algorithm and support vector machines, they are computationally more efficient and easier to be extended for handling multi-class learning problems. In particular, they can be parallelized, so that multi-core and GPU implementations can be performed to reduce run time significantly.

2.15   Multiple Instance Learning

Among the several machine learning methods, multiple-instance learning (MIL) is considered as a variation on the supervised learning. Instead of requiring an elaborate set of labeled instances, the learner receives a set of labeled bags. Each of these bags contains many instances. In the simplest case of a binary classification problem, if all the instances of a bag come from the negative class, then the bag may be labeled negative. On the other hand, if atleast one of the instances of a bag comes from the positive class, then the bag may be labeled positive. From such a varied collection of labeled bags, the learner tries to either induce a concept that will label individual instances correctly or learn how to label bags without inducing the concept.

While taking advantages of both semi-supervised learning and MIL, researchers [23, 48] developed tracking algorithms based on semi-supervised multiple instance learning. This system uses the merits of both techniques by incorporating more prior information and focusing on selection of positive samples to update the discriminative model during tracking.

2.16   Multi-label Learning

Supervised machine learning algorithms [39] are mostly seen to handle traditional single label data, i.e., any sample that belongs to one group/class only. However, looking around us for a more real scenario, it is seen that objects tend to naturally belong to more than one class at a time. Music, movies, articles, novels, news, art, etc., can be comfortably classified into more than one class. For example, while classifying a movie, it can belong to more than one genre like romantic, thriller, comedy, sci-fi, action, and drama. It would not justify to forcefully label it with only class and ignore the other possibilities. This is known as multi-label classification [17], where each sample belongs to one or more classes at a time. Multi-label learning mostly refers to the classification perspective. Researchers have developed various algorithms over the past decades to address multi-label data, focusing mostly on multi-label classification techniques. There are a few approaches recorded in literature that researchers have followed while dealing with this kind of data. The first branch of techniques is data transformation [17], which modifies the multi-label property of the data by converting it into binary/single-label data. After the transformation, an existing binary/single-label classifier is used to classify this data. Binary relevance (BR) [14], Label Powerset (LP) [7], etc., are some of the popular data transformation techniques. These methods were developed initially to handle the multi-label nature of data; however, in recent years, researchers do not prefer to lose the multi-label property of the data. Hence, they gravitate toward the second branch of techniques known as the problem adaptation [17]. This branch of multi-label classification methods concentrate on adapting existing classifiers or creating new ones to deal with multi-label data directly. Adaptations of decision trees [10, 11], k nearest neighbors [51], neural networks [20, 21, 38, 49, 50], deep learning [43], etc., are quite popular algorithms in multi-label literature. Apart from these, a combined approach that is used by researchers to improve the performance of multi-label classifiers is ensemble technique. Algorithms like classifier chains (CC) [30], random k labelsets (RakeL) [40] use a combination of data transformation and/or problem adaptation classifiers to build a stronger ensemble. In recent years, due to the explosion of social networks and other online data platforms, researchers are getting access to huge amount of multi-label data and are strongly focussing on developing multi-label classifiers that can perform efficient prediction. This stream of machine learning is increasingly gaining importance due its relevance and similarity to real world situations.

2.17 Learning Classifiers

If we have fairly little data and we are going to train a supervised classifier, then machine learning theory says that we should stick to a classifier with high bias. For example, there are theoretical and empirical results that Naive Bayes' classifier does well in such circumstances, although this effect is not necessarily observed in practice with regularized models over textual data. At any rate, a very low bias model like a nearest neighbor model is probably counter-indicated. Regardless, the quality of the model will be unfavorably affected by the limited training data. But low-bias or high-variance classifiers start to win out as the training set grows (they have lower asymptotic error), since high-bias classifiers are not powerful enough to provide accurate models. So, proper choice of classifiers is an important aspect in making the decision about any new test data. Learning Classifier Systems (LCSs) [9, 12] are a type of rule-based system having general mechanism of processing rules in parallel, for adaptive generation of new rules, and for testing the effectiveness of existing rules. So, basically they are rule-based systems that automatically create their rule set.

LCSs are one of the major families of techniques that apply evolutionary computation to machine learning tasks. They are rule-based systems, where the rules are generally in the traditional production system form of "IF state THEN action." The two main approaches to implementing and investigating the system empirically are the Pittsburgh-style [12] and the Michigan-style [12]. In Pittsburgh-style approach, there are a number of rule sets, and the learning system seeks to optimize the whole classifier by recombining and reproducing the best of the rule set. However, in the Michigan-style approach, there is only a single set of rules. The algorithm focuses to select the best classifier within the rule set.

In recent time, LCSs have shown efficient result at solving automatic classification tasks. LCSs are now contemplated as sequential decision problem-solving systems bestowed with a generalization property. Indeed, from a reinforcement learning point of view, LCSs could be seen as learning systems creating a compact representation of their problems.

2.18   Learning in Machines

In 1959, Arthur Samuel defined machine learning as a "Field of study that gives computers the ability to learn without being explicitly programmed" [36]. This puts us to a grave question "can machines have thinking ability or make decisions of their own like us?"

Machine learning is a scientific methodology that explores the theory and development of algorithms that can be learnt from data [19]. Such algorithms work by developing a model from example inputs and exploring those inputs to make predictions or decisions, rather than following strictly static program instructions. Machine learning tasks are generally classified into three main categories, depending on the nature of the learning "signal" or "feedback" available to a learning system just as the three mechanisms of. These are:

Supervised machine learning: The computer is provided with example inputs and their corresponding outputs, given by an "expert," and the aim is to learn a general rule that maps inputs to outputs.

Unsupervised machine learning: No information about data is provided to the computer, it has to find the structures in the input data by itself. Unsupervised learning can be a goal in itself (finding hidden patterns in data) or a means toward an end.

Reinforcement machine learning: A computer program interacts with a dynamic environment where it must perform a certain goal (such as driving a vehicle), without an expert explicitly telling it whether it has come near to its goal or not. In other words, there is a critic who is criticising its activities but not telling it the correct answer.

Semi-supervised machine learning   algorithms generally train using a small portion of the labeled data, which are then combined with larger amounts of unlabeled data. This basically combines supervised and unsupervised learning in situations where labeled data are difficult or expensive to procure. **Transduction** is a special case of this principle where induction and deduction are combined in a single step, and it relates with the field of semi-supervised learning where during learning, unlabeled data are used. By eliminating the need to construct an accurate model, transduction gives opportunities for achieving greater accuracy, as has been demonstrated in bioinformatics and text analysis applications where the entire set of problem instances is known during learning, except that a portion of the targets are missing.

There are other learning methods too. In **active learning**, the learner interactively chooses which data points to label. The hope of active learning is that interaction can substantially reduce the number of labels required. Most machine learning researchers focus on domain-specific learning algorithms. Among other categories of machine learning techniques in **learning to learn** [5], the machine uses labeled data from related tasks. It is also known as multi-task learning where the machine learns its own set of assumptions to predict outputs given inputs, which it has not encountered based on previous experience. **Developmental learning**, explained for robot learning, generates its own sequences (also known as curriculum) of learning situations for cumulatively acquiring a stock of plays for skills through autonomous self-exploration and social interaction with human teachers, and making use of guidance mechanisms such as maturation, motor synergies, active learning, and imitation. In brief, the design of learning algorithm is dictated by:

- What type of performance element is used.
- Which functional component is to be learned.
- How that functional component is represented.
- What kind of feedback is available.

Learning agent (machine) = performance element + learning element.

Machine learning studies computer algorithms for learning to do different actions. For instance, we may be interested in learning the process to complete a task, for making accurate predictions or to behave intelligently. The learning process is always based on some sort of observations or data, such as examples, instruction, or direct experience. So typically, machine learning is all about learning for doing better performance in the future based on what was experienced in the past. The emphasis of machine learning is on automatic methods. In other words, the goal is to build learning algorithms that learns automatically without human assistance or intervention. Many a times, there comes a specific task in mind, such as spam filtering. But, instead of programming the computer to solve the task directly, in machine learning, methods are developed by which the computer will come up with its own program based on the examples that are provided. Essentially, it is a technique of teaching computers of how to make and improve predictions or behaviors based on some data. The data, however, depend completely on the problem. It can be readings from a robot's sensors while it learns to walk or the actual output of a program based on a certain input. Another way of thinking about machine learning is that it is "**pattern recognition**" the process of teaching a program about reacting or recognizing patterns [25].

Machine learning theory, also known as Computational Learning Theory [3], aims to understand the basic principles of learning as a computational process. This field wants to understand at a precise mathematical level what capabilities and information are fundamentally required to learn different kinds of tasks successfully and to understand the basic algorithmic principles involved in making computers learn from data and improving their performance using feedbacks. The goals of this theory are both to help in the design of better automated learning methods and to understand the fundamental issues in the learning process itself. Machine learning theory draws elements from both the theory of statistics and computation and includes tasks such as:

- Developing mathematical models that capture main aspects of machine learning, where one can analyze the inherent ease or difficulty of different types of learning problems.
- Proving guarantees for algorithms (under what conditions will they succeed, amount of data, and computational time required) and developing machine learning algorithms that probably meet some desired condition.
- Mathematically analyzing general issues, like: "when can one be confident about predictions made from limited data?," "how much power does active participation add over passive observation for learning?," and "what kinds of methods can learn even in the presence of large quantities of distracting information?"

Machine learning theory is both a fundamental theory with various basic and compelling foundational questions and a topic of practical importance that helps to advance the state of the art in software by providing mathematical frameworks for designing new machine learning algorithms. It is an exciting time to work in this domain, as connections to many other areas are being explored and discovered and as new machine learning applications are bringing out new questions to be modeled and studied. It can be safely said that the actual potential of machine learning still lies beyond the frontiers of our imagination.

References

1. Aggarwal, C. C. (2014). *Data classification: Algorithms and applications* (1st ed.). Chapman and Hall/CRC.
2. Alpaydin, E. (2014). *Introduction to machine learning* (Adaptive computation and machine learning series). MIT Press.
3. Anthony, M. H. G., & Biggs, N. (1997). *Computational learning theory* (vol. 30). Cambridge University Press.
4. Arel, I., Rose, D. C., & Karnowski, T. P. (2010). Deep machine learning-a new Frontier in artificial intelligence research [research frontier]. *IEEE Computational Intelligence Magazine, 5*(4), 13–18.
5. Baxter, J. (2000). A model of inductive Bias learning. *Journal of Artificial Intelligence Research (JAIR), 12*, 149–198.
6. Bishop, C. M. (2006). *Pattern recognition and machine learning* (1st ed.). Springer.
7. Boutell, M. R., Luo, J., Shen, X., & Brown, C. M. (2004). Learning multi-label scene classification. *Pattern Recognition, 37*(9), 1757–1771.
8. Breiman, L. (2001). Random forests. *Machine Learning, 45*(1), 5–32.
9. Bull, L. (2004). *Applications of learning classifier systems* (vol. 150). Springer.
10. Clare, A., & King, R. D. (2001). Knowledge discovery in multi-label phenotype data. In *European conference on principles of data mining and knowledge discovery* (pp. 42–53).
11. De Comité, F., Gilleron, R., & Tommasi, M. (2003). Learning multi-label alternating decision trees from texts and data. In *International workshop on machine learning and data mining in pattern recognition* (pp. 35–49).
12. Drugowitsch, J. (2008). *Design and analysis of learning classifier systems: A probabilistic approach* (Studies in computational intelligence). Springer.
13. Gall, J., Razavi, N., & Van Gool, L. (2010). On-line adaption of class-specific codebooks for instance tracking. In *Proceedings of the British Machine Vision Conference (BMVC)* (pp. 1–12).
14. Gonçalves, T., & Quaresma, P. (2003). A preliminary approach to the multilabel classification problem of Portuguese juridical documents. In *Portuguese conference on artificial intelligence* (pp. 435–444).
15. Guyon, I., Dror, G., & Lemaire, V. (2013). *Unsupervised and transfer learning: Challenges in machine learning* (vol. 7). Microtome Publishing.
16. Haykin, S. (2010). *Neural networks and learning machines*. PHI Learning.
17. Herrera, F., Charte, F., Rivera, A. J., & Del Jesus, M. J. (2016). *Multilabel classification: Problem analysis, metrics and techniques*. Springer.
18. Krizhevsky, A., Sutskever, I., & Hinton, G. E. (2012). Imagenet classification with deep convolutional neural networks. *Advances in Neural Information Processing Systems, 25*, 1097–1105.
19. Langley, P., & Morgan, M. B. (1996). *Elements of machine learning* (Machine learning series). Morgan Kaufmann.
20. Law, A., Chakraborty, K., & Ghosh, A. (2017). Functional link artificial neural network for multi-label classification. In *International conference on Mining Intelligence and Knowledge Exploration (MIKE)* (pp. 1–10).
21. Law, A., & Ghosh, A. (2019). Multi-label classification using a cascade of stacked autoencoder and extreme learning machines. *Neurocomputing, 358*, 222–234.
22. Lepetit, V., & Fua, P. (2006). Keypoint recognition using randomized trees. *IEEE Transactions on Pattern Analysis and Machine Intelligence, 28*(9), 1465–1479.
23. Li, G., Huang, Q., Qin, L., & Jiang, S. (2013). SSOCBT: A robust semi-supervised online CovBoost tracker that uses samples differently. *IEEE Transactions on Circuits and Systems for Video Technology, 23*(4), 695–709.
24. Lu, H., Plataniotis, K. N., & Venetsanopoulos, A. N. (2011). A survey of multilinear subspace learning for tensor data. *Pattern Recognition, 44*(7), 1540–1551.

25. Nigrin, A. (1993). *Neural networks for pattern recognition* (Bradford books). MIT Press.
26. Ozuysal, M., Calonder, M., Lepetit, V., & Fua, P. (2010). Fast keypoint recognition using random ferns. *IEEE Transactions on Pattern Analysis and Machine Intelligence, 32*(3), 448–461.
27. Pan, S. J., Tsang, I. W., Kwok, J. T., & Yang, Q. (2010). Domain adaptation via transfer component analysis. *IEEE Transactions on Neural Networks, 22*(2), 199–210.
28. Pan, S. J., & Yang, Q. (2010). A survey on transfer learning. *IEEE Transactions on Knowledge and Data Engineering 22*(10), 1345–1359.
29. Pise, N. N., & Kulkarni, P. (2008). A survey of semi-supervised learning methods. *International Conference on Computational Intelligence and Security, 2*, 30–34.
30. Read, J., Pfahringer, B., Holmes, G., & Frank, E. (2009). Classifier chains for multi-label classification. In *Joint European conference on machine learning and knowledge discovery in databases* (pp. 254–269).
31. Reimann, P., Spada, H., & Foundation, E. S. (1996). *Learning in humans and machines: Towards an interdisciplinary learning science* (Collection of Jamie and Michael Kassler). Emerald Group Publishing Limited.
32. Ren, T., Qiu, Z., Liu, Y., Yu, T., & Bei, J. (2015). Soft-assigned bag of features for object tracking. *Multimedia Systems, 21*(2), 189–205.
33. Samudrala, S. (2019). *Machine intelligence: Demystifying machine learning, neural networks and deep learning.* Notion Press.
34. Scudder, H. (1965). Probability of error of some adaptive pattern-recognition machines. *IEEE Transactions on Information Theory, 11*(3), 363–371.
35. Shotton, J., Johnson, M., & Cipolla, R. (2008). Semantic Texton forests for image categorization and segmentation. In *IEEE conference on computer vision and pattern recognition (CVPR)* (pp. 1–8).
36. Simon, P. (2013). *Too big to ignore: The business case for big data* (Wiley and SAS business series). Wiley.
37. Strang, G. (2009). *Introduction to linear algebra* (4th ed.). Wellesley-Cambridge Press.
38. Sun, X., Xu, J., Jiang, C., Feng, J., Chen, S.-S., & He, F. (2016). Extreme learning machine for multi-label classification. *Entropy, 18*(6), 225.
39. Theodoridis, S., & Koutroumbas, K. (2008). Pattern recognition. Edn. 4, rev., Academic Press. ISBN: 0080949126, 9780080949123.
40. Tsoumakas, G., & Vlahavas, I. (2007). Random k-labelsets: An ensemble method for multilabel classification. In *European conference on machine learning* (pp. 406–417).
41. Vincent, P., Larochelle, H., Lajoie, I., Bengio, Y., & Manzagol, P. A. (2010). Stacked denoising autoencoders: Learning useful representations in a deep network with a local denoising criterion. *The Journal of Machine Learning Research, 11*, 3371–3408.
42. Wilson, R. A., & Keil, F. C. (2001). *The MIT encyclopedia of the cognitive sciences* (A Bradford book). MIT Press.
43. Wu, F., Wang, Z., Zhang, Z., Yang, Y., Luo, J., Zhu, W., & Zhuang, Y. (2015). Weakly semi-supervised deep learning for multi-label image annotation. *IEEE Transactions on Big Data, 1*(3), 109–122.
44. Yang, F., Lu, H., & Chen, Y.-W. (2010). Bag of features tracking. In *IEEE 20th international conference on pattern recognition (ICPR)* (pp. 153–156).
45. Yang, F., Lu, H., & Yang, M. H. (2013). Learning structured visual dictionary for object tracking. *Image and Vision Computing, 31*(12), 992–999.
46. Yang, F., Lu, H. H., Zhang, W., & Yang, G. M. (2012). Visual tracking via bag of features. *IET Image Processing, 6*(2), 115–128.
47. Yang, L., & Jin, R. (2006). Distance metric learning: A comprehensive survey. *Michigan State Universiy, 2*, 2.
48. Zeisl, B., Leistner, C., Saffari, A., & Bischof, H. (2010). On-line semi-supervised multiple-instance boosting. In *IEEE conference on Computer Vision and Pattern Recognition (CVPR)* (pp. 1879–1879).

49. Zhang, M. L. (2009). ML-RBF: RBF neural networks for multi-label learning. *Neural Processing Letters, 29*(2), 61–74.
50. Zhang, M. L., & Zhou, Z. H. (2006). Multilabel neural networks with applications to functional genomics and text categorization. *IEEE Transactions on Knowledge and Data Engineering, 18*(10), 1338–1351.
51. Zhang, M. L., & Zhou, Z. H. (2007). ML-KNN: A lazy learning approach to multi-label learning. *Pattern Recognition, 40*(7), 2038–2048.
52. Zhong, Q., Qingqing, Z., & Tengfei, G. (2012). Moving object tracking based on codebook and particle filter. *Procedia Engineering, 29*, 174–178.

Part II
Preparation of Data

Chapter 3
Types of Data

Data are given values to occurrence, items, and observations. Their attributes can be described. Harvard psychologist S. S. Stevens coined some terms in the mid-1940s with respect to data representations. In order to better understand a continuum of psychophysically employed methods of gradation and the quantitative techniques, the words nominal, ordinal, interval, and ratio were used to keep with the "allowed" levels [5].

Empirical or statistical calculations presume that the parameters have different reference scales (numerical values). For instance, Mr. Kim is weighing 74 kg. The mean weight can be estimated if the weights of 10 such individuals are available. However, calculating an average skin tone would not be meaningful. Mean values of a categorical factor make little sense because the category levels are not intrinsically ordered. Moreover, we would also achieve an insensitive result if we had attempted to calculate the mean academic experience as defined in Sect. 3.2.3. Since the gap between the basic academic levels is very unequal, the importance of that term is unclear. Briefly, an average includes a numerical parameter. Sometimes, we have factors that are an overlap between ordinal and interval. For instance, the five-point scale of a decision being taken is used when the options that the members may give will be, "strongly accepting," "approving," "positive," "disagree," or "strictly disagreeing." When we cannot be certain that the distances between the five points are the same, we cannot say it is an interval factor, but we would say it is an ordinal factor.

In the following sections, we will describe different types of numerical data.

3.1 Numerical Data

Numerical details are tangible knowledge. Although some other types of data seem to have numerical values, the numerical data are always collected in numbers. The amount of people who went to the film festival (say 100,000) over a period would

© The Author(s), under exclusive license to Springer Nature Singapore Pte Ltd. 2025   35
A. Ghosh, *Data Science and Cases in Sustainability*, Mathematics for Sustainable
Developments, https://doi.org/10.1007/978-981-96-8362-8_3

be an illustration of numerical data. Numerical data may be further subdivided as discrete or continuous.

3.1.1 Discrete and Continuous

Discrete numerical data are data that have a limit and can also be measured and counted. It can only have some values by nature. The number of pens in a box (say 7), for example, is discrete. Only certain values can be used for discrete data. There can be an endless number of such values, but each of these values is different, and there is no blurry line between them. Numerical data are isolated, like the number of apples in a basket (say 20). It will never be a fractional number or a negative number, but can only be whole numbers. Discrete data may be categorical too, for instance, dark or light, man or woman, true or false.

On the other hand, continuous numerical data depict values with endless opportunities, such as Mr. Kim's weight. On a particular day, he may weigh 74 kg, and on some other day, when he has lost weight, he might be 71.3 kg. In order to make it quantifiable, ongoing data are placed within specific range. They are not limited to specified values but can take any value across a continuous range. An endless number of other values can exist between any two continuous data. In essence, continuous data are always numerical.

3.1.2 Interval

The form of interval denotes the degree of disparity. The interval variable values are arranged and are spaced equally. Interval variable parameter can take on either positive or negative values and can be calculated on a linear scale. The intervals are supposed to be the same over the entire range. It not only enables us to identify the calculated objects but also to calculate and compare the discrepancies. It can be said that the temperature of 40 °C is higher than 30 °C, and a rise from 20 °C to 40 °C is double that of the rise from 30 °C to 40 °C. Counts are interval scale measurements, such as counts of years of education, publications, or citations.

3.1.3 Ratio

A ratio scale is an interval measure where differences are defined rather than compared to a rational zero. Numbers can also be compared to each other in a ratio scale as multiples. Therefore, one individual can be twice the size of another. For example, a score of 60 has a ratio of 0.6 compared to a score of 100. Ratio data can be divided and compounded because the disparity between one and two is not only the same as between three and four, but also that four is half of two. The interval and ratio data are of a quantifiable or numeric nature.

3.2 Non-numeric Data

Non-numerical information is any kind of data expressed in word or any other language. It utilizes symbols, emojis, pictures, letters (e.g. a, b, c, d,...), etc. Such statistics can be contained in a text format only. To collect non-numerical data about a new batch of students, an individual can report that they are polite, civic, environmentally conscious and has an optimistic community spirit. There are different types of non-numerical data as described in the following sections.

3.2.1 Categorical

For those types of information that distinguish between different groups and typically mention a limited number of categories, categorical data are used. This includes the type of product, color (red green), age, gender (male, female), etc. All data are obtained in numbers, but what these numbers imply are often not known to us. Such numbers are described by categorical data. We would not understand what to do with them or what to say if the numbers 4, 10, and 12 had been issued to an applicant. Though, if the submission were made categorically, they would have a clearer understanding of the group as 4 apples, 10 oranges, and 12 bananas. This is because the numbers are categorized. In categorical data, a set of data is sorted or divided into different categories, according to the attributes of the data.

3.2.2 Nominal

The term "nominal" comes from the Latin word *nomen*, which means name. Nominal information is objects identified by a simple system of naming. Only a nominal scale specifies, although not described, that the items measured are common. An example of nominal data is the id-card number of a person (237 – 184 – 596), the SAARC countries, or the name of the avatar in a game. A number may be allocated to the nominal objects for example course six meaning course in EECS in MIT. It may seem ordinary, but it is used for reference purposes only. Nominal items are usually categorical and feature unordered categories of actual response. Nominal levels without any numerical significance are used for marking variables. "Nominal" scales could be simply called "labels." All these scales are mutually exclusive, and none of them have any numerical significance. To sum up, it is a discrete classification of data, where data are neither measured nor ordered but subjects are merely assigned to distinct categories: for instance, a record of students' course choices constitutes nominal data, which could be correlated with school results.

3.2.3   Ordinal

The order of values is important and significant for ordinal scales, but the differences between each are not really known. Ordinal levels are usually non-numerical indicators such as pleasure, joy, and pain. Let's use this as an example. Suppose we ask a question to a random person "how are you feeling today?" The possible responses can be:

(i)   Heartbroken
(ii)   Sad
(iii) Indifferent
(iv)   Happy
(v)   Cheerful

Objects on an ordinal level are defined by their placement on the scale ordering in some sort of order. The third, fourth, and fifth participants in a contest, for instance, are ordinal levels of the participants. In an ordinal dataset, the correct way to identify the central tendency is to measure the mode or the median. As the set is ordinal, the mean value can not be determined. This is because, unlike normal data, the variables in an ordinal set are arranged in an order. Ordinal variables, therefore, cannot be equally divided between categories. Had the categories been equally spaced, the variable would have been an interval variable.

3.3   Presentation of Data

The way the data are presented in tables, graphs, and diagrams is an important step of the data analysis and reporting process. While the findings can be shown in the report format, the details are typically more comprehensible if viewed as a table or chart. Statistics and graphs will relay the important data points and patterns to the user easily.

3.3.1   Tables

Tables are the standard mediums of displaying subjective and categorical statistics, but quantitative details can also be presented. They are an arrangement of information, in general where the data are basically rectangularly organized in columns or rows. An example of tabular data is given in Table 3.1.

The simplest table is the two-column frequency table (Table 3.2).

The first column indicates the grouping of the data (range of weights in kg), while the second column lists the frequencies or count for each group.

3.3.2   Scatter Plots

A scatter plot (diagram or graph) [2] generally gives an idea of how bivariate (two variables) information is distributed. It provides a nice overview of the relation between the two parameters and helps describe the coefficient of correlation or the regression parameters (Fig. 3.1). It actually gives some information about the structure of the data. Scatter plots show how much one variable is affected by another (covariance). Three-dimensional scatter diagrams are also frequently used (Fig. 3.2).

Table 3.1 Tabular data

Name	Position	Office	Salary (in dollars)
Jared Andrews	Sportsperson	San Francisco	715,000
Diane Day	Miner	New York	225,250
Guy Jordan	Chief Marketing Officer	New York	615,000
Omar Tucker	Author	San Francisco	432,500
Unity Butler	Marketing Designer	San Francisco	85,675
Timothy Mooney	Office Manager	London	136,200
Tiger Nixon	System Architect	Edinburgh	320,800
Thor Walton	Developer	New York	98,540
Tatyana Fitzpatrick	Regional Director	London	385,750
Suki Burks	Developer	London	114,500

Table 3.2 Two-column frequency table for the weight of students in a school

Weight in kg	Frequency
35–40	20
40–45	55
45–50	61
50–55	11
55–60	6
60–65	25
65–70	75
70–75	8
75–80	5
80–85	1

Fig. 3.1 Example of a 2D
scatter plot

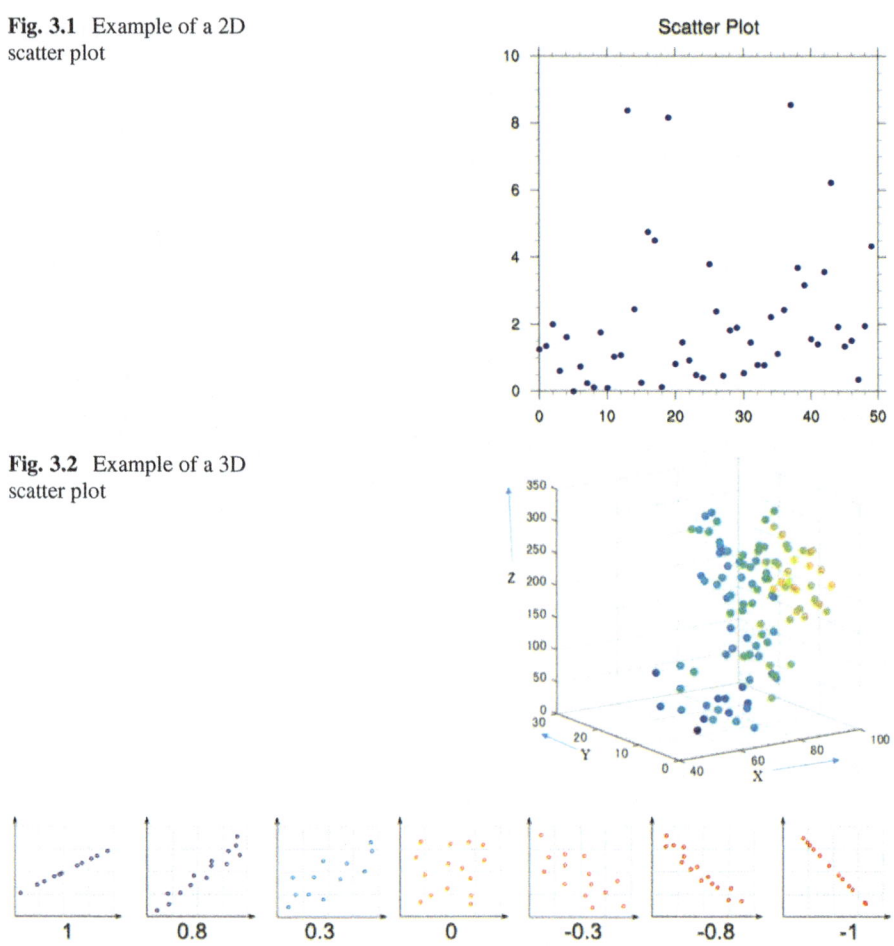

Fig. 3.2 Example of a 3D
scatter plot

Fig. 3.3 Example of scatter plots with different correlation coefficient values

Figure 3.3 shows 2D scatter plots of data with different values of correlation
coefficient ranging from 1 to −1. If the variables are highly correlated, then the
scatter plot looks like a line.

3.3.3   Graphs

Graphs are a helpful method to present quantitative data. A standard graph uses
two rectangular coordinates (called the x and y axes). The independent variable is
usually plotted on the horizontal x axis, while the response or outcome variable
is plotted on the vertical y axis (Fig. 3.4). The response variable is typically a

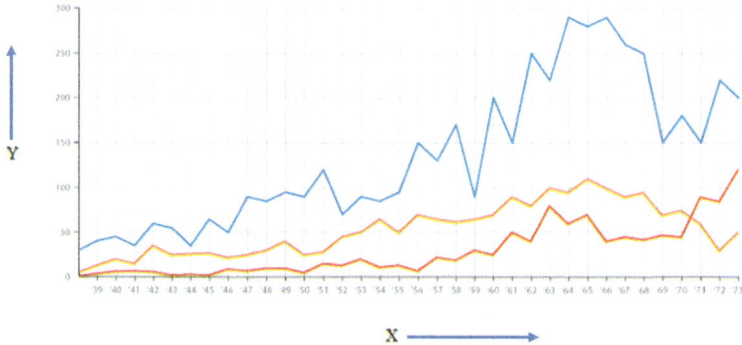

Fig. 3.4 Example of a graph

quantitative measure such as a frequency (count) or a percentage. This indicates how *y* varies with change of *x*.

Two extremely useful display charts for the quantitative (calculated or tallied) data are frequency histogram and relative frequency histogram. In particular, a frequency distribution of a quantitative variable is represented by them. Histograms or pie chart shows a single variable's distribution. A bar diagram shows the relationship between two or more factors, typically one empirical and one subjective or descriptive.

3.3.4 Histograms

A histogram is a means to describe information calculated at an interval (discreet or continuous). It is often used to display the main aspects of the distribution of information in a convenient way for exploratory data analysis. This splits the range of potential values into categories or groups for a set of data. A rectangle is built for each group with a baseline length equal to this group's value range and an area corresponding to the number of observations. The rectangles can therefore be drawn of a uniform length. It could be thought of as a representation of a frequency table also. The histogram in Fig. 3.5 depicts the weights (in kgs) of the members in a fitness center. For example, there are 35 members who weigh 71 kg, 1 member weighing 66 kg, and so on.

It is an estimate of the probability distribution of a continuous variable (quantitative variable) and was first introduced by Karl Pearson [3, 4].

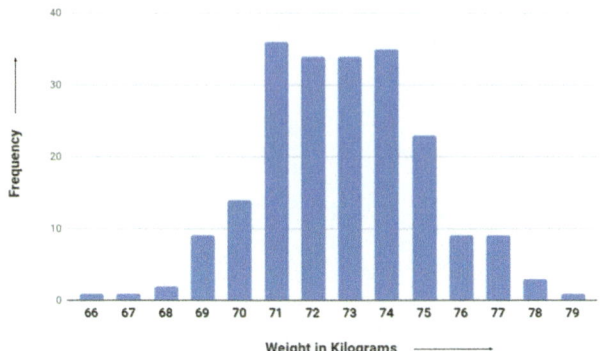

Fig. 3.5 Example of a histogram

Fig. 3.6 Example of a pie
chart

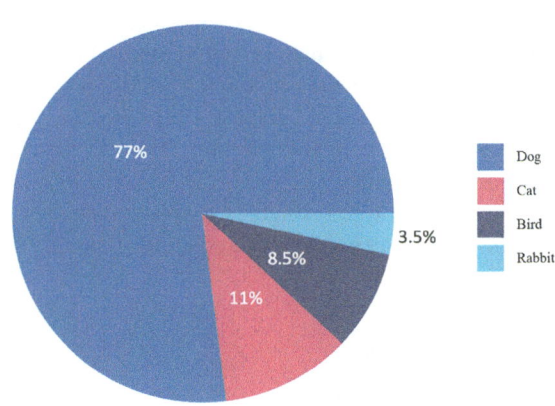

3.3.5   Pie Chart

A pie chart is a circular graphical object separated into slices to display the ratios.
In a pie chart, the longitude of the arc of each slice is proportional to the amount
specified (and therefore its central angle and area). It is referred to for its similarity
with a slit of a pie.

The whole pie constitutes the total amount equivalent to 100%, whereas each
slice denotes the percentage of the particular variable it represents. Let us consider
an example where we need to represent the types of pets owned in a locality through
a pie chart. Out of the total number of pets, 77% pet are dogs, 11% are cats, 8.5%
are birds, and 3.5% are rabbits. This is represented in a pie chart as in Fig. 3.6.

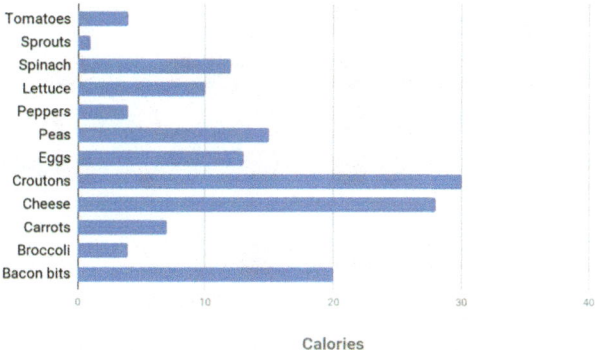

Fig. 3.7 Example of a bar chart

3.3.6   Bar Chart

A bar chart shows visually the values of various variables on a common scale using bars. The numerical values of the variables are shown by the same width bars in height or length. There's a distinct variety of bar charts. Usually, bar diagrams are scaled so that all information fits the diagram. Usually, frequency bars are arranged in chronological sequence (time). In Fig. 3.7, 12 different food items and their calories per serving are represented in the form of a bar chart. The foods are arranged randomly.

3.3.7   Structured, Semi-structured, and Unstructured Data

Data that are present in a fixed field within a record or file is called structured data (Fig. 3.8a). In this figure, data items with similar feature arrive in a regular manner, i.e., same color boxes appear in a single column. This includes data present in relational databases and spreadsheets. Handling structured data is easier, and there are some Database Management Systems (DBMSs) [1] for them.

Unstructured data (Fig. 3.8b) are all those things that cannot be so easily classified and fit into a neat box: photos and graphic images, videos, streaming data, PDF files, web pages, emails, blog entries, power point presentations, wikipedia, and some word processing documents. Dealing with unstructured data is the most difficult task.

An intersection between the former two is semi-structured data. These data are hierarchical, but neglect the rigid framework of the data model. With semi-structured data (Fig. 3.8c), tags or other types of markers are used to locate specific elements from the data, but the data do not have a rigid structure. In the figure, each column contains a combination of different color boxes, but the order of arrival is not strict. For example, the word processing program can now contain annotations

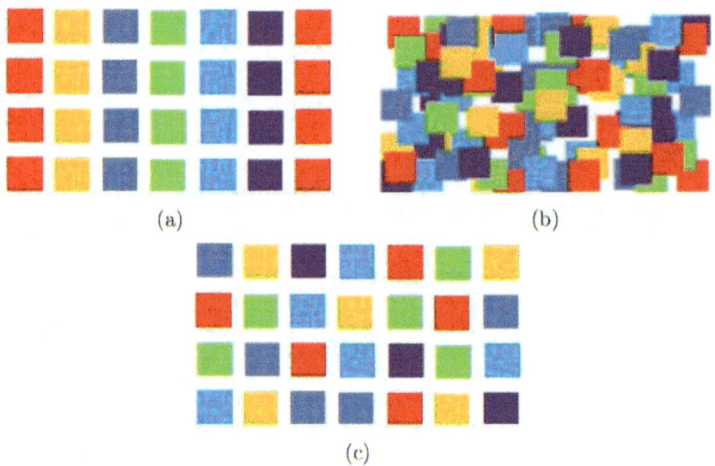

Fig. 3.8 Types of data: (**a**) structured data, (**b**) unstructured data, (**c**) semi-structured data

to indicate the name of the author and the generated date, with the rest of the text remaining essentially unstructured. Mails are linked to the unstructured information format as they have details like attachments, sender information, receiver information, location, period, and other specified fields. Keywords such as the author, time, location, and keywords can be used to mark images or other illustrations, enabling graphics to be ordered and found. Semi-structured information management is usually done by XML and other markup languages. Extraction of information from these details is a major topic of the day.

Everywhere unstructured data can be found. In fact, most individuals and organizations are forced to live around unstructured data. Similar to structured data, unstructured data can be either machine generated or human generated. Followings are few instances of sensor-generated unstructured data:

Satellite images: This includes weather data or the data captured by the Government using satellite surveillance imagery. Like Google Earth can be used to capture images.

Scientific data: Atmospheric data, Seismic imagery, and data from high-energy physics.

Photographs and video: Generated by surveillance, security, and traffic video.

Radar or sonar data: They come from vehicular, meteorological, and oceanographic seismic profiles.

References

1. Garcia-Molina, H. (2008). *Database systems: The complete book.* Pearson Education India.
2. Keim, D. A., Hao, M. C., Dayal, U., Janetzko, H., & Bak, P. (2010). Generalized scatter plots. *Information Visualization, 9*(4), 301–311.
3. Pearson, K. (1893). Contributions to the mathematical theory of evolution. *Proceedings of the Royal Society of London, 54*(326–330), 329–333.
4. Pearson, K. (1895). Contributions to the mathematical theory of evolution. II. Skew variation in homogeneous material. *Philosophical Transactions of the Royal Society of London Series A, 186*, 343–414.
5. Stevens, S. S. (1946). On the theory of scales of measurement. *Science, 103*, 2684.

Chapter 4
Pre-processing of Data

There are many aspects to the success rate of making a machine learn a job. Major factors deciding how effective the training are the quality and reliability of the example data. If an educator gives the students inaccurate or meaningless classes, the students show a poor result in the test. Likewise, if the training data contain a lot of meaningless and obsolete information or any interference and unreliability, then the learning (education) would be wrong, and the acquisition of skill would be far more complicated and inappropriate. We, therefore, need to prepare the data so that good examples can be presented as a training set to the machine. Pre-processing of information requires clearing of data, normalization, modification, extraction of features, selection of features, etc. The clean training set is the product of data preprocessing. The following sections address some of the pre-processing procedures of data.

4.1 Data Cleaning

The data being evaluated might be incomplete, chaotic, erroneous, or anomalous. Irrelevant results will always misrepresent the data and thus make the decision biased. The data cleaning procedure, also known as data cleansing or scrubbing, deals with data inspection and elimination of errors and inconsistencies to enhance data quality. During collection of data, several quality issues inevitably occur (e.g., due to incorrect registration, lack of information, or other incomplete evidence due to measurement error). If aggregation of multiple data sources is needed, from data centers, federate database systems, or global cloud-based data systems, the importance of cleaning increases significantly. This is because the sources also include duplicate information in multiple formats. Consolidation of different data representations and deletion of redundant information are required to provide access to reliable and consistent data. Cleaning of data filters the information using many

A. Ghosh, *Data Science and Cases in Sustainability*, Mathematics for Sustainable Developments, https://doi.org/10.1007/978-981-96-8362-8_4

advanced yet straightforward techniques. It reduces the chance of a system over-adapting (overfitting) to the data, when incorrect values are prevented or "cleaned" in simple terms. Several data cleaning techniques are discussed in the following subsections:

4.1.1 Missing Feature Values

There are many available datasets where some entries of the feature values might be missing (Table 4.1).

In that case, no learning with all these instances is relevant to the machines. These data points have many missing feature values. The absence of feature values during the data collection may have several reasons. Data may not be easily obtained (for instance, people refuse to provide their age and weight; the training set creator becomes careless about the features), features may not be relevant to all circumstances (e.g., kids do not earn an annual income), or the data are damaged. To handle this problem of missing values, the following methods are commonly used:

- **Ignoring Instances with Unknown Feature Values**: This is the most direct form. This simply ignores instances with at least one unspecified feature. The fourth, fifth, and seventh data instances in Table 4.1 would be ignored. So, we will be left with only 70% of the available (collected) data for training.
- **Use of Most Common Feature Value**: For all uncertain values of the feature, the most frequently occurring value of the feature is selected. In this case as the gender "M" occurs more than "F," the value at Gender of seventh instance would be given as "M," even though it is wrong.
- The value of the feature, which occurs commonly within the same class, could also be selected to be the value for all the unknown entries. Suppose, there are two classes—those who play cricket and those who do not. For those who play

Table 4.1 A database where some feature values are missing

Name	Weight	Gender	Play cricket?
Mr. Amit	68	M	Y
Mr. Anil	61	M	Y
Ms. Swati	58	F	N
Ms. Richa	–	F	Y
Mr. Steve	55	M	–
Ms. Rina	64	F	Y
Ms. Rashmi	57	–	Y
Mr. Kunal	58	M	Y
Mr. Richard	72	M	N
Ms. Mary	52	F	N

cricket, it is seen that most of them have gender "M." So, the seventh instance will be identified as "M" as it plays cricket.

- *Use of Arithmetic Mean*: When numerical values are present in features, the usual choice is to add a mean value of the feature determined from the available cases to fill the missing data values. A better solution instead of using the "general" feature mean is to use the feature mean for all samples belonging to the same class. For example, for the class that plays cricket, the mean value of weight is 64. Then, the value of weight of the fourth instance will be filled as 64.
- *Regression or Classification Methods*: Create a regression or classification model based on the available instances for a given feature, treating it as the outcome and using all other relevant features as predictors.
- *Hot Deck Imputation*: Identifying the case most similar to the present case using a missing value and substituting the feature value present in that instance. As it can be seen that the fourth and sixth instances are similar, the weight column of the fourth instance will be substituted by the value of weight at the sixth instance, i.e., 64.
- *Treating Missing Feature Values as Special Values*: Treating "unknown" itself as a new value for the features, which contain missing values. Rather than estimating an unknown feature value, it is possible to replace each missing value with an arbitrary unique value. Suppose the missing weight value in the fourth instance can be 40, a new value.

4.1.2 Duplicate Data

The information is combined or obtained from several sources to the data warehouse. The volume of information increases by combining data from multiple sources. We may have redundant data as well. For mining purposes, the data warehouse may have terabytes of data. Dataset can contain duplicate or almost duplicate data items. The same person has several email addresses, for instance. The integration of information from heterogeneous sources has become an important issue. The same file in a server can be saved by different users at several different places, or two or more non-identical files may still contain much the same information. Refer to Table 4.2.

The first and seventh entries are duplicate records. These entries have been added by a system user multiple times, for example, re-registering because a person has forgotten his details. These types of duplicates are often the most damaging for good analysis. So, removing these duplicates is essential so as to reduce the cost of storing dirty data. Replica-free repositories will improve the efficiency and save processing time.

Table 4.2 A database where
there are duplicate entries

First name	Last name	Age
Renee	Sharp	31
Bradley	Huchiston	30
Monica	Stuart	27
Bradley	Stuart	21
Steve	Jones	31
Elizabeth	McCullum	20
Renee	Sharp	31
Harris	Ford	27
Richard	Frank	41
Mary	Magdalene	36

4.1.3   Noisy Data

Noise is an inevitable and naturally occurring error. Removal of noisy items is an important objective for data cleansing because noise inhibits most forms of data analysis. Many popular techniques for data cleaning rely on noise reduction. Also, irrelevant or weakly relevant data items can also greatly obstruct the analysis of data. These data are meaningless for the model to be constructed. However, the definition of noise would depend on the domain of the problem. For example, if we are analyzing the academic records of students where the marks are given out of 100, an entry like 20 where generally most of the students have got above 60 appears to be a noisy entry. However, the same mark of 20 is not noisy if the subject was too hard to answer, and on an average, all the students had performed poor. Therefore, strategy of reducing it would be different for each domain. There can be different approaches for noise reduction.

4.1.3.1   Binning

Binning method eliminates noise by considering the neighborhood of a particular data point. The values are distributed into various buckets or bins according to some rules. Let us consider a simple example to describe the different strategies. Suppose we are analyzing the academic percentage of 10 students in a class. The data obtained after being sorted look somewhat like this:

$$30,\ 71,\ 73,\ 74,\ 79,\ 85,\ 87,\ 88,\ 92,\ 96.$$

The sorted data are first put into bins using **equal frequency bins**. If the data are divided into two bins, then

$$\textbf{Bin 1}:\ 30,\ 71,\ 73,\ 74,\ 79.$$

$$\textbf{Bin 2}:\ 85,\ 87,\ 88,\ 92,\ 96.$$

Smoothing by bin means: In this process, each value in a bin gets replaced by the mean of the data in that bin. The entry 30 appears to be a noisy input. It is replaced by the average of the first bin, i.e., by 65.

Bin 1 : 65, 65, 65, 65, 65.

Bin 2 : 90, 90, 90, 90, 90.

Smoothing by bin boundaries: The maximum and minimum values in a given bin get recognized as the boundaries. Each bin value then gets replaced by its closest boundary value.

Bin 1 : 30, 79, 79, 79, 79.

Bin 2 : 85, 85, 85, 96, 96.

The larger is the boundary width (difference between maximum and minimum boundary) of the bin, the greater is the effect of smoothing. Bins might be of equal width where the interval range of values in each bin is constant.

Now, since binning method only takes the neighbourhood values, they perform local smoothing of the data.

4.1.3.2 Regression

By fitting a function through scattered data points, for instance, regression [2], data can be smoothed. The goal of linear regression [6] is to find a best fit line for two variables in order to find relationship between the two. Consider an instance of linear single-dimensional regression. Assuming that we have data in (x, y) format and that data distribution is assumed to obey a straight line. We do assume, though, that the target values y are noisy. Now that data are supposed to be on a straight line $y = a + bx$, every noise value is far off from the line. In this scenario, the value of the noise expected by regression will smooth out. Figure 4.1 provides an physical understanding of linear regression.

Multiple linear regression [6] is an extension of linear regression, where two or more variables are used to fit into a multidimensional surface. This regression method builds a mathematical equation to fit the data and helps to smoothen out the noise.

There are also other methods like clustering and combined computer and human inspection methods that can smooth the data to eliminate the effect of noise [7].

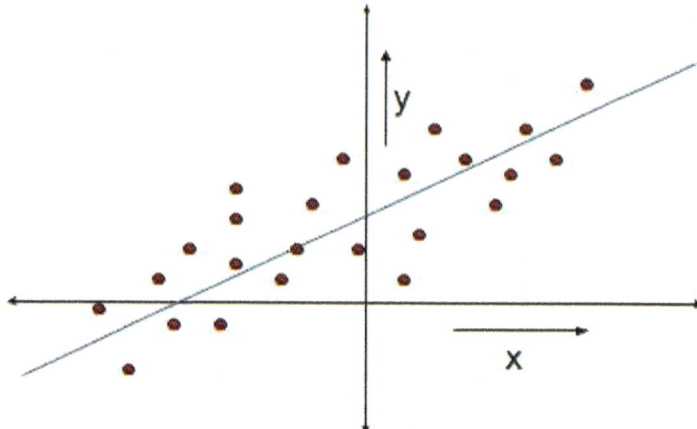

Fig. 4.1 Visualization of linear regression

4.2 Data Integration

The task of data analysis includes data aggregation, which integrates data in a coherent data storage process from multiple sources like data warehouses. The data may come from various sources including multiple databases, data cubes, flat files, etc. However, there are several problems related to the data integration process like redundancy, entity identification. Suppose a certain database stores the customer name under the heading "Customer" and another database stores the names under the heading "Cust_Name." This creates trouble in entity matching. This is called entity identification problem. Another problem that is encountered is redundancy in databases. In case of merging data from two sources, the data redundancy occurs, which is another form of duplicate data. Suppose data from two electronic stores are merged in a single database. It is pretty clear that "Television" and "TV" are the same product. But since one of the stores made the entry as "Television" and another store entered it as "TV," it may create a problem. An attribute that can be derived from another need not be stored. For example, in a given record, the date of birth and age of a person are two such attributes. If one of them is known, the other can be easily determined.

4.3 Data Normalization

Normalization means transforming the feature values by "scaling down." For a feature, the highest and the lowest values are often quite different, such as 0.01 and 1000. The difference in magnitudes becomes significantly smaller when normalization is done. Without normalization, some important features that inherently have the low value get eliminated while those that are not so important get priority only because of their high values. The two most common methods for normalizing are:

(i) **Min-max normalization:**

$$x' = \frac{(x - x_{min})}{(x_{max} - x_{min})};$$ (4.1)

(ii) **z-score normalization:**

$$x' = \frac{(x - \bar{x})}{\sigma};$$ (4.2)

where, σ is the standard deviation given by

$$\sigma = \frac{1}{N} \sum_i (x_i - \bar{x})^2$$ (4.3)

x_{max} is the maximum value of the feature, x_{min} is the minimum value of the feature, $\bar{x}$ is the expectation value of the feature, i.e., $\frac{\sum_{i=0}^{N} x_i}{N}$, is the the mean, x is the old feature value, and x' is the normalized one [1].

4.4 Instance Selection

The choice of instances is directly linked to data reduction, and in many KDD implementations, it becomes increasingly important when computational capacity or space efficiency is required. The collection of instances is used not only to control the noise but also to handle the ineffectiveness of training from very large datasets. The choice of the example is a challenge of optimization in which the mining performance is preserved while the sample size [4] is minimised. It decreases data volume and allows a training algorithm to function and to operate with large data efficiently. There are a number of sampling approaches for instances from a wide set of data. The well-known are:

- **Random sampling** that randomly selects a subset of instances from the whole data set.
- **Stratified sampling** that is used in the cases where the class labels are not uniformly distributed in the training sets.

Minority group instances are more commonly picked to even the distribution. The statistics group also supports sampling, which states that an effective computation-intensive approach, which operates on a subsample of the information that, in turn, have greater precision than a less sophisticated one that uses the entire database [3]. In fact, the rate of increase in accuracy deteriorates as the amount of data increases. The efficacy of the sampling depends on how the rate of increase in data reliability significantly declines.

In addition, the learning agents are expected to generalize themselves consistently over unknown instances of every class. The learner will do well on unknown instances for both classes in two-class (class 1 and class 2) problems. That is the ideal scenario, of course. The learning agents face imbalanced datasets in many implementations that can lead to the agent being skewed toward one class. It occurs when one of the classes is very small as opposed to the other classes of the training dataset. Recently, the machine learning group is quite interested in working with imbalanced datasets. Commonly used solutions of instance selection are:

(i) Duplicating training instances of the minority class. This is, in effect, resampling the instances and is known as over-sampling.
(ii) Removing training examples of the over represented class. This is known as downsizing. It is expected to reflect that the overall size of the dataset is smaller after balancing.

The grave concern that arises is whether every instance is equally useful for the learning process or not. So, we must have some **relevant instance selection** procedure for this [5]. For the purpose of motivation, let us study a simple example. Suppose, people were asked to vote which movies they like. It is seen that anyone who likes "Superman" always likes "Batman" with a high probability. Now, those few exception instances (people) that have not liked Superman but liked Batman or vice versa can mislead the prediction. If such instances are ignored, it might be expected that prediction time could be shortened and the accuracy might be improved. Selecting **task relevant data** is also of utmost importance. For example, while analyzing viewers of the "superhero" movie genre, we can easily eliminate users whose votes do not contain any movie from this genre. Generally, experts may delete obsolete, noisy, or redundant data by choosing specific instances. Using high-quality data will provide better outcomes and computing costs would also reduce. Moreover, if a dataset is too large, a making a machine learn all of it not be feasible. The instance selection, in this scenario, reduces data and allows a machine learning algorithm to work with massive volumes of datasets efficiently.

4.5 Data Augmentation

An abundance of good quality data is the secret to great designs of machine learning algorithms. Reliable data are not always accessible, and scarcity may prevent a model from improving its performance. An augmentation of the database with newer training samples is one way to solve this issue. Intelligent approaches to generative data augmentation will increase the training data by several folds. Indeed, the model becomes more stable (and avoids overfitting) and can be even simpler by modifying the training set. Several strategies to data augmentation are open. Adding noise and transformations to existing data are among the most basic solutions. Samples can be inserted in the sparse regions of the dataset with imputation and dimensionality reduction. More advanced approaches include simulation of data based on dynamic systems or evolutionary systems.

Some of the well-known augmentation techniques are:

- Rotation (90° clockwise)
- Horizontal flip
- Vertical flip
- Scaling (by a factor of 2)
- Adding Gaussian noise

Let us describe it with an example. Let the following 3×3 matrix be the original data.

$$\begin{bmatrix} 1 & 34 & 39 \\ 64 & 81 & 18 \\ 33 & 58 & 74 \end{bmatrix}$$

Thus the augmented data could be

- Rotated matrix (90° clockwise):

$$\begin{bmatrix} 33 & 64 & 1 \\ 58 & 81 & 34 \\ 74 & 18 & 39 \end{bmatrix}$$

- Horizontally flipped matrix:

$$\begin{bmatrix} 39 & 34 & 1 \\ 18 & 81 & 64 \\ 74 & 58 & 33 \end{bmatrix}$$

- Vertically flipped matrix:

$$\begin{bmatrix} 33 & 58 & 74 \\ 64 & 81 & 18 \\ 1 & 34 & 39 \end{bmatrix}$$

- Scaled matrix (by a factor of 2):

$$\begin{bmatrix} 2 & 68 & 78 \\ 128 & 162 & 36 \\ 66 & 116 & 148 \end{bmatrix}$$

- Matrix with Gaussian noise:

$$\begin{bmatrix} 1.03 & 33.61 & 37.86 \\ 63.30 & 83.01 & 18.36 \\ 35.40 & 57.67 & 74.98 \end{bmatrix}$$

- Normalized matrix (values scaled to [0, 1]):

$$\begin{bmatrix} 0.00 & 0.42 & 0.48 \\ 0.80 & 1.00 & 0.21 \\ 0.41 & 0.70 & 0.91 \end{bmatrix}$$

- Clipped matrix (values clipped to [10, 60]):

$$\begin{bmatrix} 10 & 34 & 39 \\ 60 & 60 & 18 \\ 33 & 58 & 60 \end{bmatrix}$$

Figure 4.2 shows the corresponding visualization of the augmented data matrices. Figure 4.2 is the visualization of the matrices.

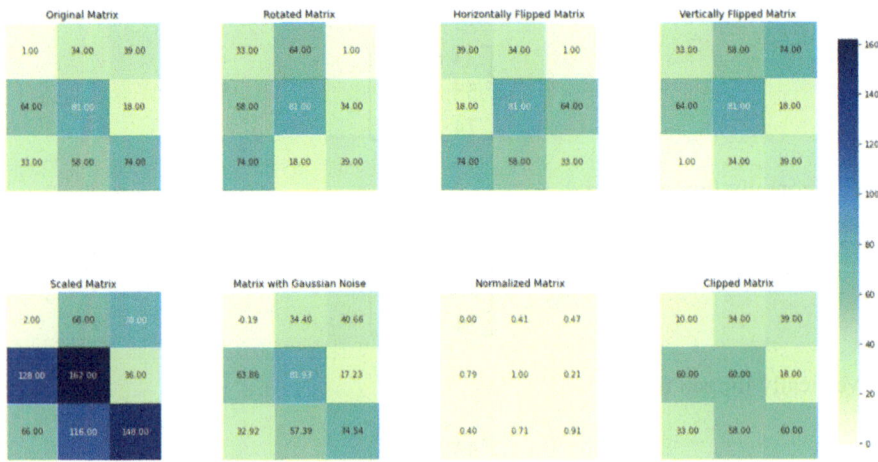

Fig. 4.2 Visualization of original and augmented matrices

References

1. Bernstein, P. A., & Goodman, N. (1980). What does Boyce-Codd normal form do? *Proceedings of the Sixth International Conference on Very Large Data Bases, 6*, 245–259.
2. Collins, M., Schapire, R. E., & Singer, Y. (2002). Logistic regression, AdaBoost and Bregman distances. *Machine Learning, 48*(1–3), 253–285.
3. Friedman, J. H. (1998). Data mining and statistics: What's the connection? *Computing Science and Statistics, 29*(1), 3–9.
4. Liu, H., & Motoda, H. (2002). On issues of instance selection. *Data Mining and Knowledge Discovery, 6*(2), 115–130.
5. Liu, H., & Motoda, H. (2013). *Instance selection and construction for data mining* (vol. 608). Springer.
6. Seber, G. A. F., & Lee, A. J. (2012). *Linear regression analysis* (vol. 936). Wiley.
7. Vaseghi, S. V. (2006). *Advanced digital signal processing and noise reduction*. Wiley.

Chapter 5
Dimensionality Reduction

5.1 Patterns and Features

Pattern is defined as a composite of attributes that are characteristic of an individual/object/entity. Depending upon the application, objects can be image, signal waveform, or any type of measurements. These objects have the generic name **patterns**. Patterns have several inherent **features** that are used to describe them. These features could be numerical, nominal, or ordinal values describing some characteristics of the object. Pattern **class** is a set of patterns having similar attribute or feature values. Suppose a set of m known object classes. Either through a description or by a set of examples for each category, they are known. For example, we have a summary of the nature of each of the men and women faces or a collection of samples of each of the two classes (men and women) for the gender classification problem. In the general case, for objects that cannot be placed into one of the known classes, a different reject category shall be provided. A **true class** is a class to which an object belongs. Each class must have a class label that distinguishes it from other classes. For example, we have 10 classes: "zero," "one," "two," "three," ..., "nine." Whenever a digit is to be identified, it will belong to one of these 10 classes.

Pattern recognition task is related to the act of taking in data and classifying it into a number of categories or classes. Some classes have a clear cut meaning, and in the simplest case, they are mutually exclusive, i.e., one pattern belonging to a class will not belong to another class. For example, in signature verification, the signature is either genuine or fraud (two classes). In some problems, it would be difficult to define the true class: for example, the classes of left-handed and right-handed people. Some people who can write with both their hands can belong to both the classes. Natural variability in data creates difficulty in defining sharp discrimination criterion between classes. Every class has its own model, that is, its own description, which is typically mathematical in nature, e.g., a probability density function (like Gaussian). So, a "hypothesis" is drawn about the class model, and accordingly, patterns are sorted into the category that suits them the best. In a

© The Author(s), under exclusive license to Springer Nature Singapore Pte Ltd. 2025
A. Ghosh, *Data Science and Cases in Sustainability*, Mathematics for Sustainable
Developments, https://doi.org/10.1007/978-981-96-8362-8_5

classification problem, a pattern is represented as a $\{\vec{x}, w\}$ pair of variables where $\vec{x}$ represents a set (feature vector) of observations or characteristics, and w is the term behind the observation (class label). The value of a pattern or a feature vector is associated to its ability to differentiate instances from different classes. Instances of the same class must have equivalent feature values, although samples belonging to different class should have differing feature values. If there are two classes "pen" and "pencil," objects belonging to the "pen" class have nib and ink as features, whereas objects in the "pencil" class should have lead as a feature.

Features are measurable quantities that can be used to define a pattern and distinguish between two pattern class. Features in real world are nothing but attributes of the objects. For example, suppose there are two kinds of pens—Fountain pens and Ball-point pens. We may describe these two classes of pens through their features as illustrated in Table 5.1.

We need to know which features can optimally distinguish these two classes of pens. As we can see, features x_3, x_4 and x_5 are common to both. So, they are irrelevant for consideration. x_1 and x_2 can be used for classification. Suppose we take a number of pens out of which some are fountain and some are ball-point. We need to classify these patterns (each pen represents a pattern) based on these two features. A pattern is represented as a point in l-dimensional feature space (l being number of features used—here $l = 2$). To represent graphically, we have a two-dimensional feature space with x_1 and x_2 as the two coordinate axes, and we plot the patterns on the two-dimensional *feature-space* (Fig. 5.1). Circles represent ball pens, and triangles represent fountain pens.

Table 5.1 Features of pens

Features	Feature symbol	Fountain pens	Ball pens
Nib width	x_1	3–6 mm	0.5–1 mm
Density of ink	x_2	1.11–1.3 g/cc	1.12 1.4 g/cc
Nib material	x_3	Steel	Steel
Body material	x_4	Plastic	Plastic
Presence of cap	x_5	Yes	Yes

Fig. 5.1 Representations of patterns in feature space

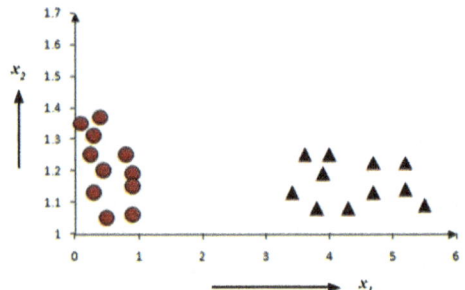

A pattern is thus described by a set of features (and its class label). The set of features, as for this case it is $\{x_1, x_2\}$, is often called the *feature vector*. For an l-dimensional feature space, a feature vector can be represented as $\{x_1, x_2, x_3, \ldots x_l\}^T$ or

$$\left\{ \begin{array}{c} x_1 \\ x_2 \\ x_3 \\ . \\ . \\ . \\ x_l \end{array} \right\} . \tag{5.1}$$

Feature Domain refers to the set of values a particular feature may have. In operations or while adding new objects to a dataset, the feature values might be checked for the defined domain. For example, the salary of workers may range between \$3000 and \$5000. This is the domain of the salary (feature) of workers. Now, if a new entry is made to the dataset that has a feature of \$6000, he must not be a worker. He may be an outlier from the class of workers. So, specification of feature domain is important in cases of outliers.

Features can be Quantitative (Numerical) or Qualitative (Categorical) [17]. The features that can have large number of possible discrete values are treated as quantitative features. Qualitative features on the other hand have limited values, gradations, or levels.

Depending on how features may classify the objects, they can be treated as good features or bad features. The standard of a feature vector is associated with its ability to identify instances from different classes (Fig. 5.2). Example patterns from one class should have similar feature values, while example patterns from different classes should have different feature values.

Features must be invariant [25] to:

- Translation
- Rotation
- Scale
- Noise
- Projection to other spaces

Features invariant to translation do not change if the object undergoes a translation from one point to another, e.g., the length of a rectangle (Fig. 5.3). Figures

Fig. 5.2 An example of good and bad features

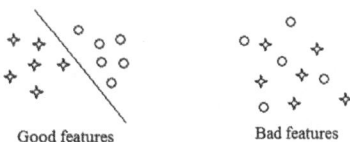

Good features   Bad features

Fig. 5.3 Length of rectangles does not change on translation

Fig. 5.4 Square rotated
through 90°

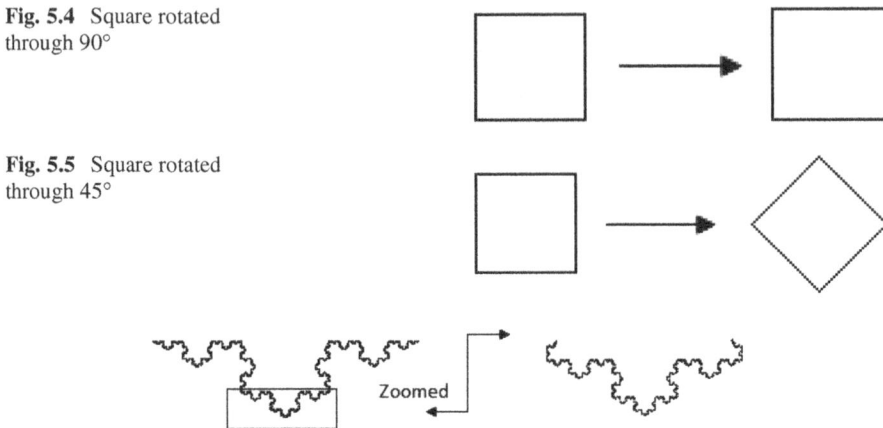

Fig. 5.5 Square rotated
through 45°

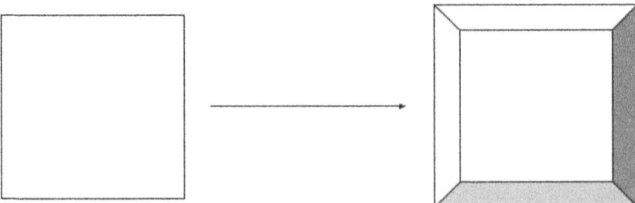

Fig. 5.6 Part of the fractal is zoomed in

invariant to rotation do not change if the object gets rotated by an angle, e.g., a
square rotated by 90° (Fig. 5.4). But if the square is rotated by 45°, then it is no
longer invariant (Fig. 5.5).

Features invariant to change in scale of measurements are called scale-invariant
features. As, for example, after zooming out one part of the fractal in Fig. 5.6, its
features remain the same.

Features invariant to noise do not alter their values in the presence of noise or
unwanted data. For example, a man will have the same fingerprint irrespective of
some cuts and bruises on his hand. Features invariant to projection remain the same
even if they are projected on some other space. For example, a shadow of man's
right palm will have five fingers as the man himself though the shadow is projected
somewhere on a plane (Fig. 5.7).

Fig. 5.7 Projected on higher-dimensional space

5.2 Peaking Phenomena—the Curse of Dimensionality

One common occurrence is that the overall error of a built classifier will initially reduce and subsequently increase while the feature numbers increase. Experimentally, if the number of training samples used while developing the classifier is less than the number of features, the additional features will actually reduce the quality of the classifier. This paradoxical behavior is referred to as "peaking phenomena." In fact, it has been seen that the number of training data points a classifier needs for its training is an exponential function of the number of features. Thus, dimensionality reduction becomes of utmost importance to reduce computational time.

Dimensionality reduction basically takes the data points to a different space. If on that space an extra dimension is added or removed, the points will be transformed to some other space where they might not be separable. In order to develop a classifier having good generalization capability, the number of training points N must be large enough with respect to the number of features d.

The curse of dimensionality is a grave issue in data analysis. More features may certainly give a detailed description of the data, but that may produce noise to degrade the performance. The curse of dimensionality refers to how certain learning algorithms may perform poorly in higher-dimensional data. The "curse of dimensionality" is the fact that when the input space has many dimensions, there is a big difference between the performance on the training set and the performance on unseen data (test set).

There is a 100 m long straight path, for example, and somebody has dropped a coin on it. If he travels along the path, it will take 2 minutes and would not be very difficult to find it. Now if it is a $100 \, m^2$ on each side and he lost the coin somewhere, it would then be quite hard to look for it. It would be like searching in an entire soccer field. It might take several hours or even a day. Now consider the situation when the coin is lost in a $100 \, m^3$. This would seem like looking for a coin in a 30-story building.

The generalizing ability of the model decreases in case of high-dimensional data. This causes an over fitting condition and does not perform well on unseen datasets.

When dimensionality increases, i.e., an instance has multiple feature values, data become increasingly sparse in the space that it occupies (Fig. 5.8). Definitions of density and distance between points, which is very important in clustering and outlier detection, become less important in that case. The purpose of dimensionality reduction [27] has thus gained significance. The question then is, how to reduce the dimension?

In Fig. 5.8, as we can see that the sample data points in two dimensions are represented in the form of $\{x_1, x_2\}$. The average distance between these points is more, which means there are less number of neighbors within the same radius. If the dimensionality of the data points is decreased somehow to 1, the decision-making process, which depends on finding the similarity within data points, becomes

Fig. 5.8 Data sparsification
in high dimensions

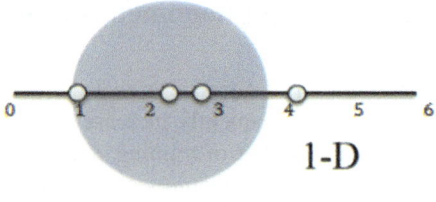

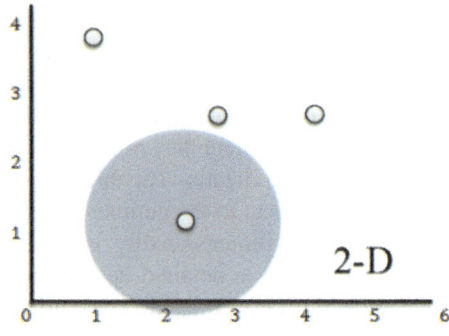

Fig. 5.9 Performance of a
classifier model may increase
upto a certain optimum
choice of dimensions and
then degrades as the
dimension increases (Peaking
phenomenon)

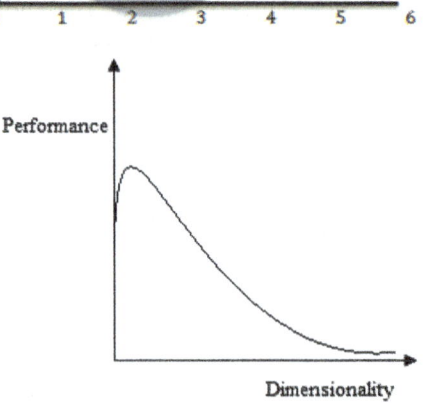

easier. In most of the cases, the information lost by discarding some features is
compensated by a more accurate mapping in lower-dimensional space. However,
we cannot decrease the number of features to a very low number also. In that
case, the decision-making system will not have enough information about the
data. So, the performance of the decision model reaches the maximum when the
number of dimensions is optimum, but then it decreases rapidly (Fig. 5.9), peaking
phenomenon.

Fig. 5.10 Feature selection

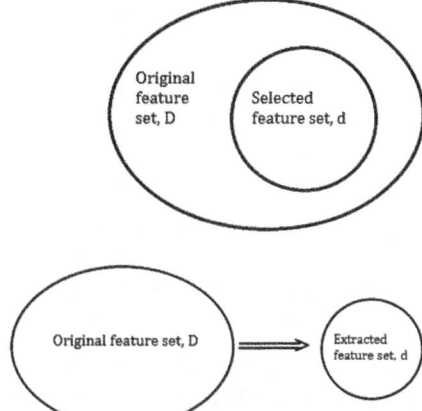

Fig. 5.11 Feature extraction

Dimensionality reduction methods can be broadly grouped into *feature extraction* and *feature selection*. To understand the difference between the two, let us take data points described by D-features given by

$$\mathbf{x} = \left\{ \begin{array}{c} x_1 \\ x_2 \\ x_3 \\ . \\ . \\ . \\ x_D \end{array} \right\} .$$

The feature selection method selects a subset of existing features without a transformation (Fig. 5.10).

$$\left\{ \begin{array}{c} x_1 \\ x_2 \\ x_3 \\ . \\ . \\ . \\ x_D \end{array} \right\} \implies \left\{ \begin{array}{c} x_1 \\ x_2 \\ \vdots \\ x_d \end{array} \right\} . \tag{5.2}$$

On the other hand, feature extraction techniques generate new features by transforming the existing feature space into a lower-dimensional one (Fig. 5.11).

$$
\begin{Bmatrix} x_1 \\ x_2 \\ x_3 \\ . \\ . \\ . \\ x_D \end{Bmatrix} \implies \begin{Bmatrix} y_1 \\ y_2 \\ \vdots \\ y_d \end{Bmatrix} = f \left(\begin{Bmatrix} x_1 \\ x_2 \\ x_3 \\ . \\ . \\ . \\ x_D \end{Bmatrix} \right) . \tag{5.3}
$$

Criterion for feature selection or feature extraction can be different based on different problem settings:

- **Unsupervised setting:** To minimize the information loss.
- **Supervised setting:** To maximize the class discrimination.

Although feature selection can be looked into as a special case of feature extraction, in practice, they are quite different and have their own sets of methodologies.

5.3 Separability Measures

The objective of a feature selection algorithm is to identify the relevant feature subset. The relative values assigned to different subsets reflect their greater or lesser relevance to the objective function. There are several approaches to calculate the goodness measure $J(.)$ of a feature subset d. For example, the feature subset that gives the minimum classification error is the ideal feature subset. Selection of feature f_1 & f_2 does not generate good separability measure between two classes (Fig. 5.12). Instead of that, selection of feature f_3 & f_4 provides good measure to separate those two classes (Fig. 5.13). However, it is generally difficult to estimate the error rate. Alternatively, class separability may reflect the discriminative power of the feature.

Let $J: d \subseteq D \to \mathbb{R}$ be an evaluation measure to be maximized, where d is a feature subset. Following are some of the ways this can be achieved.

5.3.1 Probability of Error

If the final goal is to develop a classifier that is able to label the instances correctly that are generated by the same probability distribution, we need to minimize the total Bayesian error probability of the classifier. Therefore, the total probability of error is a clear choice for J.

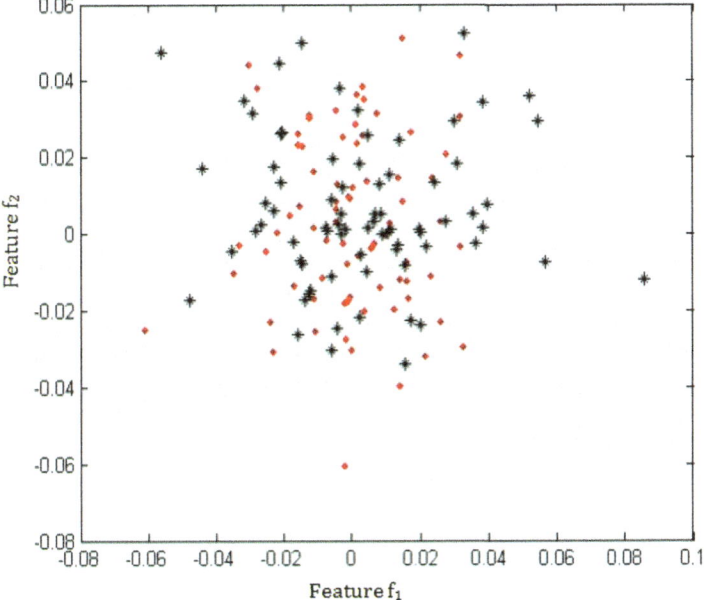

Fig. 5.12 Scatter plot of patterns with features f_1 & f_2 (overlapping classes)

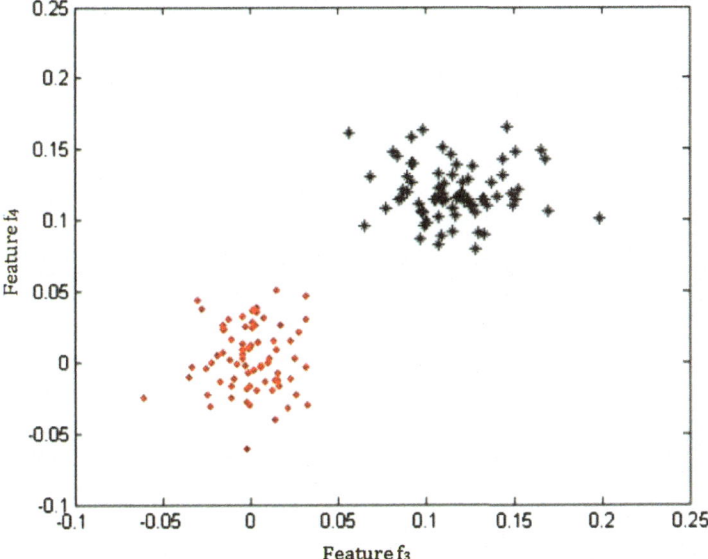

Fig. 5.13 Scatter plot of patterns with features f_3 & f_4 (more class separability)

Let $\mathbf{x} \in \mathbb{R}^n$ represents the unlabeled instances, and $\Omega = \{\omega_1, \omega_2, \omega_3, \ldots \omega_m\}$ be a set of classes. Then the probability of error is defined as [5]:

$$P_e = \int [1 - \max_i p(\omega_i|\mathbf{x})] p(\mathbf{x}) dx; \qquad (5.4)$$

where, $p(\mathbf{x}) = \sum_{i=1}^{m} p(\mathbf{x}|\omega_i) P(\omega_i)$ is the probability distribution of the instances, and $p(\omega_i|\mathbf{x})$ is the posterior probability of ω_i being the class of $\mathbf{x}$. The use of this probability by means of the construction of a classifier, using a sample (training) dataset, is the base for the *wrapper* methods [14] of feature selection.

Although the error probability is the ideal measure from the concept of theory, but from practical point of view, it is not easy to evaluate this error value. Therefore, a number of alternative methods have been suggested.

5.3.2   Pearson's Correlation Coefficient

Dependence measures or correlation measures qualify the ability to predict the value of one variable from that of the other. This signifies the amount of variation in one feature affected by the variation in another.

The correlation r between two variables x and y is given by

$$r = \frac{\sum_{i=1}^{n} (x_i - \bar{x})(y_i - \bar{y})}{\sqrt{\sum_{i=1}^{n} (x_i - \bar{x})^2} \sqrt{\sum_{i=1}^{n} (y_i - \bar{y})^2}}; \qquad (5.5)$$

where, $\bar{x} = \frac{1}{n} \sum_{i=1}^{n} x_i$ and $\bar{y} = \frac{1}{n} \sum_{i=1}^{n} y_i$. The coefficient value can range from -1 to $+1$. If the coefficient value is in the negative range, then the variables are negatively correlated. In other words, if the value of one variable increases, the value of the other decreases (Fig. 5.14).

If the value is in the positive range, i.e., positive correlation, then this means both values increase or decrease together. For negative correlation, the value will be in the negative range, i.e., when a value of one variable increases, the other variable's value decreases. No correlation occurs when there is no linear dependency between the variables. The correlation coefficient can also be used to bring out the correlation between a feature and the class label of the data we are using. If there is a strong correlation, then that feature is likely to be good at separating the data and will be useful for classification.

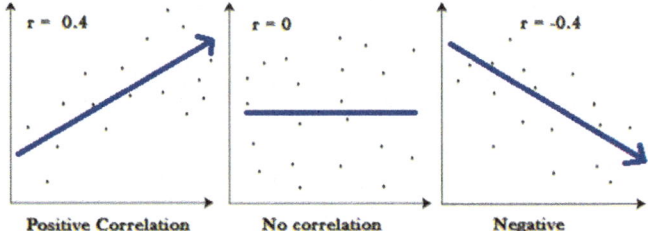

Fig. 5.14 An illustrative example of positive correlation, no correlation and negative correlation between two variables

5.3.3 Information Measures

Usually, these indicators specify the information gain (for example, entropy) of a feature. The difference between the prior uncertainty and estimated posterior uncertainty using feature f is known as the information gain of f. Feature f_1 is preferred to feature f_2 if the information gain from feature f_1 is higher than that from feature f_2. Entropy is generally used as information measure, which characterizes the inhomogeneity of an arbitrary collection of examples. The entropy measure is considered to be a measure of the system's unpredictability and is defined as:

$$H(Y) = -\sum_{y \in Y} p(y) \log(p(y)) \tag{5.6}$$

where $p(y)$ is the marginal probability density function of the random variable Y modeling the system. If the observed values of Y in the training dataset are distributed according to the values of a second random variable X, and the entropy of Y with respect to the partitions induced by X is less than the entropy of Y prior to partitioning, then there is a relationship between variables Y and X. The entropy of Y after observing X is then:

$$H(Y|X) = -\sum_{x \in X} p(x) \sum_{y \in Y} p(y|x) \log(p(y|x)) \tag{5.7}$$

where $p(y|x)$ is the conditional probability of y given x.

Another such measure called "mutual information" indicates the extent of shared information between two variables. In other words, it measures how a certain variable is if another variable is known. If there is a lot of shared information between a feature and a class of labeled data, then that particular feature is important for classification purposes. This works in a similar way to Pearson's correlation coefficient.

If the random variables X and Y are discrete, then mutual information is:

$$I(X;Y) = \sum_{y \in Y} \sum_{x \in X} p(x,y) \log(\frac{p(x,y)}{p(x)p(y)}). \tag{5.8}$$

If the random variables X and Y are continuous, then mutual information is:

$$I(X;Y) = \int_y \int_x p(x,y) \log(\frac{p(x,y)}{p(x)p(y)}) dx dy; \tag{5.9}$$

where $p(x,y)$ is joint probability function of X and Y.

5.3.4 Divergence

These measures indicate a probabilistic distance or divergence among the class-conditional probability densities $p(\mathbf{x}|\omega_i)$ using the given formula:

$$J = \int f[p(\mathbf{x}|\omega_i), p(\mathbf{x}|\omega_j)] d\mathbf{x}; \tag{5.10}$$

where $f(\cdot)$ represents a function.

This measure is valid if J satisfies the following conditions:

- $J \geq 0$.
- $J = 0$ only when the $p(\mathbf{x}|\omega_i)$ are equal $\forall \omega_i$.
- J is maximum when the classes are non-overlapping.

If the features used in a solution $d \subset D$ are good, the divergence among the conditional probabilities will be high. Poor features will result in very similar probabilities and lower divergence. Some classical choices of this measure are:

5.3.4.1 Chernoff Bound [4]

$$J_{\text{che}} = -ln \int [\{p(\mathbf{x}|\omega_i)\}^s \{p(\mathbf{x}|\omega_j)\}^{1-s}] d\mathbf{x}, s \in [0,1]. \tag{5.11}$$

5.3.4.2 Bhattacharya Distance [2]

$$J_{\text{bha}} = -ln \int [\sqrt{p(\mathbf{x}|\omega_i) \cdot p(\mathbf{x}|\omega_j)}] d\mathbf{x}. \tag{5.12}$$

5.3.4.3 Kullback–Liebler Divergence [15, 16]

$$J_{KL} = \int (p(\mathbf{x}|\omega_i) - p(\mathbf{x}|\omega_j))(ln\{p(\mathbf{x}|\omega_i)\} - ln\{p(\mathbf{x}|\omega_j)\})d\mathbf{x}. \qquad (5.13)$$

5.3.4.4 Kolmogorov Distance

$$J_{\text{kol}} = \frac{1}{2}E\{|P(\omega_i|\mathbf{x}) - P(\omega_j|\mathbf{x})|\}. \qquad (5.14)$$

5.3.4.5 Matusita Distance [21]

$$J_{\text{mat}} = \sqrt{[\int (\sqrt{p(\mathbf{x}|\omega_i)} - \sqrt{p(\mathbf{x}|\omega_j)})^2 d\mathbf{x}]}. \qquad (5.15)$$

5.3.4.6 Jeffries–Matusita Distance

$$J_{JM} = [\int (\sqrt{p(\mathbf{x}|\omega_i)} - \sqrt{p(\mathbf{x}|\omega_j)})^2 d\mathbf{x}]. \qquad (5.16)$$

5.3.4.7 Patrick–Fisher Distance [1]

$$J_{PF} = \sqrt{[\int (p(\mathbf{x}|\omega_i) - p(\mathbf{x}|\omega_j))^2 d\mathbf{x}]}. \qquad (5.17)$$

These measures follow the assumption that instances of different classes are distant in the input space. These measures do not need the modeling of any density function and their relation to the probability of error can be very loose. These distance metrics are explained in detail in Sect. 6.3.

5.3.5 Fisher's Separability Measure

The main drawback of the aforementioned class separability criterion involving the probability density functions (*pdfs*) of the data with respect to different classes is

Fig. 5.15 The two features x_i and x_j can produce a good separability of the three classes

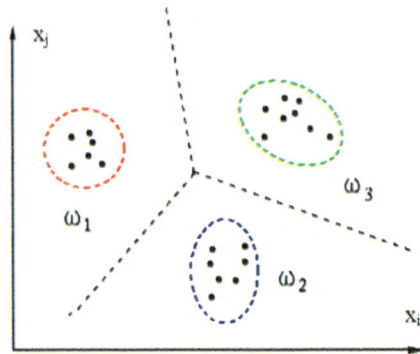

that they become very difficult to calculate for *pdfs* other than Gaussian types. Such criteria involving the *pdfs* of feature vectors of data may contain much information about the separability of different classes; but it is always desirable to have some easily computable criteria. Here lies the necessity of Fisher's Discriminant Analysis, which is based on the scatter of the feature vector in the feature space. In feature selection, we need to evaluate how separable a set of classes are in a d-dimensional feature space by some criterion (Fig. 5.15).

Suppose, we have C number of classes. So, the total number of samples

$$K = \sum_{i=1}^{C} K_i \tag{5.18}$$

where K_i is the number of samples in class ω_i. Overall mean vector:

$$\mathbf{m} = \frac{1}{K} \sum_{\mathbf{x}} \mathbf{x} = \sum_{i=1}^{C} \frac{K_i}{K} \mathbf{m}_i = \sum_{i=1}^{C} P_i \mathbf{m}_i \tag{5.19}$$

where $P_i = K_i/K$ is the a priori probability of class ω_i and m_i is mean vector of each class ω_i.

Scatter matrix of class ω_i (same as the covariance matrix of the class):

$$\mathbf{S}_i = \frac{1}{K_i} \sum_{\mathbf{x}_i \in \omega_i} (\mathbf{x}_i - \mathbf{m}_i)(\mathbf{x}_i - \mathbf{m}_i)^T, \qquad (i = 1, 2, \cdots, C). \tag{5.20}$$

5.3.5.1 Within-Class Scatter Matrix

$$\mathbf{S}_W = \sum_{i=1}^{C} P_i \mathbf{S}_i = \frac{1}{K} \sum_{i=1}^{C} \sum_{\mathbf{x}_i \in \omega_i} (\mathbf{x}_i - \mathbf{m}_i)(\mathbf{x}_i - \mathbf{m}_i)^T. \tag{5.21}$$

5.3.5.2 Between-Class Scatter Matrix

$$S_B = \sum_{i=1}^{C} P_i(\mathbf{m}_i - \mathbf{m})(\mathbf{m}_i - \mathbf{m})^T. \tag{5.22}$$

5.3.5.3 Total Scatter Matrix

$$S_T = \frac{1}{K} \sum_{\mathbf{x}} (\mathbf{x} - \mathbf{m})(\mathbf{x} - \mathbf{m})^T. \tag{5.23}$$

It can be shown that $S_T = S_W + S_B$, i.e., the total scatteredness is the sum of within-class scatteredness and between-class scatteredness.

We can use the traces of these scatter matrices as some scalar criterion to measure the separability between the classes:

$$J_B = tr S_B; \tag{5.24}$$

and

$$J_W = tr S_W. \tag{5.25}$$

One can show that J_B is the weighted average of the Euclidean distances between $\mathbf{m}_i$ and $\mathbf{m}$ for all C classes, and J_W is the average variance over all C classes.

It is obviously desirable to maximize J_B and at the same time also minimize J_W, so that the classes are maximally separated for the classification to be most effectively carried out.

5.4 Feature Selection

Feature selection means that every possible irrelevant and redundant function is identified and removed to help reduce the dimensionality curse. It achieves the advantages of reduced dimensions (such as a significant decrease in time and memory and in many situations eliminates irrelevant characteristics or noise reduction) that data mining algorithms require, which allows data to be interpreted more easily. It eliminates redundant features that contain an insignificant amount of extra information than several other attributes present. For example, a consumer purchase price and the sales tax charged are correlated. So, it is sufficient to keep any one of the two to get the full insight. Feature selection also removes irrelevant features, which has no useful information related to the data mining task at hand.

For instance, students' ID is often not relevant to the task of predicting students' performance.

Feature selection is a common term used in the analysis of data to characterize the tools and techniques used to minimize processing and analyze input to a manageable size. For a successful analysis of the data, the potential for the selection of the features is vital since features can be very costly to obtain and give valuable information to form the classifier. Moreover, having plenty of non-informative features will overburden the classifier and result in high misclassification errors. The selection of features helps to solve the two issues when we have too much of insignificant data, we have too little of significant data. Feature selection techniques can be categorized in two ways:

- **Feature ranking**: Ranking features based on some criterion and those above a certain threshold (criterion value) gets selected.
- **Subset selection**: A subset of features are selected, which has good predictive capability.

There are many ways to implement feature selection algorithms depending on the type of data and the algorithm chosen for analysis. Methods of this kind include sequential selection techniques like forward and backward. A traditional feature selection method comprises four steps named as:

- subset generation
- subset evaluation
- stopping criterion
- validation

Feature subset generation (Fig. 5.16) is a search technique that generates candidate feature subsets for evaluation based on a certain search strategy. Each candidate subset gets evaluated and compared with the previous best one according to certain objective measures. If the new subset appears to be better, it replaces the best one. This process gets repeated until a given stopping criterion is satisfied. The main aim of feature selection is therefore to reduce the number of features while maintaining high performance by discarding those features that are least significant.

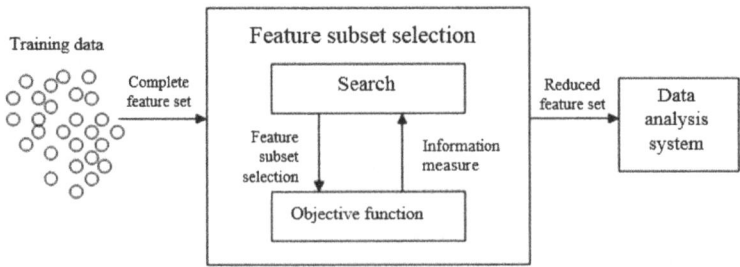

Fig. 5.16 A diagrammatic representation of the feature selection procedure

Search Strategies: If we want to create a feature subset of size d from a complete feature set of size D, we would have to do an exhaustive evaluation of feature subsets involving $^D C_d$ combinations for a fixed value of D and 2^d combinations if D needs to be optimized as well. Even for moderate values of D and d, the number of combinations is too big. So, a search strategy is needed to direct the feature subset selection process to get a reasonably good subset without exploring all possible combinations of features.

Objective Function: The objective function evaluates candidate subsets and returns a measure of their importance in the model construction. A feedback is used by the search strategy to select new candidate subsets. Objective functions for feature selection can be categorized into two broad categories, namely *Filters* and *Wrappers*.

5.4.1 Filter Methods

Filter methods [18] apply some ranking over features. The ranking denotes how useful each feature individually is for learning. The features are ranked according to some univariate metric. The objective function evaluates feature subsets by their information content, typically interclass distance, statistical dependence, or information theoretic measures. Some examples are Fisher score [7, 20], information gain [23], etc. Suppose, there are some features f_1, f_2, f_3 and f_4, and we will have to maximize a certain ranking criterion function $J(X)$ such that the features obey

$$J(f_1) \geq J(f_2) \geq J(f_3) \geq J(f_4).$$

The features are then ranked in the order f_1, f_2, f_3 and f_4.

In filter approaches to the subset feature selection, biases of induction algorithms are not taken into account. They select functional subsets independent of the induction algorithms. In some cases, algorithm-specific measures can be identified and calculated effectively.

Once this ranking is computed, the first d features are selected to make the feature subset. These are easily scalable to very high-dimensional datasets. It is also computationally fast and simple and independent of classification algorithm. There are, however, certain limitations as they ignore the interaction with the classifier and also interaction between features. Each feature is considered independently. This may lead to worse classification performance when compared to other types of feature selection techniques [10]. There may be a case where

$$J(f_2, f_3) \geq J(f_1) \geq J(f_2) \geq J(f_4) \geq J(f_3)$$

and

$$J(f_2, f_3) \geq J(f_1, f_2).$$

In that case, the first two features may not be the best, if used together. Thus, different subsets are analyzed for ranking measures. Several searching techniques make it easy for the method to work efficiently.

5.4.2 Wrapper Methods

In the case of the wrapper method [13], choice of the subset function is performed by means of the induction algorithm as a black box. This means that the information about the classification model used is needed. With the induction algorithm itself, the selection algorithm performs a search for a best subset according to the evaluation function. Using accuracy estimation techniques, the accuracy of the induced classifier is estimated.

These methods assess subsets of variables according to their usefulness to a given predictor. The objective function is a pattern classifier, which evaluates feature subsets by their predictive accuracy. The method conducts a search for a good subset using the learning itself as a part of the evaluation function. Wrapper methods [19] are called so because they wrap up a classifier in feature selection algorithm. The problem with this approach is that feature space is vast, and finding out the accuracy for every possible combination is time-consuming.

Wrappers generally achieve better recognition rates than filters since they are tuned to the specific interactions between classifier and the dataset. They also avoid any chance of overfitting by using cross-validation measures of predictive accuracy. However, as the classifier needs to be trained for each subset, the method can become computationally difficult. Also, the solutions lack generality since the optimal feature subset is specific to the classifier under consideration.

5.5 Search Strategies

Exhaustive search is computationally prohibitive. So, heuristics have been used to avoid exhaustive search. There is a large number of search strategies that can be categorized into three groups:

- **Sequential Algorithms**: These algorithms sequentially add or eliminate features but have a tendency to be trapped in a local minimum. Some of the examples of this series are Sequential Forward Selection and Sequential Backward Selection
- **Exponential Algorithms**: These algorithms evaluate a number of subsets that grow exponentially with the dimensionality of the search space. The most

representative algorithms under this class are Exhaustive Search, Branch and Bound search, etc.

- **Randomized Algorithms**: There is a randomness in the search procedure of such algorithms so as to avoid local minima. Some examples include Random Generation and Genetic Algorithms

We will discuss only some algorithms in this context.

5.5.1 Sequential Forward Selection

Sequential forward selection (SFS) algorithm [12] is the simplest greedy search method. It starts with an empty set and then repeatedly adds the most significant feature with respect to the set that has already been selected. The most significant feature is the one that has the highest objective function $J(F_k \cup f)$, where f is the feature that is to be added, and $F_k = Y$ is the set of features that has already been selected. The algorithm SFS can be represented as follows:

- Start with the empty set $F_0 = \phi$. The SFS search algorithm takes the whole feature set as input, starts with an empty set, and goes on adding the features one by one.
- Select the next best feature satisfying the condition

$$\arg\max_f [J(F_k \cup f)]. \tag{5.26}$$

with the condition that f is not a subset of F_k. To put it simply, we go through the feature space and look for the feature f, which maximizes our criterion if we add it to the feature subset (where J is the criterion function).
- Update $F_{k+1} = F_k \cup f$.
- Go to Step 2 and run the loop d times to get a subset of size d. We add features to the new feature subset F_k until we reach the number of specified features for our final subset. For example, if our desired number of features is 5 and we start with the "null set," we would add features to the subset until it contains five features.

The sequential forward search has a bottom-up approach of searching. To illustrate, let us consider the objective function as

$$J(X) = -2f_1f_2 + 3f_1 + 5f_2 - 2f_1f_2f_3 + 7f_3 + 4f_4 - 2f_1f_2f_3f_4.$$

Fig. 5.17 Stepwise
representation of a sequential
forward selection search
example

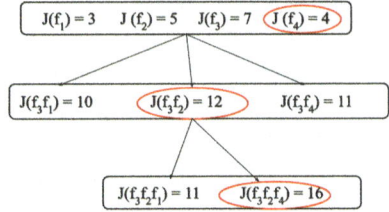

Here, for any selected k^{th} feature, the indicator variable is $f_k = 1$ and $f_{j \neq k} = 0$. The sequential search will be done as shown in Fig. 5.17. Let $d = 3$. The sequential forward search algorithm may generate weaker subsets because the importance of the variables is not assessed in the presence of other variables not included yet. Moreover, the features that are no longer important in the context to other newly added features cannot be removed once they are included.

5.5.2   *Sequential Backward Elimination*

The sequential backward selection (SBS) algorithm is the opposite to the sequential forward selection. The only difference is that we start with the complete set of features instead of the "null set" and go on removing features sequentially until we reach the number of desired features f^d. The biggest downside of this algorithm is that the features once removed can never be added back.

The algorithm can be represented as follows:

- Start with a subset of all features $Y_0 = X = \{f_1, f_2, \ldots, f_n\}$. The SBS search algorithm takes the whole feature set as input and goes on eliminating the features one by one. If our feature space consists of 10 dimensions ($D = 10$), it will start with all the 10 features.
- Select the less informative feature that does not improve prediction accuracy for the target variable, i.e., satisfying the condition,

$$\arg\max_{f}[J(Y_k - f)] \tag{5.27}$$

with the condition that f belongs to a subset of Y_k. Removing a feature may actually increase the objective function $J(Y_k - f) > J(Y_k)$; such functions are said to be non-monotonic.
- Update $Y_{k+1} = Y_k - f$.

(continued)

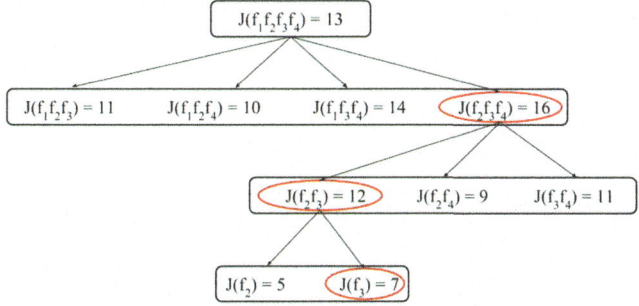

Fig. 5.18 Stepwise representation of a sequential backward search example

---

- Go to step 2 and run the loop till a subset of size d is obtained. We remove features from the existing feature subset X_k until we reach the number of specified features for our final subset. For example, if our desired number of features is 5 and we start with the subset of size 10, we would remove features from the subset until it contains only five features.

To illustrate, let us consider the following objective function

$$J(X) = -2f_1f_2 + 3f_1 + 5f_2 - 2f_1f_2f_3 + 7f_3 + 4f_4 - 2f_1f_2f_3f_4.$$

Here, for any selected k^{th} feature, the indicator variable is $f_k = 1$ and $f_{j \neq k} = 0$. The sequential elimination will be done as shown in Fig. 5.18

Sequential backward elimination works the best when the optimal feature subset becomes large, because most of its time is spent visiting large subsets. The main constraint of it is that it is unable to reevaluate the importance of a feature after it has been discarded.

5.5.3 *Generalized l-r Selection*

The plus-l minus-r selection [9, 18] method is a generalization of the SFS and SBS. The nesting problem encountered in the aforementioned two algorithms, where inclusion or removal of a feature cannot be undone once the operation is finished, is dealt here. This algorithm was introduced by Stearns (1976). If l > r, this method consists of applying the SFS l-times followed by r-steps of SBS. On the other hand, if l < r, this method consists of applying the SBS l-times followed by r-steps of SFS. The cycle is repeated until the required number of features is reached. It is a forward method if l > r; and is a backward method if l < r, and the algorithm can be stated as:

- If l>r, then begin with an empty set $Y_0 = \phi$, else begin with the full set $Y_0 = X = \{f_1, f_2, \ldots, f_n\}$ and go to Step 3.
- Repeat l-times finding

$$f^{\dagger} = \arg\max_{f}[J(Y_k \cup f)],\tag{5.28}$$

where $f^{\dagger} \in X - Y_k$
- Update $Y_{k+1} = Y_k \cup f^{\dagger}$
- Repeat r-times checking

$$f = \arg\max_{f}[J(Y_k - f)],\tag{5.29}$$

where $f \in Y_k$
- Update $Y_{k+1} = Y_k - f$.
- Go to Step 2 and run the loop till a subset of size d is obtained.

Plus-l minus-r selection attempts to stay away from the nesting problem by performing l forward selection steps followed by r backward selections and looping until the desired number of features are found. It attempts to compensate for weaknesses in SFS and SBS by backtracking. The drawback of this approach is that the optimal choices for l and r are difficult.

5.5.4   Branch and Bound Algorithm

The branch and bound algorithm [8] takes help of a tree to select a set of features. We assume here that the criterion function J is known, and without loss of generality, we need to maximize it. If we want to get the optimal feature subset of size d from the original feature set of size D, we need to run an exhaustive search over $^{D}C_d$ subsets. In case of $D = 100$ and $d = 10$, we would need to evaluate about 2^{12} possible subsets. Without doing the exhaustive search, we may find the optimal subset if the criterion function J satisfies the following property.

Let

$$S = \{f_1, f_2, \ldots, f_D\}\tag{5.30}$$

and

$$A_1 \subseteq A_2 \subseteq S; \tag{5.31}$$

then we will have

$$J(A_1) \leq J(A_2) \forall A_1, A_2. \tag{5.32}$$

In other words, it is a monotonically non-decreasing function on the number of features. It means more features have more information(at least equal). This property, however, may not always hold good for all criterion functions. In cases where this property holds, the branch and bound algorithm select the optimal feature subset.

In this algorithm, a tree (with each node denoting a subset of features) is constructed from right to left and is asymmetric in nature. The basic thought behind it is to reduce the search space. So, effectively, the search for optimal subset of features moves from the right side (narrower portion) of the tree to the left side (wider portion) of the tree. If a leaf node at any point of time gives a better (comparison is done using the function J(.)) subset of features than that of the remaining unexpanded nodes to its left, then the rest of the tree need not be constructed or selected.

To illustrate this algorithm, let us consider that there are only six features f_1, f_2, f_3, f_4, f_5, and f_6, and we would like to select only two out of them, i.e., $D = 6$ and $d = 2$. So, there are $^6C_2 = 15$ possible subsets. We make a tree (Fig. 5.19) where the number of nodes in the last level (leaf nodes) is the number of possible subsets. The number of features to be preserved is the number of branches that have not been expanded yet. Each node denotes a subset of the original feature set. The tree is constructed such that

No. of branches at a node + No. of features to be preserved at that node = $d + 1$.
$$\tag{5.33}$$

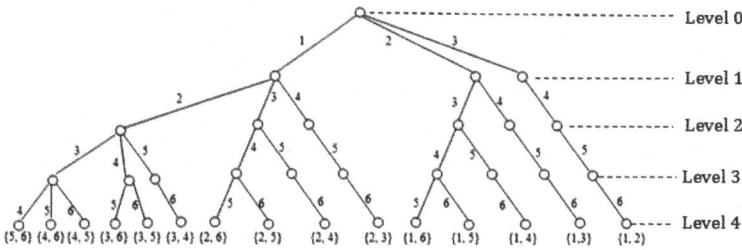

Fig. 5.19 Tree construction for branch and bound algorithm where f_1, f_2, f_3, f_4, f_5 and f_6 are represented by 1, 2, 3, 4, 5, and 6 respectively

We begin at Level 0: the number of branches from the root node is $2 + 1 - 0 = 3$. Without loss of generality, we assume that at the first level the features f_1, f_2 and f_3 are removed. There are many ways to choose this set of features to be removed at a level. One of the ways is to choose them randomly. Let us consider the following case: $J(S - \{f_1\}) <= J(S - \{f_2\}) <= J(S - \{f_3\})$. This means that removal of f_1 reduces the criterion function by the largest amount. This means that removal of f_3 reduces the criterion function by the least of the amount. So, the features f_1 and f_2 are more important than feature f_3. The three branches are labeled as f_1, f_2 and f_3 from left to right. At every node, J is computed. After each level, a particular feature is excluded so that finally we get all possible combinations(subsets). However, the features that are once excluded in the previous nodes in the same level of the same parent node will not be excluded from the remaining nodes as well as its child nodes. So, there are $(D - d + 1) = 6 - 2 + 1 = 5$ levels.

Now, we will start from the right most node (denoting feature subset $S - \{f_3\}$). In the subtree below this node, the features f_1 and f_2 are to be preserved. So, the number of branches coming out of this node is $2 + 1 - 2 = 1$. At this point, f_4 is deleted to get a new node (denoting feature subset $S - \{f_3, f_4\}$). Then features to be preserved turn out to be 2 (f_1 and f_2). So, the branches coming out of this node are $2 + 1 - 2 = 1$. At this point, f_5 is deleted to get a new node (denoting feature subset $S - \{f_3, f_4, f_5\}$). Then features to be preserved turn out to be 2 (f_1 and f_2). So, the branches coming out of this node are $2 + 1 - 2 = 1$. At this point, f_6 is deleted to get a new node (denoting feature subset $S - \{f_3, f_4, f_5, f_6\} = \{f_1, f_2\}$). The feature subset $\{f_1, f_2\}$ is the right-most leaf node obtained. This leaf node is compared with the nodes denoting feature subsets $S - \{f_2\}$ and $S - \{f_1\}$ by comparing $J(f_1, f_2)$ with $J(S - \{f_2\})$ and $J(S - \{f_1\})$. This comparison guarantees all the comparisons with other subsets under these two nodes. If $J(f_1, f_2)$ turns out to be greater, then $\{f_1, f_2\}$ will be the best and optimum feature subset (considering the monotonicity property of the function J).

5.6 Feature Extraction

As the dimensionality of a domain widens, the number of features D increases. Detecting an optimal feature subset becomes unmanageable. Feature extraction methods create a set of new features by combining the existing features. Feature extraction is the technique of defining a new set of features, which will most efficiently or meaningfully represent the original information that is critical for analysis and classification. Some of the popular feature extraction methods are discussed further.

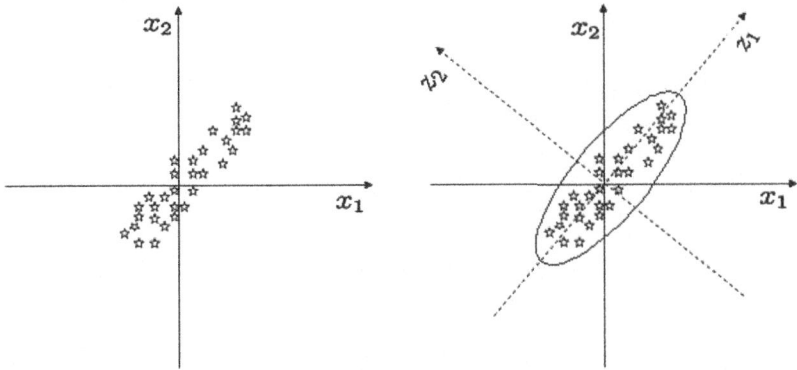

Fig. 5.20 Two principal components z_1 and z_2

5.6.1 Principal Component Analysis

Principal Component Analysis(PCA) [6] is a technique to identify the underlying, orthogonal dimensions that explain relations between variables in the dataset. It [24] takes a large number of correlated (interrelated) variables and transform them into a decreased number of uncorrelated variables (principal components) while retaining maximal amount of variation. We may easily understand the geometric picture of principal components (PCs) (Fig. 5.20).

Intuition: This basically includes finding the axis that shows the greatest variation and projecting all points onto that axis [11]. The 1st principal component z_1 is a maximum variance fit to a line in the feature space. The 2nd principal component z_2 is a line in the plane perpendicular to the 1st principal component (Fig. 5.20).

PCA steps (to reduce dimensionality from D to d)

- Normalize the dataset by moving the origin to the center of the dataset.
- Find the eigenvectors of the covariance matrix.
- Sort the eigenvalues in "goodness" order (non-increasing) that is highest to lowest. This gives the components in order of significance.
- The eigenvectors with the largest eigenvalues correspond to the principle components of the dataset.
- If the dimensions are highly correlated, there will be a small number of eigenvectors with large eigenvalues and d will be much smaller than D.

A covariance matrix Σ represents the variance and covariance of data features. Eigenvectors of Σ point to directions in feature space, and eigenvalues indicate the variance along those directions.

Proof:

- Let Σ be the covariance matrix and v be an eigenvector with eigenvalue λ.
- By definition of eigenvectors: $\Sigma * v = \lambda * v$.
- $\Sigma * v$ scales v by λ, meaning v aligns with a principal axis of the data, and λ measures variance along this direction.
- The eigenvector corresponding to the largest eigenvalue indicates the direction of maximum variance.

In Fig. 5.21, suppose all our data points are "inside" the blue cuboid.

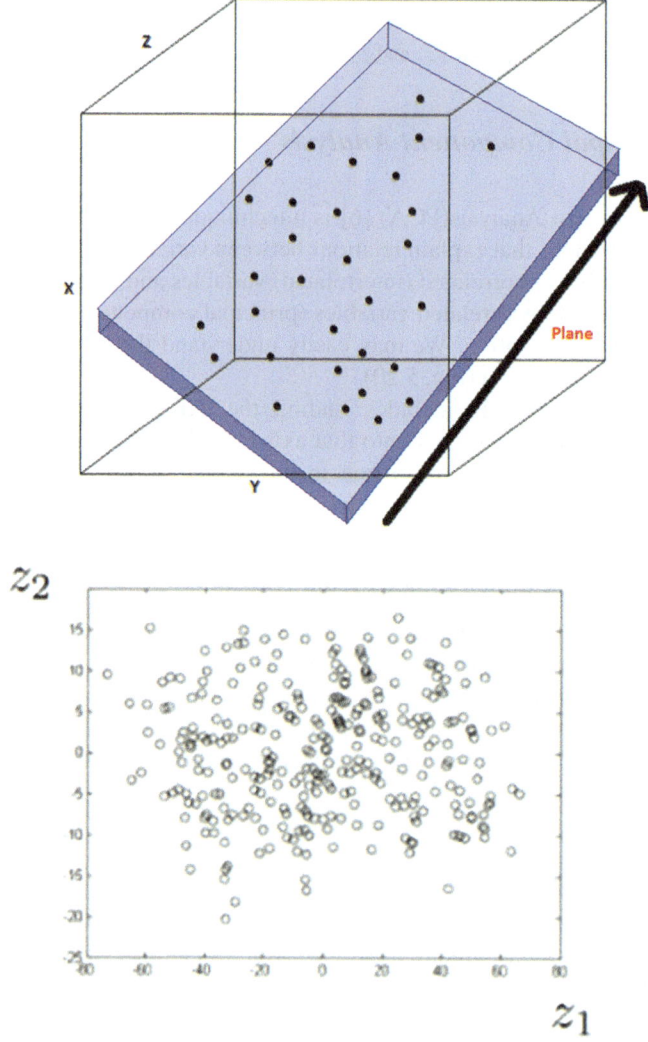

Fig. 5.21 Two principal components z_1 and z_2

As they are all in this comparatively shallow region, we can basically ignore any one of the two dimensions. So, we can now draw two new lines (z_1 and z_2) along the box's x and y planes, and we plot the positions in that box, i.e., lose the information in our z-dimension of the "shallow box." Thus, our 3D map has now been simplified to a 2D map. The problem essentially consists of rotating the ellipsoid of data, such that the direction of the data variance becomes the first component (Fig. 5.22).

Advantages of PCA: PCA according to modern data analysis is a standard tool since it is a straightforward parametric process to extract useful information from complex datasets. In addition, PCA offers a road map with a negligible effort to reduce complex datasets to a lower-dimensional space and expose the hidden and simplistic structure, which often underlies it.

Limitations of PCA: PCA works in an unsupervised manner. So, if we apply this in classification problem, it may increase the misclassification error because it disturbs the decision boundary between classes. If the dimensions are highly uncorrelated, then PCA will not be much useful.

In PCA, the main concept is to represent the dataset in some other space to obtain the relevant information by decreasing the redundancy. Now this works fine as long as the relation between the subsets is fairly linear. If we have a situation as shown in Fig. 5.23, we have a problem.

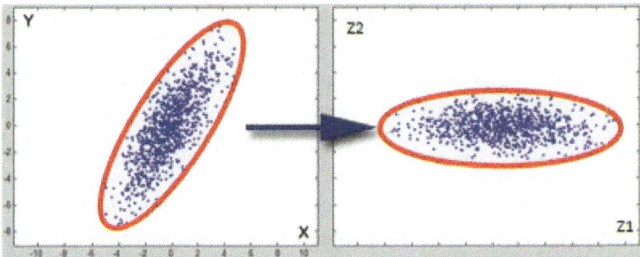

Fig. 5.22 Diagrammatic representation of simplified PCA

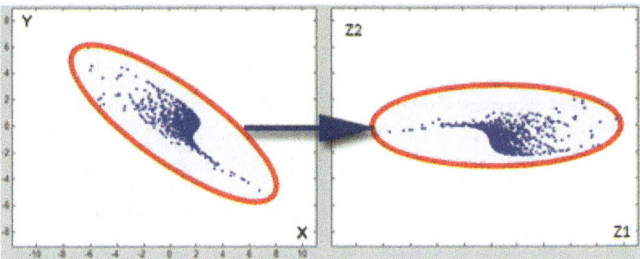

Fig. 5.23 Nonlinear problem: PCA gives bad results

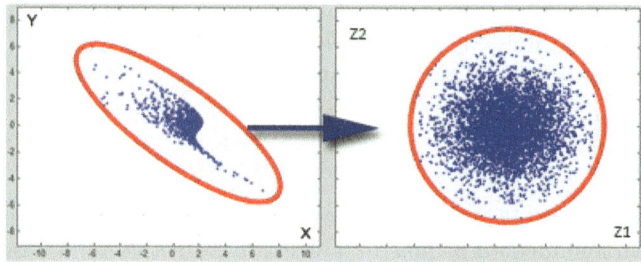

Fig. 5.24 Nonlinear problem: Results by neural network consideration

The PCA tries to produce features by finding out the maximum variance. It fails as the largest variance is not along a single direction, but along a nonlinear path (Fig. 5.24).

5.6.2   Linear Discriminant Analysis

The objective of Linear Discriminant Analysis (LDA) is to perform dimensionality reduction as well as to preserve as much of the class discriminatory information as possible [3, 26]. LDA classifies objects into one of the several groups based on a set of features that describe the objects [22].

For example, let us consider a pattern classification scenario, where C classes, e.g., salmon, seabass, and tuna are present. Each class has N_i number of d-dimensional samples, where $i = 1, 2, \ldots, C$. Hence, we have a set of d-dimensional samples $x_1, x_2, \ldots, x_{N_i}$ belonging to class ω_i. Stacking these samples from different classes into one matrix X such that each column represents one sample, we seek to obtain a transformation of X to Y by projecting the samples in X onto a hyperplane with dimension $C - 1$ (Fig. 5.25).

In order to find a good projection vector, a measure of separation between the projections is needed to be defined. Fisher Linear Discriminant projects on to a

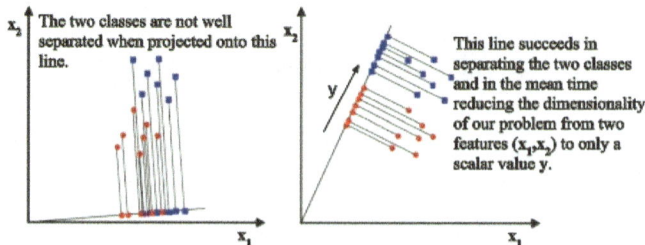

Fig. 5.25 An example to illustrate LDA: Of all possible lines, it selects the one that maximizes the separability of the classes

Fig. 5.26 Data for linear discriminant analysis

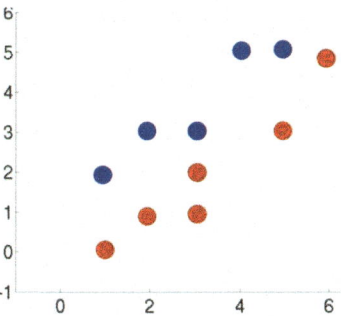

line, which preserves the direction useful for data classification. (PCA finds the most accurate data representation in a lower-dimensional space, by projecting data in the directions of maximum variance. The directions might be not useful for classification.) Let us consider an example (Fig. 5.26):

Class 1 has five samples $w_1 = [(1, 2), (2, 3), (3, 3), (4, 5), (5, 5)]$

Class 2 has six samples $w_2 = [(1, 0), (2, 1), (3, 1), (3, 2), (5, 3), (6, 5)]$

PCA performs very faultily on these data since the direction of largest variance is not useful for classification. So, we apply LDA. For this, we perform the following steps:

- First obtain the mean for each class

$$\mu_1 = [3 \quad 3.6];$$

$$\mu_2 = [3.3 \quad 2].$$

- Compute scatter matrices S_1 and S_2 for each class

$$S_1 = \sum_{x_i \in w_1} (x_i - \mu_1)(x_i - \mu_1)^T \qquad (5.34)$$

$$= \begin{bmatrix} 10 & 8 \\ 8 & 7.2 \end{bmatrix} \qquad (5.35)$$

$$S_2 = \sum_{x_i \in w_2} (x_i - \mu_2)(x_i - \mu_2)^T \qquad (5.36)$$

$$= \begin{bmatrix} 17.3 & 16 \\ 16 & 16 \end{bmatrix} \qquad (5.37)$$

- Define between class scatter matrix

$$S_B = (\mu_1 - \mu_2)(\mu_1 - \mu_2)^T; \qquad (5.38)$$

which measures the separation between the means of two classes (before projection).

- Now define the within class scatter matrix

$$S_W = S_1 + S_2 \tag{5.39}$$

$$= \begin{bmatrix} 27.3 & 24 \\ 24 & 23.2 \end{bmatrix}. \tag{5.40}$$

- Let the direction of the line be given by a unit vector v where

$$v = S_W^{-1}(\mu_1 - \mu_2); \tag{5.41}$$

if S_W has a full rank. If not then, v can be found by the generalized eigen value using

$$S_B v = \lambda S_W v. \tag{5.42}$$

Thus,

$$y = v^T x_i \tag{5.43}$$

is the projection of x_i onto a one-dimensional subspace.

- Scatter for projected samples of class 1 (ω_1) is

$$S_1' = \sum_{y_i \in \omega_1} (y_i - \mu_1')^2; \tag{5.44}$$

and for projected sample of class 2 (ω_2) is

$$S_2' = \sum_{y_i \in \omega_2} (y_i - \mu_2')^2 \tag{5.45}$$

where

$$\mu_1' = v^T \mu_1; \tag{5.46}$$

$$\mu_2' = v^T \mu_2. \tag{5.47}$$

Thus, Fisher linear discriminant is to project on a line in the direction of v, which maximizes

$$J(v) = \frac{(\mu_1' - \mu_2')^2}{S_1' + S_2'}. \tag{5.48}$$

Fig. 5.27 A representative
dataset to illustrate LDA

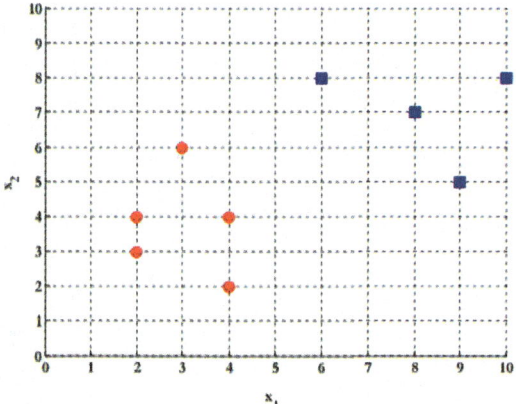

If we find v, which makes $J(v)$ large, we are guaranteed that the classes are well
separated.

Let us take an example. We take a sample space of two classes (Fig. 5.27):

Samples for class ω_1 : $\mathbf{X}_1 = (\mathbf{x}_1, \mathbf{x}_2) = (4, 2), (2, 4), (2, 3), (3, 6), (4, 4)$.

Sample for class ω_2 : $\mathbf{X}_2 = (\mathbf{x}_1, \mathbf{x}_2) = (9, 10), (6, 8), (9, 5), (8, 7), (10, 8)$.

For this dataset, the eigenvalues came out to be 8.88×10^{-16} and 12.2,
respectively.

Now, we see how a choice of the two eigenvectors for the data affects separability
between the classes (Figs. 5.28 and 5.29). The eigenvector with highest eigenvalue
will lead to good separability of the classes. If we have a class distribution as in the
following figure (Fig. 5.30), LDA is not going to be effective.

LDA is a parametric method, i.e., it makes the assumption of unimodal Gaussian
likelihood distribution. If the distributions are significantly non-Gaussian, the
LDA projections may not preserve complex structure in the data required for
classification. Also, when the discriminatory information is not in the mean but lies
in the variance of the data, LDA is going to become unsuccessful (Fig. 5.30).

5.6.3   Independent Component Analysis

Independent component analysis (ICA) is a computational method for separating
a multivariate signal into additive subcomponents. This is achieved by assuming
that the subcomponents of the considered signal are non-Gaussian in nature and
that they are statistically independent from each other. In the model, the data
variables are assumed to be linear mixtures of some unknown latent variables,
and the mixing system is also unknown. The latent variables are called the
independent components of the observed data. These independent components,
also called sources or factors, can be found by ICA. ICA can also considered as

Fig. 5.28 Projection onto
this vector leads to poor
separability between the
classes

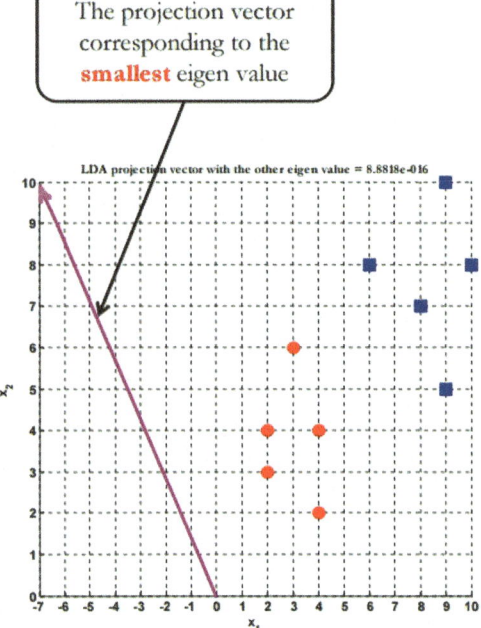

Fig. 5.29 Projection onto
this vector leads to good
separability between the
classes

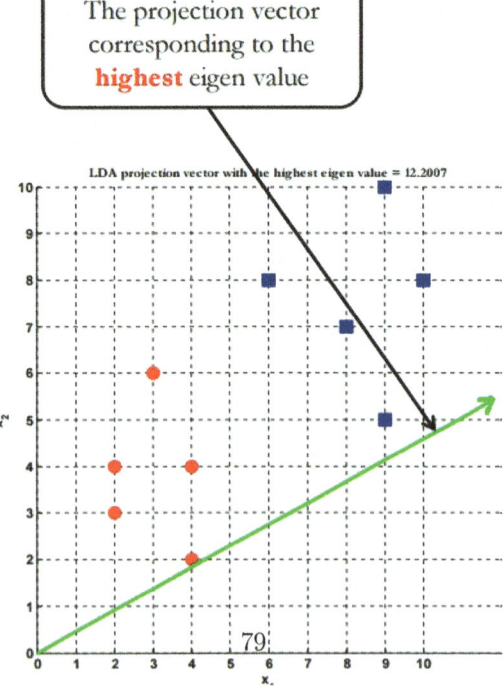

Fig. 5.30 LDA will fail as the discriminatory information is in the variance of the data

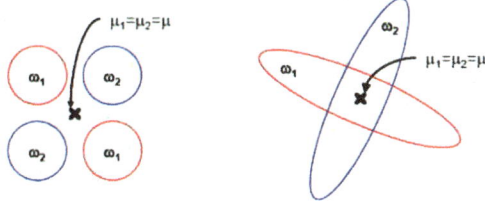

a special case of blind source separation. Principal Component Analysis (PCA) and Independent Component Analysis (ICA) are although similar techniques, but are different approaches and can perform different tasks. Specifically, PCA is often used to compress the information like dimensionality reduction. Whereas the ICA aims to separate information by transforming the input space, which can divide into the maximally independent basis. However, one of the common things in both approaches is it requires the input data to be autoscaled. The independent components generated by the ICA are assumed to be statistically independent of each other. The independent components generated by the ICA must have non-Gaussian distribution. The number of independent components generated by the ICA is equal to the number of observed mixtures. The data analyzed by ICA could originate from many different kinds of application fields, including digital images, document databases, economic indicators, and psychometric measurements. In many cases, the measurements are given as a set of parallel signals or time series; the term blind source separation is used to characterize this problem. Typical examples are mixtures of simultaneous speech signals that have been picked up by several microphones, brain waves recorded by multiple sensors, interfering radio signals arriving at a mobile phone, or parallel time series obtained from some industrial process.

Basic Steps of ICA:

- Centering: Subtract the mean of the dataset X to center it around zero.
- Whitening: Transform X into a whitened form Z by decorrelating the features (e.g., using PCA).
- Decomposition: Maximize non-Gaussian form of the whitened signals Z to extract independent components: $X = W \cdot S$, where W is mixing matrix and S is independent components.

5.6.4 Correlated Component Analysis

The Correlated Component Analysis (CCA) exploits a second-order statistics to estimate the mixing matrix from the statistics of data and noise. It is also known as canonical-correlation analysis (CCA). It is a statistical way of making sense of cross-covariance matrices. If two vectors $\mathbf{X} = (X_1, \ldots, X_n)$ and $\mathbf{Y} = (Y_1, \ldots, Y_m)$ of random variables are present, and there are correlations among the variables, then

Fig. 5.31 Correlation

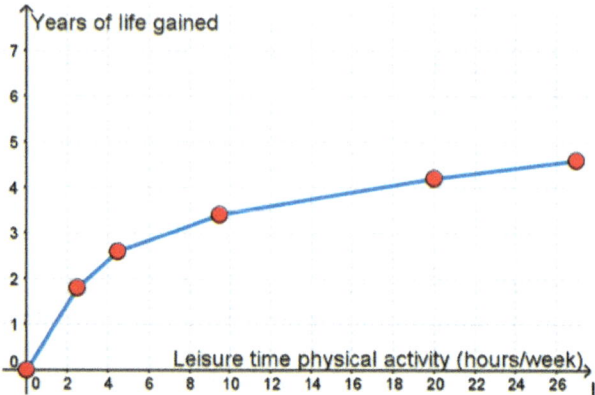

CCA will find linear combinations of X_i and Y_j, which have maximum correlation with each other.

Consider as an example the relation between duration of physical activity and years of life gained (Fig. 5.31). We may consider physical activity parameters, such as the climbing pace on a staircase, how fast one can walk. However, there may also be health factors, such as blood pressure, blood sugar levels, body mass index, and cholesterol. Therefore, there are two kinds of variables, and the interaction between the physical activity variables and the health variables must be investigated.

An approach to study the relationship of two groups of variables is to use CCA. It describes how the first collection of variables is related to the second collection of variables. One collection of the variables is not necessarily considered to be autonomous and the other to be dependent on it. They are assumed to be correlated, and CCA finds that relationship.

Correlation measures the relationship between variables:

- Gather data: Collect paired observations (X, Y).
- Calculate means: Compute the mean of X and Y.
- Compute covariance: measure how X and Y vary together.

$$\text{Cov}(X, Y) = \frac{1}{n} \sum_{i=1}^{n} (X_i - \bar{X})(Y_i - \bar{Y}).$$

- Standardize: Normalize by standard deviations of X and Y.

$$r = \frac{\text{Cov}(X, Y)}{\sigma_X \sigma_Y};$$

where r is the Pearson correlation coefficient, indicating strength and direction of the relationship.

5.6.5 *Factor Analysis*

Factor analysis (FA) is a valuable tool to research parameter interactions in complex concepts like socioeconomic status, dietary patterns, or psychology. It enables researchers to explore notions that are not easily measured by breaking many variables into several interpretable underlying factors. The key concept in factor analysis is that many factors found have common reaction trends, both related to a latent variable (i.e., not explicitly measured).

Individuals may, for instance, react similarly to queries related to employment, education, and work, which are all linked to the latent socioeconomic parameter status.

The same number of factors exists as variables in each factor analysis. Every factor reports a certain amount of the total variation in the measured variables. The eigenvalue is a metric of how much of the variability of the measured variables a factor describes. Factors having an eigenvalue ≥ 1 describes more variance than a single measured variable.

So, if the factor for socioeconomic status had an eigenvalue of 2.3, it would explain as much variance as 2.3 of the three variables. This factor, which captures most of the variance in those three variables, could then be applied in other analyses.

Factor analysis is a technique to investigate whether a number of variables of interest $Y_1, Y_2, \ldots, Y_l$ are linearly related to a smaller number of unobservable factors $F_1, F_2, \ldots, F_k$.

Factor analysis can be explained using the context of a simple example. Candidates applying for a teaching post must pass three tests: written test, interview, and classroom management. Let Y1, Y2, and Y3, respectively, represent a candidate's score in these tests. The available data comprise the grades of five candidates (in a 10-point numerical scale), as shown in Table 5.2.

It has been suggested that these scores are functions of two underlying factors, F_1 and F_2, tentatively and rather loosely described as quantitative ability and verbal ability, respectively. Assumption is made that each Y variable is linearly related to the two factors in the following manner:

$$Y_1 = \beta_{11} F_1 + \beta_{12} F_2 + e_1 \qquad (5.49)$$

$$Y_2 = \beta_{21} F_1 + \beta_{22} F_2 + e_2 \qquad (5.50)$$

$$Y_3 = \beta_{31} F_1 + \beta_{32} F_2 + e_3 \qquad (5.51)$$

Table 5.2 Candidate grades

Candidate no.	Y_1	Y_2	Y_3
1	3	6	5
2	7	3	3
3	10	9	8
4	3	9	7
5	10	6	5

Table 5.3 Value of loadings

Variable Y_i	F_1, β_{i1}	F_2, β_{i2}
Written test Y_1	0	0.7
Interview Y_2	0.65	0.11
Classroom management Y_3	0.38	0

The error terms e1, e2, and e3 explain that the hypothesized relationships are not exact. It is crucial in understanding factor analysis to remember that F stands for a function of variables and not a variable.

In the special vocabulary of factor analysis, the parameters β_{ij} are referred to as loadings. For example, β_{12} is called the loading of variable Y_1 on factor F_2.

Interview and classroom management scores are highly dependent on verbal ability, while written test needs a strong quantitative skill. Quantitative skills should help a student in written and interview, but not in classroom management. Verbal skills should be useful in interview and classroom management; but not in written test. How each variable is related to the underlying factor is explained by the so-called factor loading. In other words, it is expected that the loadings roughly follow the given structure (Table 5.3):

Since factor loadings could be explained like standardized regression coefficients, it can be also said that the variable interview has a correlation of 0.65 with factor 1. This would be considered a strong association for a factor analysis in most research fields.

There are, however, some assumptions in factor analysis:

- There are no outliers in the dataset.
- Adequate sample size, i.e., the data should have more variables than factors. Also, each variable must also have more data values than factors.
- Each variable must be unique and cannot be derived from the other.
- Each of the variables should be linear in nature.
- The data are interval scaled. Nominal and ordinal data do not work with factor analysis.

Factor analysis is a statistical method that can be used to interpret the interrelationships among a large number of variables, and these variables can be explained in terms of their common underlying dimensions (factors). It includes finding a process to condense the information present in a number of original variables into a smaller set of dimensions (factors) with a minimum loss of information. FA and PCA are not much different than canonical correlation in terms of generating canonical variate from linear combinations of variables.

The following steps apply in case of factor analysis:

- The correlation matrix of the input variables is computed as

$$\begin{bmatrix} \alpha_{11} & \alpha_{12} & \alpha_{13} \\ \alpha_{21} & \alpha_{22} & \alpha_{23} \\ \alpha_{31} & \alpha_{32} & \alpha_{33} \end{bmatrix}. \tag{5.52}$$

- The orientation of the factors is computed based on the eigenvectors and eigenvalues of the correlation matrix.
- The vectors in each column are normalized. The normalized values are the factor analysis coefficients. Looking at three variables, the factor analysis coefficients matrix looks like:

$$\begin{bmatrix} \beta_{11} & \beta_{12} \\ \beta_{21} & \beta_{22} \\ \beta_{31} & \beta_{32} \end{bmatrix} \tag{5.53}$$

where β_{ij} represents the factor loadings.

The factor analysis model can be algebraically written as follows. If l variables $Y_1, Y_2, \ldots, Y_l$ measured on a sample of N subjects are taken, then variable Y_i could be explained as a linear combination of k factors $F_1, F_2, \ldots, F_k$ where, $k < l$. Thus,

$$Y_i = \beta_{i1}F1 + \beta_{i2}F2 + \ldots + \beta_{ik}F_k + e_i; \tag{5.54}$$

where the β_is are the factor loadings (or scores) for variable Y_i and e_i represents the part of variable Y_i that cannot be "explained" by the factors. Following are the three important steps in a factor analysis:

(i) **Calculate initial factor loadings**: It could be done via various different ways; the two most commonly used methods are described briefly here:

- Principal component method: As suggested by the name, this follows the same principle by which a principal components analysis is carried out. But, the obtained factors would not actually represent the principal components (although the loadings for the jth factor will be proportional to the coefficients of the jth principal component) where $j = 1, 2, \ldots, k$.
- Principal axis factoring: This is a technique that attempts to find the lowest number of factors that take accountability for the variability in the original feature that is associated with these factors (this is in contrast to the principal components method, which looks for a set of factors that can account for the total variability in the original variables).

These two techniques will appear to provide similar results when the variables are quite highly correlated and/or the number of original variables is quite high. Irrespective of the used method, the resulting factors at this stage will be uncorrelated.

(ii) **Factor rotation**: The factors are rotated once the initial loadings of the factor are calculated. This is done to find easier to understand factors. If there are clusters of variables, i.e., subgroups of variables that are closely inter-related, a rotation is carried out to try and ensure that the loading of the remaining variables in a subgroup score is as large (positive/negative) as feasible on a given parameter. In another way, the purpose of the rotation is to ensure that only one factor contains high loads for all variables. The orthogonal and oblique rotations are two types of rotation strategies. The rotated variables stay uncorrelated in orthogonal rotation, whereas the resultant factors for oblique rotation are correlated. There are numerous rotation approaches of each kind. Varimax rotation is the most common orthogonal type.

(iii) **Calculation of factor scores**: While calculating the final factor scores (the values of the k factors, $F_1, F_2, \ldots, F_k$, for each observation), a decision is needed to be made as to how many factors should get included. This gets typically done using one of the following methods:

- Choose k in such a way that the factors account for a particular percentage (e.g., 75%) of the total variability in the original variables.
- Choose k to be equal to the number of eigenvalues over 1 (if you are using the correlation matrix). A different criterion must be used if you use the covariance matrix.
- Use the scree plot of the eigenvalues. This will specify whether there is an obvious cutoff between large and small eigenvalues.

References

1. Aladjem, M. E. (1996). Two-class pattern discrimination via recursive optimization of Patrick-Fisher distance. In *Proceedings of the 13th international conference on pattern recognition* (vol. 2, pp. 60–64).
2. Bhattacharyya, A. (1946). On a measure of divergence between two multinomial populations. *Sankhyā: The Indian Journal of Statistics*, 401–406.
3. Burges, C. J. C. (2010). *Dimension reduction: A guided tour* (Foundations and trends in machine learning). Now Publishers.
4. Chernoff, H. (1952). A measure of asymptotic efficiency for tests of a hypothesis based on the sum of observations. *The Annals of Mathematical Statistics*, 493–507.
5. Devyver, P. A., & Kittler, J. (1982). *Pattern recognition: A statistical approach*. Prentice-Hall.
6. Dunteman, G. H. (1989). *Principal components analysis*. No. 69 in a Sage Publications. SAGE.
7. Gu, Q., Li, Z., & Han, J. (2011). Generalized Fisher score for feature selection. In *Proceedings of the twenty-seventh conference on uncertainty in artificial intelligence* (pp. 266–273), Barcelona, Spain.

8. Horst, R., & Romeijn, H. E. (2002). *Handbook of global optimization* (vol. 2). Springer.
9. Hu, W. C. (2013). *Big data management, technologies, and applications* (Advances in data mining and database management). Information Science Reference.
10. John, G. H., Kohavi, R., Peger, K., et al. (1994). Irrelevant features and the subset selection problem. In *Proceedings of the eleventh international conference on machine learning* (pp. 121–129).
11. Jolliffe, I. T. (2013). *Principal component analysis* (Springer series in statistics). Springer.
12. Kittler, J. (1986). Feature selection and extraction. In Young, T. Y. & Fu, K. S. (King Sun) (eds.), *Handbook of pattern recognition and image processing* (pp. 59–83), Academic Press.
13. Kohavi, R., & John, G. H. (1997). Wrappers for feature subset selection. *Artificial Intelligence, 97*(1), 273–324.
14. Kotsiantis, S. B., Kanellopoulos, D., & Pintelas, P. E. (2006). Data preprocessing for supervised learning. *International Journal of Computer Science, 1*(2), 580–585.
15. Kullback, S. (1997). *Information theory and statistics.* Courier Corporation.
16. Kullback, S., & Leibler, R. A. (1951). On information and sufficiency. *The Annals of Mathematical Statistics, 22*(1), 79–86.
17. Kuncheva, L. I. (2004). *Combining pattern classifiers: Methods and algorithms.* Wiley.
18. Liu, H., & Motoda, H. (2007). *Computational methods of feature selection* (Chapman and Hall/CRC data mining and knowledge discovery series). CRC Press.
19. Liu, H., & Motoda, H. (2012). *Feature selection for knowledge discovery and data mining* (The Springer international series in engineering and computer science). Springer US.
20. Longford, N. T. (1987). A fast scoring algorithm for maximum likelihood estimation in unbalanced mixed models with nested random effects. *Biometrika, 74*(4), 817–827.
21. Matusita, K. (1955). Decision rules, based on the distance, for problems of fit, two samples, and estimation. *The Annals of Mathematical Statistics*, 631–640.
22. McLachlan, G. (2004). *Discriminant analysis and statistical pattern recognition* (Wiley series in probability and statistics). Wiley.
23. Mitchell, T. M. (1997). *Machine learning* (1st ed.). McGraw-Hill.
24. Schólkopf, B., Smola, A., & Múller, K.-R. (1997). Kernel principal component analysis. In *Artificial Neural Networks-ICANN'97* (pp. 583–588), Lausanne, Switzerland.
25. Theodoridis, S., Pikrakis, A., Koutroumbas, K., & Cavouras, D. (2010). *Introduction to pattern recognition: A matlab approach.* Academic.
26. Tzanakou, E. M. (1999). *Supervised and unsupervised pattern recognition: Feature extraction and computational intelligence* (Industrial electronics). Taylor and Francis.
27. Yan, J., Zhang, B., Liu, N., Yan, S., Cheng, Q., Fan, W., Yang, Q., Xi, W., & Chen, Z. (2006). Effective and efficient dimensionality reduction for large-scale and streaming data preprocessing. *IEEE Transactions on Knowledge and Data Engineering, 18*(3), 320–333.

Part III
Pattern Recognition

Chapter 6
Pattern Recognition Systems

Brains are programmed to see patterns in nature, be it humans or any other living creature. For example, the nature of recognizing patterns is present even in a child. Young children can identify and absorb the characteristics of their native language by finding trends in the words and phrases constantly spoken around them. Humans are, by nature, wonderful pattern recognition systems. They have the ability to recognize various types of patterns and then transform these bits of information into actionable, concrete steps. As a toddler learns words and concepts, he starts to recognize patterns for differentiating between objects. Kurzweil [6] describes a series of thought experiments to suggest that the brain contains a hierarchy of pattern recognizers. Intelligence, therefore, is a matter of being able to retain more patterns than others. Ray Kurzweil was among the first few to recognize how the link between human intelligence and pattern recognition could be utilized to develop the next generation of artificially intelligent systems. In the past, when IBM was able to build machines that could recognize as many chessboard patterns as a chess grandmaster, the machines were termed "smarter" than humans.

Looking at the history, pattern recognition systems have come a long way. Previously, it was mostly restricted to theoretical research in the field of statistics for deriving models out of a huge amount of available data. With the advent in technology, number of practical applications have increased drastically, which have lead to further theoretical evolution. Currently, pattern recognition has become an integral part of any machine learning model that possesses decision-making capabilities, and they use various mathematical techniques [15]. Pattern discovery in data is a basic topic of research and ways back to have a long and successful history. Pattern recognition mostly revolves around the automatic discovery of regularities in data making use of algorithms and using these regularities to take actions like classification of the data into multiple categories [5]. Pattern recognition has been a field of growing interest since 1960. However, looking back into the past tens of millions of years, humans have been doing pattern recognition with much ease resulting in an evolution of a developed neural and cognitive system in human brain.

© The Author(s), under exclusive license to Springer Nature Singapore Pte Ltd. 2025   101
A. Ghosh, *Data Science and Cases in Sustainability*, Mathematics for Sustainable Developments, https://doi.org/10.1007/978-981-96-8362-8_6

Fig. 6.1 The digit 9 written
by three different people

With the advent of computers, humans are trying to make machines do that job too. The machines now need to be intelligent enough to recognize face, understand speech, read characters, understand the condition of the nature based on specific parameters and many more complex processes, which underlie this act of pattern recognition.

Pattern recognition is mostly focused with the development of systems that can identify some patterns in data. The purpose of a pattern recognition system is to analyze a scene in the real world and to generate a description of the scene, which is useful for the accomplishment of some specific task. These real-world observations are acquired through some sensors, and the pattern recognition system classifies or describes these observations. A feature extraction mechanism then computes numeric or symbolic information from these observations. Finally, these extracted features are used to categorize the acquired data.

An example can be stated in the context of classification of geometric figures. Suppose we have three persons who write the digit 9 in three different ways (Fig. 6.1). Then the mechanism by which a computer will know that it is 9, is through geometric analysis of patterns, which are inherent to the digit 9.

Now, the question is what these patterns mean or how does a computer recognize a pattern? Humans have sense organs like eyes, ears, and nose to recognize things around them. Computers on the contrary work on only two states—a high state (usually denoted as 1) or a low state (usually denoted as 0). So, recognizing an object and then making some decision based on that analysis might become impossible for a nonliving entity like a computer. Well, there are certain procedures that actually have given that power to a machine to recognize objects.

The field of pattern recognition mainly focuses on the description and analysis of physical and logical patterns. It acquires raw data and takes actions depending on the "class" of the patterns identified in the data. In the past, this was a specialized subject due to the high price of the hardware, which was used to acquire the data and find out the answers. The fast developments in computer technology and resources enhanced the possibility of numerous applications of this field, which in turn gave rise to the need for further theoretical developments.

Mathematical pattern recognition can be defined as: the categorization of input data into identifiable classes, via the extraction of significant features or attributes of the data, using mathematical techniques [15]. The design of a pattern recognition system essentially involves three aspects: Representation of data, Categorization, and Prototyping. The problem domain decides the choice of sensors, pre-processing methods, representation scheme, and the classification model.

Patterns that are generated from raw data mostly depend on the nature of the data involved. For instance, patterns can be generated based on the statistical features of the data. In some cases, the inherent structure of the data determines the kind of the pattern that will be generated. However, in some other situations, neither of the aforementioned two exists. Then, a system is created and trained for the required responses. Thus, for a said problem, any of these approaches can be employed to reach the desired solution. Hence, to obtain the fruitful output for a pattern recognition system, there are many different mathematical techniques. The four well-known methodologies for pattern recognition are:

- Template matching
- Statistical analysis
- Syntactic matching
- Neural computing

In template matching, a prototype of the desired pattern to be recognized is compared against the actual pattern. In the statistical approach, the patterns are represented as random variables, from which the density of the classes can be inferred. Classification is done based on the statistical modeling of data. The syntactic approach sees a pattern to be composed of simpler sub-patterns, which are also in turn built from yet simpler sub-patterns, and so on. The simplest sub-patterns are the primitives. These primitive patterns are interrelated so as to create more complex patterns. Finally, a neural computing approach to pattern recognition is based on human nervous system and is strongly related to the statistical methods, since they can be regarded as parametric models with their own (neural) learning scheme.

6.1   Block Diagram of a Pattern Recognition System

We now need to know how to make machines decide which class a pattern belongs to. There are several steps in pattern recognition task. We take the raw data, preprocess it, select the relevant features, then categorize the patterns according to their feature description, do some postprocessing, and then make a decision (Fig. 6.2).

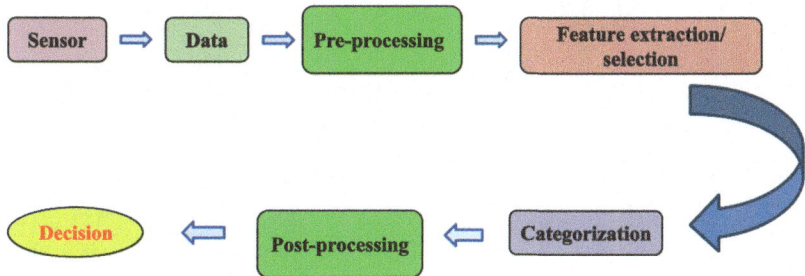

Fig. 6.2 A pattern recognition system

The job of the system doing pattern recognitions is to categorize an object into the correct class based on the measurements about the object. The possible classes are generally well-defined before the pattern recognition system is in action. Most of the systems doing pattern recognition can be imagined of consisting of the following phases:

- Sensing (measurement)
- Pre-processing and segmentation
- Feature extraction/selection
- Categorization
- Post-processing
- Decision

Each of these stages are described as follows.

Sensing [10] is referred as doing measurement or inspection about the objects to be categorized. To explain this task, let us consider a simple example. There are two different types of leaves (Fig. 6.3) whose features (like, thickness, slimness, roundness, and regularity) are measured by various sensors (as shown in Table 6.1).

Similarly, we can consider other examples, where the input data consist of an image of the earth, and the sensing equipment is a satellite camera. Often, a single observation (for instance, image) comprises information about several objects to be categorized. For example, we wish to identify the forests, swamps, croplands, and grasslands in a particular region of interest from an image. We must then classify several shades of the green color to recognize whether it is a tree from the forest

Fig. 6.3 Two types of leaves

Table 6.1 Features of the two types of leaves measured by the sensors

Features	Leaf 1	Leaf 2
Thickness (mm)	0.5591	0.3727
Slimness	0.0024	0.0169
Roundness	0.0412	0.0848
Regularity	0.1809	0.0655

or a grass from the grassland. The data here are probably an image of the area that includes the sub-images of all the shades of green to be classified and some soil background.

Pre-processing is referred to filtering the raw data to suppress noise information and other operations (on the raw data) for improving their quality.

In **segmentation**, the measured data are divided to represent exactly one object to be classified. For instance, in a location recognition, an image of the whole area of land is required to be partitioned into sub-images that represents just one type of land (e.g., forest, water, and concrete)

Feature extraction is needed especially when dealing with pictorial information where there can be an enormous amount of data per object. A high-resolution photograph can comprise a large number of pixels. Then each pattern vector can have over a million components. The most part of these data is non-informative or irrelevant. In feature extraction, we search for the optimum number of features that can efficiently characterize the data for categorization. Result of this feature extraction phase is known as a feature vector (a pattern) as discussed in the previous section.

The **classifier** takes the feature vector obtained from the object as input. It then matches the feature vector (i.e., the object) to the most appropriate class. The classifier might be viewed as a mapping of the feature space to the array of possible classes. However, the classifier cannot differentiate between two objects having the same feature vector. The key interest of pattern recognition task lies with the classifier design. The abstraction offered by the concept of the feature vector allows for the creation of a general, application-independent theory for designing the classifiers.

Postprocessing mainly is concerned about enhancing the accuracy of the classification task. Different classification actions can also have different costs associated with them. This information could be included in the classifier design. The system can then use the information about the other classification results to make correction to a probable misclassification. There are other considerations that need to be addressed in addition to classification accuracy when developing a pattern recognition system. For instance, the time spent to classify an object could be crucial. The cost of the system could also be a crucial factor.

The main job of the pattern recognition system is to give a final **decision** upon an overall action based on the result of classification.

The process of designing a pattern recognition system is iterative in nature. If the system designed is not well enough, we go back and try to improve some or all of the five stages of the system. After that, the system is repeatedly tested and improved till it does meet the requirements.

The classifier component of a pattern recognition system needs to be taught to recognize certain feature vectors belonging to a certain class. Now, we shall have to take two aspects into consideration:

(i) The job of the system is to assign a previously unseen object to a correct class.
(ii) It is not possible to define the correct classes for all the possible feature vectors.

For instance, if we take a 16×16 binary image where each pixel represents a feature, which has a value of either 0 or 1, we will have $2^{256} \approx 10^{77}$ possible feature vectors. In simple words, there is going to be more feature vectors in the feature space than there are atoms in this galaxy. The question now stands is: how we can ensure whether the object is put into its true class or not. In this context we need to have an idea about the classifiers. It is now well understood that "similar patterns" are put in one category and dissimilar patterns in different categories. In the upcoming sections, some famous similarity and dissimilarity measures used for identifying similar patterns have been discussed.

6.2   Similarity Measures

Measuring similarity between two data points is a core requirement for several pattern recognition tasks. Finding out similarity between patterns having real-valued features/attributes is relatively easy. The study of similarity between data objects with categorical variables is, however, relatively difficult. In the late 1800s, Pearson proposed a Chi-square statistics that are often used to test independence between categorical variables in a contingency table [2]. Some of the more common similarity measures used are Dice Coefficient [3], Jaccard Similarity Index [4], Cosine Similarity [1, 14], and Overlap Coefficient [9].

6.2.1   Angle Measures

Let us begin with the definition of the dot product for two vectors: $\mathbf{x} = (x_1, x_2, x_3, \ldots)$ and $\mathbf{y} = (y_1, y_2, y_3, \ldots)$, where x_i and y_i are the ith components of the vectors and n is the dimension of the vectors (Fig. 6.4). We know that

$$\mathbf{x} \cdot \mathbf{y} = \|\mathbf{x}\| \|\mathbf{y}\| \cos \theta. \tag{6.1}$$

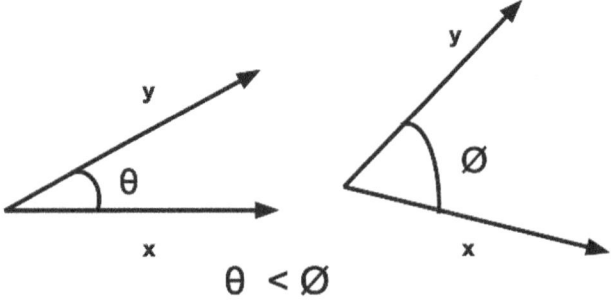

Fig. 6.4 Angle measures

Here, θ is the angle between **x** and **y**. Thus,

$$\cos\theta = \frac{\mathbf{x}\cdot\mathbf{y}}{\|\mathbf{x}\|\|\mathbf{y}\|}. \tag{6.2}$$

6.2.2 Cosine Similarity

The cosine similarity between two vectors (or patterns) is a measure that computes the cosine of the angle between them. This metric is a measurement of orientation and not magnitude that determines the relation between feature vectors with respect to the angle instead of magnitude. Cosine similarity is a commonly employed similarity measure for real-valued vectors, used in information retrieval to score the similarity of documents in the vector space model.

$$\mathbf{Cosineh}(\mathbf{x}, \mathbf{y}) = \frac{2\sum\limits_{h=1}^{n} x_h y_h}{\sum\limits_{i=1}^{n} x_i^2 \sum\limits_{i=1}^{n} y_i^2}. \tag{6.3}$$

6.2.3 Tanimoto Measures

In some cases, each attribute is binary such that each bit reflects the absence or presence of a characteristic. Therefore, it is easier to determine the similarity via overlap or intersection of the sets. Tanimoto Coefficient uses the ratio of the common bits (logical AND) to the union of bits of two vectors as the measure of similarity between them represented by

$$T(\mathbf{x}, \mathbf{y}) = \frac{N_x \cap N_y}{(N_x \cup N_y) - (N_x \cap N_y)}. \tag{6.4}$$

Here N_x represents the number of attributes in vector **x** and N_y in vector **y**. The Tanimoto similarity is only applicable for a binary variable, and it ranges from 0 to +1 (where +1 is the highest similarity).

6.2.4 Distance-Based Similarity Measures

One of the oldest and most influential theoretical assumptions is that perceived similarity is inversely related to distance. The two most popular distance measures are Euclidean distance and city-block distance. If the features of a pattern **x** in an

Fig. 6.5 Euclidean and
City-Block distance between
two points

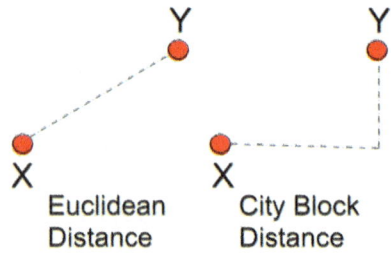

Euclidean   City Block
Distance   Distance

d-dimensional space are $(x_1, x_2, \ldots, x_d)$, then the Euclidean distance from $\mathbf{x}$ to another pattern $\mathbf{y}$ is

$$d(\mathbf{x}, \mathbf{y}) = \sqrt{\sum_{i=1}^{d}(x_i - y_i)^2}. \tag{6.5}$$

The city-block distance between these two vector is defined as (Fig. 6.5):

$$d(\mathbf{x}, \mathbf{y}) = \sum_{i=1}^{d} ||(x_i - y_i)||. \tag{6.6}$$

6.2.5 Probabilistic Similarity Measures

In scenarios where the difference between distributions cannot be adequately captured by geometric distances, probabilistic similarity measures offer more nuanced approach. Unlike traditional distance-based measures, which compare test patterns to single pattern from each group individually, probabilistic measures take into account the distribution of a set of points. This results in a more comprehensive assessment of similarity. This is crucial as distances based on single patterns from each group at a time may yield vastly different results. One prominent example is the Bhattacharyya coeeficient. The Bhattacharyya distance between two points $\mathbf{x}$ and $\mathbf{y}$ in d-dimensional space is defined as:

$$\mathbf{BC}(\mathbf{x}, \mathbf{y}) = -\log(\sum_{i=1}^{d} \sqrt{x_i\, y_i}). \tag{6.7}$$

Here, x_i and y_i represent the ith components of vector $\mathbf{x}$ and $\mathbf{y}$, respectively. The sum is taken over all dimensions, d, in the vectors.

This coefficient quantifies the Bhattacharyya distance between two probability distributions represented by the points **x** and **y**. Another notable measure is the Kullback–Leibler (KL) divergence, expressed as:

$$\mathbf{D_{KL}(x||y)} = \sum_{i=1}^{d} x_i \log(\frac{x_i}{y_i}). \tag{6.8}$$

Here, x_i and y_i represent the ith dimensions of points **x** and **y**, respectively. This equation quantifies the difference between the two multivariate probability distributions represented by **x** and **y**. It is crucial to note that KL divergence is not symmetric, meaning $\mathbf{D_{KL}(x||y)} \neq \mathbf{D_{KL}(y||x)}$.

In addition to Bhattacharyya distance and KL Divergence, other prominent probabilistic similarity measures include the Jensen–Shannon Divergence [8], Hellinger Distance [7], Earth Mover's Distance [12], and the Wasserstein Distance [13]. These metrics play crucial roles in various domains, providing versatile tools to quantify the divergence or similarity between probability distributions, each tailored for specific applications and types of data.

6.2.6   Dice Coefficient

The Dice Coefficient [3], also known as the Sørensen Dice Index, is a similarity measure used to assess the degree of overlap or similarity between two sets or patterens, typically in the context of binary data or categorical variables. The concept behind the Dice coefficient is intuitively related to the idea of measuring the agreement or intersection between two sets. In the context of binary data, the sets can be thought of as the presence or absence of certain elements or features in two patterns.

Mathematically, the Dice Coefficient is defined as:

$$\mathbf{Dice\ h(x, y)} = \frac{2\sum_{i=1}^{d} x_i y_i}{\sum_{i=1}^{d} x_i^2 + \sum_{i=1}^{d} y_i^2}. \tag{6.9}$$

The coefficient yields values between 0 and 1, where 0 indicates no overlap or disimmilarity, and 1 signifies complete overlap or similarity. A Dice Coefficient 1 implies that the two patterns are identical, while a coefficient of 0 suggests no common elements between them.

6.2.7 Jaccard Similarity Index

The Jaccard Similarity Index is a statistical measure used to assess the similarity between two vectors, $\mathbf{x}$ and $\mathbf{y}$. It is derived from the Jaccard Index, which is often used for comparing sets. In the context of vectors, the Jaccard Similarity Index quantifies the overlap between the two vectors by taking into account the dot product relative to their respective magnitudes. Mathematically, it is represented as:

$$\mathbf{Jaccard}(\mathbf{x}, \mathbf{y}) = \frac{\sum_{i=1}^{d} x_i y_i}{\sum_{i=1}^{d} x_i^2 + \sum_{i=1}^{d} y_i^2} - \sum_{i=1}^{d} x_i y_i. \tag{6.10}$$

6.2.8 Overlap Coefficient

The overlap coefficient is a metric used to quantify the similarity between two vectors, $\mathbf{x}$ and $\mathbf{y}$, often employed in fields like biology and data mining. It measures the proportion of overlap between non-zero elements in both vectors, emphasizing the commonality in their active components. The formula, as defined, computes the overlap by doubling the sum of element-wise products of the vectors and normalizing it with the minimum sum of squares of individual vector elements. This normalization accounts for variations in vector lengths, making the Overlap coefficient robust for vectors of differing magnitudes.

$$\mathbf{Overlap}(\mathbf{x}, \mathbf{y}) = \frac{2 \sum_{i=1}^{d} x_i y_i}{\min(\sum_{i=1}^{d} x_i^2, \sum_{i=1}^{d} y_i^2)}. \tag{6.11}$$

6.3 Distance Measures

A general distance metric should obey the following rules:

$$d(\mathbf{x}, \mathbf{x}) = 0. \tag{6.12}$$

This property means one point is at zero distance from itself.

$$d(\mathbf{x}, \mathbf{y}) = d(\mathbf{y}, \mathbf{x}). \tag{6.13}$$

This property ensures the symmetry of space.

$$d(\mathbf{x}, \mathbf{y}) \leq d(\mathbf{x}, \mathbf{z}) + d(\mathbf{z}, \mathbf{y}). \tag{6.14}$$

Assuming the points $\mathbf{x}$,$\mathbf{y}$ and $\mathbf{z}$ to be at the vertices of a triangle, this property can be understood as the triangular inequality where sum of two sides of a triangle is always greater than the third side.

The following distance measures [11] are often used for pattern recognition:

6.3.1 Minkowski Distance

The Minkowski distance is a metric in a normed vector space that can be contemplated as a generalization of both the Euclidean and the Manhattan distances. The Minkowski distance of order p between two points $\mathbf{x}(x_1, x_2, x_3, \ldots, x_d)$ and $\mathbf{y}(y_1, y_2, y_3, \ldots, y_d)$ is given by

$$D(\mathbf{x}, \mathbf{y}) = (\sum_{i=1}^{d} |x_i - y_i|^p)^{1/p}. \tag{6.15}$$

For $p \geq 1$, the Minkowski distance is a metric. When $p < 1$, the distance between points $(0,0)$ and $(1,1)$ is $2^{1/p} > 2$, which violates triangle inequality as the point $(0,1)$ is at a distance 1 from both of these points. Minkowski distance is called L_p norm. For, $p = 1$, $p = 2$, the Minkowski distance becomes equal to the Manhattan (L_1 norm) and Euclidean (L_2 norm) distance, respectively. In the limiting case where p reaches infinity, the Chebychev distance (L_α) is obtained.

6.3.2 Manhattan Distance

Manhattan distance (or city-block distance) between two points is the addition of the absolute differences of their cartesian coordinates. More formally, it is the sum of the projections of the line segment between the points onto the coordinate axes (Fig. 6.6). Considering two points $\mathbf{x}(x_1, x_2, x_3, \ldots x_d)$ and $\mathbf{y}(y_1, y_2, y_3, \ldots y_d)$, the Manhattan distance between these two points is given by:

$$d(\mathbf{x}, \mathbf{y}) = \sum_{i=1}^{n} |x_i - y_i| = |x_1 - y_1| + |x_2 - y_2| + \ldots + |x_d - y_d|. \tag{6.16}$$

Fig. 6.6 Manhattan distance
between points A and B

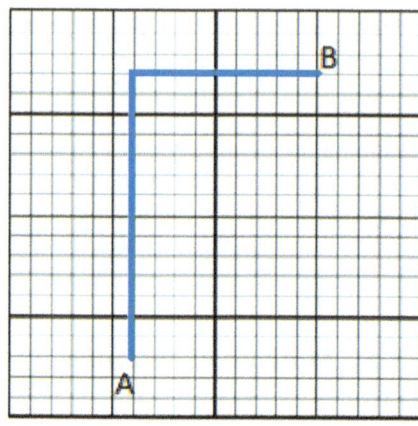

Fig. 6.7 Euclidean distance
between points A and B

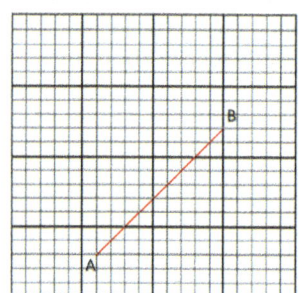

6.3.3 Euclidean Distance

From Mathematical viewpoint, the Euclidean distance between two points is the
straight line distance between two points in the Euclidean space (Fig. 6.7). A
generalized term for the Euclidean norm is the L_2 norm or L_2 distance. It is the
length of the line segment joining the two points. In two dimensions, the Euclidean
distance between two points $\mathbf{x}(x_1, x_2)$ and $\mathbf{y}(y_1, y_2)$ is given by

$$d(\mathbf{x}, \mathbf{y}) = \sqrt{[(x_1 - y_1)^2 + (x_2 - y_2)^2]}. \tag{6.17}$$

For n-dimensions, the Euclidean distance between two points $\mathbf{x}(x_1, x_2, x_3, \ldots x_n)$
and $\mathbf{y}(y_1, y_2, y_3, \ldots y_n)$ generalizes to

$$d(\mathbf{x}, \mathbf{y}) = [\sum_{i=1}^{n}(x_i - y_i)^2]^{1/2} = \sqrt{(x_1 - y_1)^2 + (x_2 - y_2)^2 + \ldots (x_n - y_n)^2}.$$

$$\tag{6.18}$$

6.3.4 Chebychev Distance

Chebychev distance represnts a special case of Minkowski distance (L_p norm) when $p \to \infty$. It is known as the L_∞-norm or Chessboard distance. It is expressed as:

$$d_\infty(\mathbf{x}, \mathbf{y}) = Lt_{p \to \infty} \left(\sum_{i=1}^{n} |x_i - y_i|^p \right)^{1/p} = ||x - y||_\infty = \max_{i=1}^{n} |x_i - y_i|.$$

(6.19)

At $p = \infty$, the expression of L_p norm becomes equivalent to only that difference in corresponding dimensions, which gives the maximum difference.

6.3.5 Hamming Distance

The Hamming distance between two strings having exact same length is the number of positions of mismatched symbols. It can be interpreted as the number of bits that need to be changed (corrupted) in order to transform one string into another. Sometimes instead of bits, characters are used. Hamming distance can be seen as Manhattan distance between bit vectors. In mathematical terms,

$$d_H(x, y) = \sum_{i=1}^{d} \delta(x - i, y_i);$$

(6.20)

where x and y are the binary strings, d is the length of the strings, x_i and y_i represent their respective elements at position i, and δ is the Kronecker delta function defined as,

$$\delta(a, b) = \begin{cases} 1 \; if \; a \neq b \\ 0 \; if \; a = b \end{cases}.$$

(6.21)

For example, let us consider the binary strings $x = 1010110$ and $y = 1001101$. The Hamming Distance between them would be:

$$d_H(x, y) = \delta(1, 1) + \delta(0, 0) + \delta(1, 0) + \delta(0, 1) + \delta(1, 1) + \delta(1, 0) + \delta(0, 1) = 3.$$

(6.22)

This implies that there are three positions where the bits in the two strings differ.

6.3.6   *Mahalanobis Distance*

Mahalanobis distance provides a way to measure how similar a set of points is to a known set of points. First, we consider a special case of weighted L_2 norm between two points x and y, where $\mathbf{B}$ is a symmetric, positive definite matrix, which is derived from the dataset under consideration.

$$D(\mathbf{x}, \mathbf{y}) = \sqrt{(\mathbf{x} - \mathbf{y})^T \mathbf{B}(\mathbf{x} - \mathbf{y})} \qquad (6.23)$$

The aforementioned weighted L_2 norm distance becomes the Mahalanobis distance when $\mathbf{B}$ becomes the inverse of covariance matrix (C) of the vectors $\mathbf{x}$ and $\mathbf{y}$. Mahalanobis distance D is then given by:

$$D(\mathbf{x}, \mathbf{y}) = \sqrt{(\mathbf{x} - \mathbf{y})^T \mathbf{C}^{-1}(\mathbf{x} - \mathbf{y})}. \qquad (6.24)$$

It takes the covariance among variables into consideration (Fig. 6.8). The distance between a vector x and a set of points having a mean m can be also expressed like

$$\mathbf{D} = \sqrt{(\mathbf{x} - \overline{\mathbf{m}})^T \mathbf{C}^{-1}(\mathbf{x} - \overline{\mathbf{m}})} \qquad (6.25)$$

where

$\mathbf{D}^2$ = Mahalanobis distance, $\mathbf{x}$ = input point
$\mathbf{m}$ = Mean value of the given set of points
$\mathbf{C}^{-1}$ = Inverse of variance-covariance matrix of x and $\overline{m}$

Fig. 6.8 Mahalanobis distance between points x and $\bar{\text{m}}$

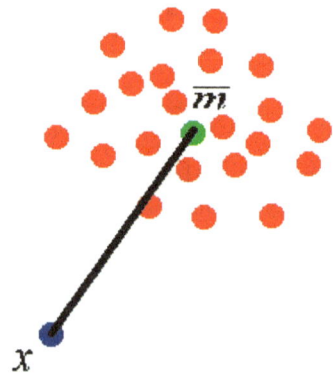

References

1. Bhattacharyya, A. (1946). On a measure of divergence between two multinomial populations. *Sankhyā: The Indian Journal of Statistics, 7*, 401–406.
2. Boriah, S., Chandola, V., & Kumar, V. (2008). Similarity measures for categorical data: A comparative evaluation. *Red, 30*(2), 3.
3. Dice, L. R. (1945). Measures of the amount of ecologic association between species. *Ecology, 26*(3), 297–302.
4. Jaccard, P. (1901). Etude comparative de la distribuition florale dans une portion des Alpes et des Jura. *Bulletin del la Société Vaudoise des Sciences Naturelles, 37*, 547–579.
5. Jordan, M., Kleinberg, J., & Schólkopf, B. (2005). *Information science and statistics*. Springer.
6. Kurzweil, R. (2012). *How to create a mind: The secret of human thought revealed*. Penguin Publishing Group.
7. Lindsay, B. G. (1994). Efficiency versus robustness: The case for minimum hellinger distance and related methods. *The Annals of Statistics, 22*(2), 1081–1114.
8. Menéndez, M. L., Pardo, J., Pardo, L., & Pardo, M. (1997). The Jensen-Shannon divergence. *Journal of the Franklin Institute, 334*(2), 307–318.
9. Mishra, S. N., Shah, A. K., & Lefante, J. J. (1986). Overlapping coefficient: The generalized approach. *Communications in Statistics-Theory and Methods, 15*(1), 123–128.
10. Moraru, A., Pesko, M., Porcius, M., Fortuna, C., & Mladenic, D. (2010). Using machine learning on sensor data. *Journal of Computing and Information Technology(CIT), 18*(4), 341–347.
11. Perlibakas, V. (2004). Distance measures for PCA-based face recognition. *Pattern Recognition Letters, 25*(6), 711–724.
12. Rubner, Y., Tomasi, C., & Guibas, L. J. (2000). The earth mover's distance as a metric for image retrieval. *International Journal of Computer Vision, 40*, 99–121.
13. Rüschendorf, L. (1985). The wasserstein distance and approximation theorems. *Probability Theory and Related Fields, 70*(1), 117–129.
14. Salton, G., & McGill, M. J. (1986.) *Introduction to modern information retrieval*. McGraw-Hill, Inc.
15. Tou, J. T. L., & Gonzalez, R. C. (1979). *Pattern recognition principles*. Addison-Wesley.

Chapter 7
Classification

The goal of a classifier is to partition the feature space into class-labeled decision regions. The decision boundaries do this job. A decision boundary separates points belonging to one class from points of other classes. The nature of the decision boundary is decided by the discriminant function used for decision. It is a function of the features. For a two-class problem, the simplest possible decision boundary is a line (Fig. 7.1a). If the number of dimensions is three, this will be a plane. The decision regions become hyper-volumes if the number of dimensions increases to more than three; and the decision boundaries become hyper-planes. For complex problems, decision boundaries are nonlinear curves (2D) (Fig. 7.1b), surfaces (3D), and hyper-surfaces for high dimensions. The final goal is to find out optimal decision boundaries in order to classify the incoming data into appropriate categories.

7.1 Training Set and Testing Set

A typical classification[6] problem comprises a training set, a validation set, and a test dataset. The process of using training data to determine the best optimal set of parameters (decision boundaries) for a classifier is called *training* the classifier [5]. It is like learning from examples. Usually, we need a dataset to design the classifier model. If we have a set of data whose true class labels are known, we divide the dataset into three parts: the training set for developing the model (determine its parameters), the validation set to checking the robustness of the built model, and the test set to check its performance (holding the parameters constant). We need a validation set (samples having known class labels but not used for training) for tuning the model (for instance, pruning a decision tree). The validation set cannot be used while testing. All three datasets need to have representative samples of the data that the model will be applied to. If the available data are sufficient in number, we can easily take multiple independent samples; one each for training, validation,

117
A. Ghosh, *Data Science and Cases in Sustainability*, Mathematics for Sustainable Developments, https://doi.org/10.1007/978-981-96-8362-8_7

Fig. 7.1 (**a**) Linear boundary
(**b**) nonlinear boundary

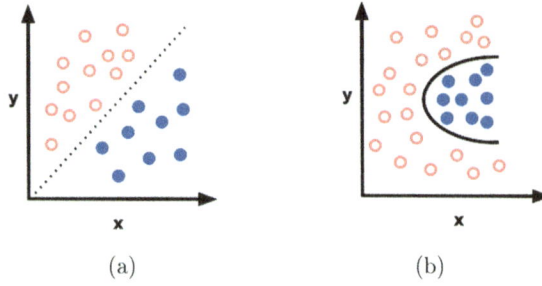

(a)   (b)

and testing. The more the training, the better is the classifier model. The more the test data, the more accurate is the error estimate. The concern is that procuring labeled data often become a costly and time-consuming process. So, we can obtain a limited dataset and use a holdout procedure, which involves setting aside a portion of the data for training and testing. To put it in the most straightforward way, the data are randomly splitted into test, validation, and training sets.

7.2 Overfitting, Underfitting, and Generalization

If we have a training set of patterns, we can design a classifier that can optimally classify the patterns. Now, there are a number of problems regarding designing an efficient classifiers. By optimal classification, we mean that the error should be minimum and the accuracy should be maximum. There are several other aspects to look at this optimality also.

7.2.1 Overfitting and Overlearning

The first is that we have a set of finite number of available labeled patterns to design the classifier. If we design a classifier that exactly fits the boundary of training points, it would be very susceptible to noise in the design set, i.e., it would classify correctly even the points deviate from the training set even by a small amount, thereby causing errors. This is the case of *over-fitting* (Fig. 7.2b) since we are unaware of what the noise is. In the overfitting case, the classifier is good at classifying training patterns; but the performance is poor for unknown data. Sometimes, it may so happen that not all training sets have the inputs classified correctly. This may cause problems if the algorithm used is effective enough to memorize even the seemingly "special cases" that do not fit to the more general principles. This can lead to overfitting. It is quite a challenge to find algorithms that are both efficient enough to learn complex functions and stable enough to produce generalizable results. For example, if a student learns how to add, multiply, and

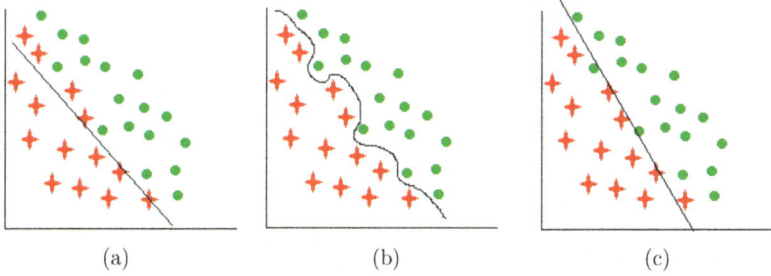

Fig. 7.2 Classifier designs: (**a**) Underfitting, (**b**) overfitting, (**c**) optimal-fitting

divide, she must have a generalization ability to learn subtraction on her own. When a model learns the training data too closely and does not generalize, this is called overlearning. The result is poor performance on data other than the training dataset. This is also called over fitting.

7.2.2   Underfitting and Underlearning

If the classifier is not complex enough, it may fail to capture the true structure of the training data. This is the case of *under-fitting* (Fig. 7.2a). In the underfit case, the classifier is not even good for the training patterns. Thus, a good classifier should provide an optimal performance on the dataset (Fig. 7.2c) for both training and testing data. In the underfit case, the classifier is not even good for the training patterns. It may be possible to achieve a 100% accuracy in classification for the test data in an overfit case, but a classifier with higher generalization capacity is much better and less susceptible to points that deviate from the training set. The other issue is how to optimize the performance of a classifier? There are several ways to measure classification performance. For example, a classifier may be trained using an optimizing criterion and tested using another. When a model has not learned enough about the structure of the input data because the learning process was terminated early, this is called underlearning. The result is poor performance on all data, including the training dataset. This is also called underfitting.

7.2.3   Generalization

Generalization is required because the model that is prepared by a machine learning algorithm needs to make predictions or decisions based on specific data instances that were not seen during training.

7.3   Decision Boundaries

Suppose we have a set of measurements (features), represented as a pattern vector $\mathbf{x} = \{x_1, x_2, x_3, \ldots, x_n\}$. We wish to know to which one (w_j) of C possible classes this pattern belongs to. A decision rule will then divide the measurement space into C regions $R_1, R_2, \ldots, R_C$ as shown in Fig. 7.3a. If a pattern $\mathbf{x}$ is in a particular region R_j, then it will belong to the class w_j. Each class may be made of several disjoint regions too as shown in Fig. 7.3b. The boundaries among regions are the decision boundaries or decision surfaces [11].

Let us assume that we have a two-dimensional data—height and weight of people from different regions (Fig. 7.4). There are two types of people in the dataset—people belonging to Asian origin and people belonging to the Caucasian origin. The circles are examples (training patterns) of Class 1 (Asians), and the diamonds are examples (labeled training samples) of Class 2 (Caucasians). Given a new data point (star), which type of race will it correspond to? The Asians usually are short in height and have varying weights, while the Caucasians are tall in stature. The decision boundary is the curve (black) separating the two classes. The boundary could be linear and nonlinear. The methods by which we can draw a straight line (a hyperplane in higher dimensions) between classes are called linear classifiers. Nonlinear classifiers design nonlinear boundaries, as in Fig. 7.5.

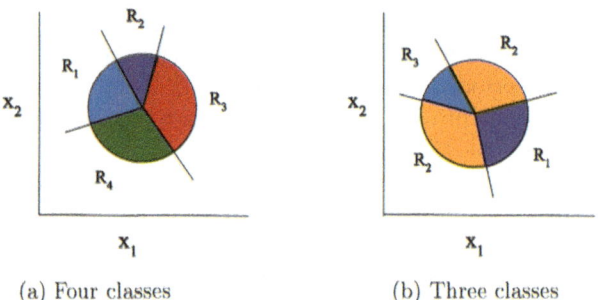

(a) Four classes   (b) Three classes

Fig. 7.3 Decision regions

Fig. 7.4 Training pattern distribution of people according to their heights and weights

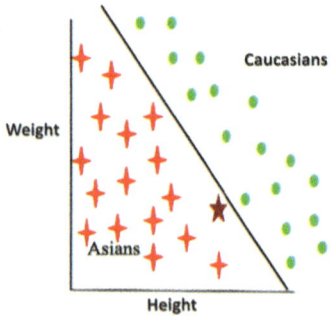

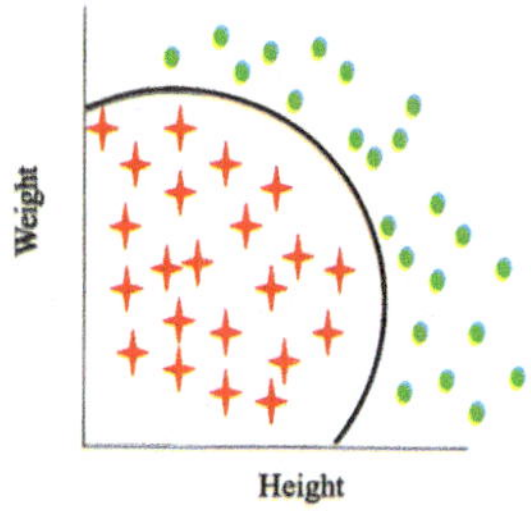

Fig. 7.5 Non-linear decision
boundary

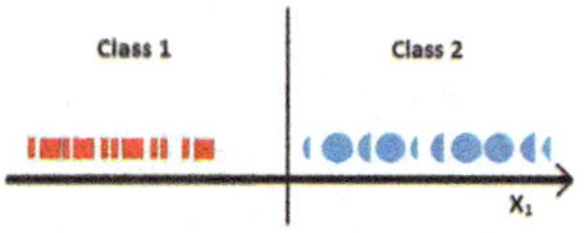

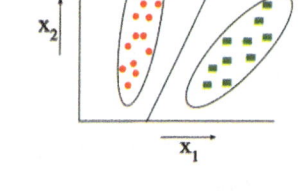

Fig. 7.6 One-dimensional
decision boundary

Fig. 7.7 Two-dimensional
decision boundary

In one dimension, refer to Fig. 7.6, the decision boundary can be described with the equation $x = c$, where c is a constant. If the input value is less than c, we assign it to "squares" (Class w_1); if it is more than c, we assign it to "circles" (Class w_2).

In 2D, a linear boundary is at $x_1 a_1 + x_2 a_2 + a_3 = 0$ (Fig. 7.7). In three dimensions, the corresponding decision boundary is a plane (in higher dimensions a hyperplane). The decision boundary for an n-dimensional data distribution will always be a hyperplane in $(n-1)$ dimensions. It can be expressed in vector notation as $\mathbf{a}^T \mathbf{x} = 0$. Here, $\mathbf{a}$ is called the augmented weight vector [3] and $\mathbf{x}$ is the augmented pattern vector.

$$a = \begin{Bmatrix} a_1 \\ a_2 \\ a_3 \\ \cdot \\ \cdot \\ \cdot \\ a_n \\ a_{n+1} \end{Bmatrix} \tag{7.1}$$

$$\mathbf{x} = \left\{ \begin{array}{c} x_1 \\ x_2 \\ x_3 \\ . \\ . \\ . \\ x_n \\ 1 \end{array} \right\} \qquad\qquad (7.2)$$

The scalar term a_{n+1} is added to the weight vector for coordinate translation purposes [3]. Nonlinear decision boundaries could be surface and hyper-surface.

7.4   Vapnik–Chervonenkis (VC) Dimension

Let us start with the representation of a learning problem. Let us assume we are looking at a classification task with two labels: "+" and "−." The data points are plotted in an d-dimensional space, and in doing a classification, what we are essentially doing is finding out a surface that has only points with the "+" label on one side of it and points with "−" labels on the other side.

So when a new data point comes, you want to

- Find out which side of this surface it falls.
- Announce the label of the new data point to be the label for that side.

VC dimension enables us to conduct our search in a principled way. For a family of surfaces, or to be precise, a family of functions—the VC dimension gives a number on which we can peg its capability to separate labels. The general idea is that the VC dimension points to a reasonable family of functions to inspect for boundaries. We pick a specific member within this family based on the exact dataset at hand.

As an example, a family of functions could be a set of hyperplanes (a plane in d-dimensions). Each member of this family is a specific hyperplane uniquely identified by its perpendicular distance from the origin and the direction of this perpendicular with respect to the origin.

It should be noted that the VC dimension is an estimate, and there are cases where this number might not be very helpful or can actually be counterintuitive.

If there are n data points, there can be 2^n possible labeling. For each of these labelings, if we can draw a function from the family of functions that separates the data, then this set of n points is said to have been shattered by the family of functions. The maximum number of points n that one can shatter is the VC dimension, h, of the family of functions.

Thus, the VC dimension gives a measure of confidence for the "separating capability" of the family of functions by looking at how many points one can actually separate knowing nothing about the distribution of labels.

There is a catch here though. A function with a VC dimension h, in general, will not be able to shatter all possible sets of h points; all that is guaranteed is that there is some set of h points that can be shattered.

7.5 Discriminant Analysis

A discriminant function $d(x)$ defines the decision hypersurface. As shown in Fig. 7.8, $d_k(x)$ and $d_j(x)$ are values of the discriminant functions for a pattern **x** respectively in classes k and j. $d_k(\vec{x}) - d_j(\vec{x}) = 0$ will then be the equation defining the surface that separates the classes w_k and w_j. The points for which $d_k(\vec{x}) > d_j(\vec{x})$ will belong to the class w_k; whereas those for which $d_k(\vec{x}) < d_j(\vec{x})$ will belong to the class w_j.

In discriminant function analysis, one or more continuous or binary independent variables (called predictor variables) are used to predict a categorical dependent variable (called a grouping variable). Discriminant analysis is a classification problem, where two or more groups or populations are identified a priori and one or more new observations are categorized into one of the known populations on the basis of the measured characteristics. For example, we have two populations of bank notes: genuine and counterfeit. Let us consider six measures on each note:

- Length
- Right-hand width
- Left-hand width
- Top margin
- Bottom margin
- Diagonal across the printed area

Fig. 7.8 Two-dimensional pattern space

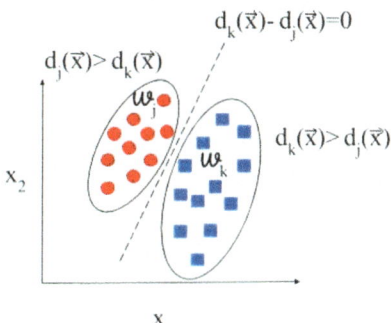

A bank note of unknown origin can be analyzed, and it can be determined just from these six measurements whether it is real or not. A note scanner can measure the notes automatically and makes a decision.

7.6 Multiclass Classification

Traditional multiclass classification problem can be mathematically defined as follows. Let us assume an object that could be represented by a vector $\vec{x} = \{x_1, x_2, \ldots, x_n\}$. This is an observation vector of the values of n features from the space X. A classification problem can be defined as a problem of assigning the object $\vec{x}$ to a class w_i, where $w_i, i = 1, 2, \ldots, C$ represents the label of class i.

The training dataset DS_T consists of observation vectors along with the corresponding class labels: $DS_T = \{\vec{x}, w_j\} | j = 1, 2, \ldots, C$, where w_j is the class label. In turn, the test dataset denoted as DS_C comprises observation vectors of feature values without class labels, to be classified and is denoted as $DS_C = \{\vec{x}\}$.

Using a training dataset, the classifier can learn a set of parameters Θ that constitute a model of the said classifier. When provided with an unlabeled observation vector $\mathbf{x}$, the classifier gives an output $u(x, \Theta)$ as result. Possible values of the output could be $u(x, \Theta) \in \{w_1, w_2, \ldots, w_C\}$, i.e., the classifier can assign the object to one of the C class labels.

Alternatively, for each class w_i, the classifier may implement a real-valued discriminant function $u_{w_i}(\vec{x}, \Theta)$, such that the greater values of this function correspond to the higher probability of class membership $p(x \in w_i | X) = p(w_i | x)$ [2]. In this case, the class with the highest value of discriminant function gets chosen:

$$\gamma(\vec{x}, \Theta) = \underset{i=1,2,\ldots C}{\arg\max}\, u_{w_i}(\vec{x}, \Theta), \tag{7.3}$$

where γ is a mapping function. For two-class classification problem, a single discriminant function in a form $u(\vec{x}, \Theta) = p(w_1 | \vec{x}) - p(w_2 | \vec{x})$ is sufficient for implementing the classification task as

$$\gamma(x, \Theta) = \begin{cases} w_1, & \text{if } u(x, \Theta) \geq 0 \\ w_2, & \text{if } u(x, \Theta) < 0. \end{cases} \tag{7.4}$$

7.7 Types of Classifiers

The block diagrammatic representation of a classifier is shown in Fig. 7.9. A classifier assigns each point of the input space to one class. Thus, the input space

Fig. 7.9 Black box
representation of a classifier

is divided into disjoint subsets, called decision regions, each of which is associated with an individual class.

Classifiers can mainly be grouped into two categories—Discriminative and Generative [4]. **Discriminative** classifiers are used for modeling the dependence of an unobserved variable w on an observed variable $\vec{x}$. In probabilistic terms, it is done by modeling the conditional probability $p(w|\vec{x})$ where w can be predicted from the knowledge of $\vec{x}$. A **Generative** classifier is used for randomly generating observable data, typically given some hidden parameters. It specifies a joint probability distribution $p(\vec{x}, w)$ over observation and label sequences. Generative classifiers are a contrast to the discriminative ones. A generative model is a complete probabilistic model of all variables, whereas a discriminative model includes a model based on the target variable(s) conditional on the observed variables. In other words, a discriminative classifier does not care about how the data were generated, it just categorizes that. On the other hand, a generative classifier models how the data were generated in order to categorize it. It looks for an answer to the question—"based on my generation assumptions, which category is most likely to generate this data?"

7.7.1 Generative Classifiers

A generative classifier tries to learn the model that generated the data behind the scenes by estimating the distributions of the data. It then uses this to predict unknown data, as it assumes that the model it learned represents the real model. As we will see, this is often not true. An example is the Naive Bayes' classifier.

To get $p(w|\vec{x})$, generative classifiers do the following:

- Estimate $p(\mathbf{x}|w)$ and $P(w)$ from the data. $P(w)$ is the apriori probability of category w.
- Multiply $p(\mathbf{x}|w)$ and $P(w)$ according to Bayes' theorem, $P(x)$ is a constant.
- Get $p(w|\mathbf{x})$.

To continue with an example, let us consider a person who lives in Delhi and earns over Rs 1,00,000. In this scenario, we can assume two events: a person living in Delhi (event-x) and a person earning over Rs 1,00,000 (event-w). A generative classifier first estimates, from the data, what is the probability that a person lives in Delhi given he earns over Rs 1,00,000, i.e., together it means $p(x|w)$. Then, it estimates how many people earn over Rs 1,00,000, regardless of if he is in Delhi or not, i.e., $P(w)$. Based on these two probabilities, it estimates the probability that given a person lives in Delhi (x), he makes over Rs 1,00,000 as salary (w), i.e., $p(w|x)$.

7.7.2 Discriminative Classifiers

A discriminative classifier tries to model by just depending on the observed data. It makes fewer assumptions on the distributions but depends heavily on the quality of the data. For example, logistic regression.

To get $p(w|\vec{\mathbf{x}})$, discriminative classifiers estimate $p(w|\vec{\mathbf{x}})$ from the data. As with our example, a discriminative classifier estimates the probability of a man making over Rs 1,00,000, given he lives in Delhi, i.e., $p(w|x)$.

The discriminative models are preferred over the generative ones. If it comes to understand speech, generative approach is to learn each language and determine as to which language the speech belongs to. Discriminative approach is to determine the linguistic differences without learning any language—a much easier task. Discriminative algorithms model the decision boundary between classes. Generative algorithms model the actual distribution of the classes. In practice, the generalization performance of generative models is often found to be poorer than that of discriminative models due to differences between the model and the true distribution of the data. However, generative models can exploit unlabeled data in addition to labeled data [10].

7.8 Evaluating the Performance of a Classifier

Creating classifiers is a multistage process. After generating a classifier from the given training dataset, validation and evaluation on the test examples should be done and the model should then be used for new examples. There are several classifier evaluation criteria.

- **Predictive (classification) accuracy:** The ability of the model by which it correctly predicts the class label of new or previously unseen data:

  accuracy = % of the test set that is correctly classified by the trained classifier.

- **Speed:** The computational costs involved in developing and using the model.
- **Robustness:** The ability of the model to make correct predictions if the given data is noisy or has several missing values.
- **Scalability:** The ability to construct the model efficiently, provided with a huge amount of data.
- **Interpretability:** The level of understanding and insight of the model.

Measuring the model's output on the test set is often helpful to the performance of the model because such a metric gives an unbiased approximation of its generalization error. The general paradigm is "train and test." The dataset is partitioned into train and test parts based on different techniques:

7.8.1   Cross-Validation

Cross-validation [1] avoids overlapping test sets. It is a method for testing and not model development. Cross-validation is a statistical method to evaluate and compare learning mechanisms by separating the data into two portions: one used during learning or training of a model and the other used during validation of the model. In typical cross-validation, the training and validation sets must cross-over in subsequent rounds to ensure that each data point gets a chance of being validated against. The following are some of the popular cross-validation techniques.

7.8.1.1   Hold-Out

Hold-out validation aims to avoid the overlap between training data and validation data, thus leading toward a more accurate estimate for the generalizing capability of the algorithm. Here, an independent test set is preferred to avoid overfitting. A natural technique is to distribute the available data into two non-overlapping sections: first for training and the second for testing. The test data are kept separately and are not used while training.

This method is very common in estimating the classifier accuracy. The entire dataset in this case is split into two independent datasets. The training set (say, for example) comprises 2/3 of the data, and the validation set comprises 1/3 of the data. It is important that the validation data are not used in any way to create the classifier. A classification model is then developed from the training set, and its performance is evaluated on the test set. This method, however, has certain limitations.

- Fewer instances are available for training the classifier as some examples are withheld for testing.
- The model will also not be independent of the training and the test data composition.
- In case of smaller data size, there will be a larger variance of the model. On the other hand, if the dataset is too large, the estimated accuracy will be low.
- The training and testing datasets are not actually independent as they are subsets of the original data only.

The hold-out method may be repeated several times to improve the classifier's performance estimation. The process is still not optimum because the different test sets usually overlap (difficulties from statistical point of view).

7.8.1.2 k-Fold

In k-fold cross-validation, the data are first divided into k equally (or almost equally) sized segments or folds. Thereafter, k iterations are done for training and validation. Within each of these iterations, a different fold of the data is held out for validation while the remaining $k - 1$ folds are used for learning. Data are usually stratified before splitting it into k folds. Stratification is the method by which data are rearranged for ensuring that each fold is a good representative of the whole sample. For instance, in a binary classification problem where each class consists of 50% of the data, it is better to arrange the data such that in every fold, each class contains around half of its instances.

This method could be efficient for datasets with moderate size. To demonstrate this method, let us distribute a dataset into two equal-sized subsets. At first, we select one of the subsets for training and the other for testing. Then in the next step, the data used earlier for testing are used for training, and the one used earlier for training is now used for testing. The total error is given by the sum of the errors in both runs.

A large number of estimates are always preferred to obtain reliable performance estimation or comparison. In k-fold cross-validation, only k estimates are procured. A widely known method of increasing the number of estimates is to perform multiple runs of the k-fold cross-validation. Before each round, the data should be reshuffled and re-stratified.

7.8.1.3   Leave-One-Out

A special case of k-fold method known as the *leave-one-out* approach where k equals total size of the dataset, and the validation set contains only one data in each run. In other words, almost all the data are used for training in each iteration except a single observation and that left-out observation is used to test the model. This method can be repeated N times so as to utilize the maximum data available. This approach has the disadvantage that the procedure has very large variance.

7.8.1.4   Bootstrapping

Bootstrap [7] method is used for small data size. Unlike the previous methods, this process relies on random sampling with replacement. It means that data already drawn for training are kept back into the original set of data so that it is equally probable to be chosen again. A dataset of N instances is sampled N times with replacement to create a new dataset of N instances. These data are used as the training set. If the original set has N data, an instance has a probability of $1 - 1/N$ of not being picked. Thus, its probability ending up in the dataset becomes $(1 - 1/N)^N \approx e^{-1} = 0.368$. This means that the training data will contain approximately 63.2% of the instances. Data that are not part of the bootstrap sample get included in the test set. The error estimate on the test data will be very pessimistic, since training was on just about 63% of the instances. Therefore, the total error becomes:

$$err = 0.632error_{test} + 0.368error_{train}. \qquad (7.5)$$

The process is repeated for several times with different replacement samples and then the results are averaged. It is probably the best way to estimate performance for very small datasets.

7.9 Confusion Matrix

A confusion matrix [9] contains information about the actual classes of the data and the predicted classes of the data by a classification system. Performance of such a system is commonly evaluated using the data in the matrix. Let us take an example of 3 classes: 15 cats, 14 dogs, and 16 rabbits (Table 7.1).

The following notations are commonly used (with respect to the class "Cats"):

Cats predicted as cats = 13 → **True Positives.**
Some other animal predicted as cats = 2 → **False Positives.**
Cats predicted as some other animals = 2 → **False Negatives.**
Some other animal predicted as other animals = 28 → **True Negatives.**

Thus, the confusion matrix for class cats stands as (Table 7.2):

Table 7.1 General confusion matrix

Actual class	Predicted class		
	Cats	Dogs	Rabbits
Cats	13	1	1
Dogs	1	12	1
Rabbits	1	2	13

Table 7.2 Confusion matrix for the class "Cats"

True Positives (23)	**False Negatives** (2)
False Positives (2)	**True Negatives** (28)

7.10   Evaluation Measures

Several standard terms have been defined to determine the performance of the classifier; and they are described in the following sections in brief.

7.10.1   Accuracy

Accuracy (AC) is the proportion of the total number of predictions that were correct. It is calculated from the following equation:

$$AC = \frac{TP + TN}{TP + FP + FN + TN}. \tag{7.6}$$

7.10.2   Precision

Precision (P) is the proportion of the predicted positive cases that were correct, computed as:

$$P = \frac{TP}{TP + FP}. \tag{7.7}$$

7.10.3   Recall

Recall (R) is the proportion of positive cases that were correctly identified, calculated using the equation:

$$R = \frac{TP}{TP + FN}. \tag{7.8}$$

It is also known as True Positive rate or True Positive fraction.

7.10.4   f-Measure

Accuracy (AC) may not be an adequate performance measure if the number of negative cases exceeds the number of positive cases. Suppose there are 1000 cases, 995 among them are negative cases, and the remaining 5 are positive cases. If the system classifies all of them as negative, then accuracy would be 99.5%, even though the classifier missed all positive cases. So, another performance measure called the f-measure [8] accounts for this by including R in a product form. It is determined using the equation:

$$f - \text{measure} = \frac{(\beta^2 + 1)PR}{\beta^2 P + R}.$$ (7.9)

It assesses the Precision and Recall trade-off by doing a weighted harmonic mean. Here, β is a positive real weight to give importance to the two measures—Precision and Recall and provides a single measurement for a system. The balanced F1-measure is often used as a useful measure by taking the harmonic mean only, i.e., with $\beta = 1$. Thus,

$$\text{F1-measure} = \frac{2PR}{P + R}.$$ (7.10)

7.11 ROC Curve

A receiver operating characteristics (ROC) graph is a methodology for visualizing, organizing, and selecting classifiers based on their performance. ROC graphs are two-dimensional graphs in which true positive (TP) rate (or fraction) $\left(\frac{TP}{TP + FN} \right)$ is plotted on the Y axis and false positive (FP) rate (or fraction) $\left(\frac{FP}{FP + TN} \right)$ is plotted on the X axis. An ROC graph depicts relative trade-offs between benefits (true positives) and costs (false positives) (Fig. 7.10).

Assume two categories, where our aim is to detect a single object against all others (a binary classifier). Let us define the following terms.

"Hit" —correctly classifying the object (true positive).
"False Alarm" —incorrectly detecting an object when it is not there (false positive).

Fig. 7.10 ROC curve

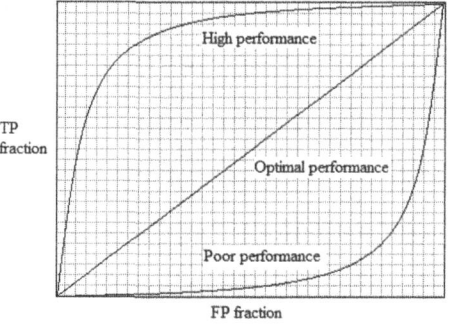

TP
fraction

High performance

Optimal performance

Poor performance

FP fraction

"Miss" —finding no object when it is there (false negative).
Probability of "hit" vs the probability of "false alarm" is called an ROC curve.

An ROC curve is usually used to compare the classifiers. Higher the value of TP
for a fixed FP, better is the performance. We are therefore aiming at maximizing the
area under the curve (AUC).

7.12   Precision–Recall Curve

In most situations when some test performances are evaluated, it is seen that the
data obtained are heavily skewed or imbalanced. For example, when performing
a particular biomedical test, it is seen that most of the subjects do not have the
disease or medical condition tested for. This happens because typical prevalence
of common diseases are in the range of   10%. It means that only   10% of
the patients with symptoms that suggest a particular disease will ultimately be
diagnosed as having that disease, and   90% will not have that disease. A receiver
operating characteristics (ROC) curve is often used when the clinical performance
of a biochemical test is evaluated. The ROC curve shows the connection/trade-off
between clinical sensitivity and specificity for every possible cutoff for a test or a
combination of tests in a graphical way. The area under the ROC curve gives an idea
about the benefit of using the test in question.

Nonetheless, it may be misleading to have a visual interpretation and comparisons of ROC curves based on imbalanced datasets. A precision–recall curve (PRC)
is an alternative to a ROC curve. It is used less often than ROC curves, but PRC
could prove to be a better choice for imbalanced datasets. Precision and Recall are
usually inversely related, i.e., as precision increases, recall drops and vice versa. The
system needs to achieve a balance between these two, and to accomplish this and
to compare performance, the precision–recall curves come in handy. Figure 7.11
shows an example of a precision–recall curve.

Fig. 7.11 Precision–recall
curve

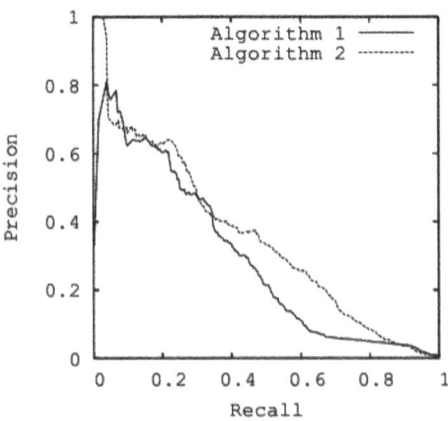

Let us consider an example. Let there be a database with 100 documents, out of which 60 are relevant to a particular keyword. If the information retrieval system returns a total of 50 documents, out of which 40 are relevant, the precision for this system is $40/50 = 0.8$ and the recall is $40/60 = 0.66$. If instead there is another system that returns only 10 documents, chances are that atleast 9 of them are relevant. This would increase my precision to 0.9 but decrease its recall to just 0.15. Thus, the aforementioned two systems need to be analyzed and compared with respect to both precision and recall. One would choose the appropriate system depending on the need (high precision, or more data with false positives allowed). This trade-off between precision and recall can be observed using the precision–recall curve, and an appropriate balance between the two obtained.

References

1. Bengio, Y., & Grandvalet, Y. Bias in estimating the variance of k-fold cross-validation. In: Duchesne, P., RÉMillard, B. (eds) *Statistical modeling and analysis for complex data problems* (pp. 75–95). Springer, Boston, MA. https://doi.org/10.1007/0-387-24555-3_5.
2. Bishop, C. M. (1995). *Neural networks for pattern recognition*. Oxford University Press.
3. Bow, S. T. (2002). *Pattern recognition and image preprocessing* (Signal processing and communications). CRC Press.
4. Dietterich, T. G., Becker, S., & Ghahramani, Z. (2002). *Advances in neural information processing systems 14. Proceedings of the 2002 conference*. MIT Press.
5. Dougherty, G. (2012). *Pattern recognition and classification: An introduction*. SpringerLink : Búcher. Springer.
6. Duda, R. O., Hart, P. E., & Stork, D. G. (2012). *Pattern classification*. Wiley.
7. Jiang, W., & Simon, R. (2007). A comparison of bootstrap methods and an adjusted bootstrap approach for estimating the prediction error in microarray classification. *Statistics in Medicine, 26*(29), 5320–5334.
8. Lewis, D. D., Schapire, R. E., Callan, J. P., & Papka, R. (1996). Training algorithms for linear text classifiers. In *Proceedings of the 19th annual international ACM SIGIR conference on research and development in information retrieval* (pp. 298–306).
9. Provost, F., & Kohavi, R. (1998). Guest editors' introduction: On applied research in machine learning. *Machine Learning, 30*(2), 127–132.
10. Vapnik, V. N., & Vapnik, V. (1998). *Statistical learning theory* (vol. 1). Wiley.
11. Webb, A. R. (2003). *Statistical pattern recognition* (Wiley InterScience electronic collection). Wiley.

Chapter 8
Classifiers

The information to design a classifier is usually represented by a labeled dataset of N patterns

$$\mathbf{X} = \{\mathbf{x}_1, \mathbf{x}_2, \ldots, \mathbf{x}_N\}, \mathbf{x}_j \in R^n. \tag{8.1}$$

The class label of pattern $\mathbf{x}_j$ is denoted by $l(\mathbf{x}_j) \in \Omega$, $j = 1, \ldots, N$ where Ω refers to the set of C classes. A classifier is defined by a function from an n-dimensional space of real numbers to a set of C classes, i.e., any function:

$$F : R^n \to \Omega = \{\omega_1, \omega_2, \ldots, \omega_c\}. \tag{8.2}$$

The classifier F maps the data points in R^n to one of the classes in Ω.

Classification pertains to known number of groups, and the objective is to assign new data points to one of these groups. We know exactly how many groups exist. We have some data that are known to come from different classes. We have new data points whose class labels are not known, and the task is to assign label to this data. Based on the information of the labeled samples, we can build classification rules to classify the new data points into one of the available classes. We must decide the way of distributing the feature space such that when we are provided with the feature vector of a test object, we can determine, quantitatively, to which of the C classes it belongs. A classification system (Fig. 8.1) has three phases:

Training: Input a training set of examples $\mathbf{X} = \{(\mathbf{x}_1, l(\mathbf{x}_1)), (\mathbf{x}_2, l(\mathbf{x}_2)), \ldots, (\mathbf{x}_N, l(\mathbf{x}_N))\}$, where ($\mathbf{x}_i$ is an observed input data and $l(\mathbf{x}_i)$ is the corresponding output label). We are given a set of training data from which the classifier is expected to learn. Based on these training data, the system will generate a classifier (function) that will map each data sample to a corresponding class label.

Validation: In this phase, a part of the labeled dataset (unseen to the classifier during training) is used to evaluate the performance of the trained classifier.

Fig. 8.1 Block diagram of a
classifier system

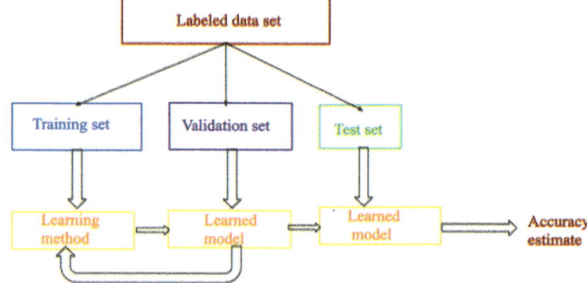

Depending on this performance, it is decided if the classifier needs to be retrained.
This step may be omitted if there is a scarcity of labeled data.

Testing: Testing of the classifier is done by analyzing whether it produces the
correct output $l(\mathbf{x}_t)$ for future examples $\mathbf{x}_t$.

In the following sections, we will describe the working principles of some of the
popular classifiers.

8.1   Minimum Distance Classifier

One of the most simple (but not the most computationally efficient) techniques
is implementing a supervised, distribution-free approach known as the minimum
distance classifier. The minimum distance classifier is used to classify unknown
new data to classes, which minimizes the distance between the test data and the
class mean. The distance is defined as an index of dissimilarity so that the minimum
distance is identical to the maximum similarity. Each class is represented by a
prototype (or mean) vector in the multidimensional feature space. For any class
w_j (where $j = 1, 2, 3, \ldots C$) having N_j number of training patterns represented by
$\mathbf{x}_i$ (where $i = 1, 2, 3, \ldots, N_j$) the mean vector can be represented as:

$$\mathbf{m}_j = \frac{1}{N_j} \sum_{i=1}^{N_j} \mathbf{x}_i; \tag{8.3}$$

where $\mathbf{x}_i \in w_j$.

For an example, let us take two classes—green circles and red diamonds. We
represent them in a two-dimensional feature space with features x_1 and x_2 as shown
in Fig. 8.2.

The pattern $\mathbf{x}_i$ is represented as $\mathbf{x}_i = \{x_1, x_2\}^T$. A mean for each class is
represented by m_j ($j = 1, 2, 3, \ldots C$). The minimum distance classifier basically
measures the distance of each of the patterns from all of the means. Then, it assigns
the pattern to that class that has the minimum distance from the mean. Suppose
distance between m_j and pattern vector $\mathbf{x}_i$ is less than distance between m_k and $\mathbf{x}_i$.

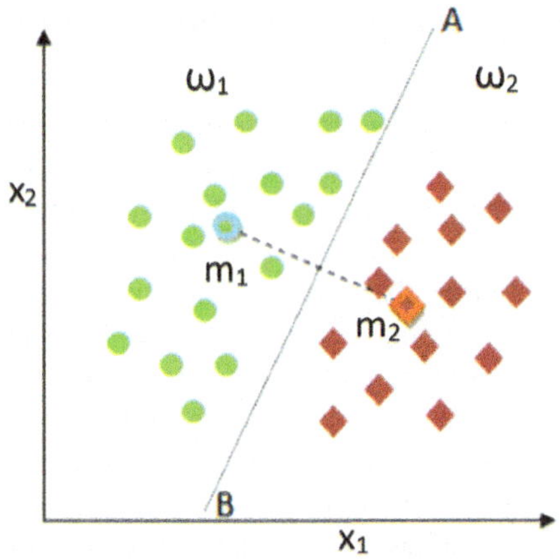

Fig. 8.2 Minimum distance classifier

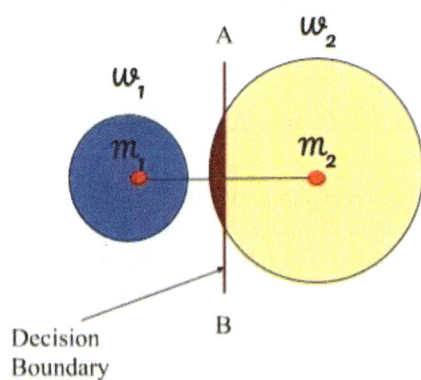

Fig. 8.3 Minimum distance classifier is insensitive to differences in variance among categories

In that case, the pattern x_i will be classified as class w_j. If the case arises such that the distance from both the means is the same for a pattern, then it cannot be decided to which class it will belong.

$$x_i \in m_j, \text{dist}(x_i, m_j) \text{ is minimum } \forall j. \tag{8.4}$$

If we draw the locus of all points, which are of equal distance from the two means, the line that comes out is the perpendicular bisector of the line joining the two means. This becomes the decision boundary whose either side represents the two class regions. In our example, if a new pattern falls on the left side of the line $\overleftrightarrow{AB}$, then it belongs to class w_1, and if it falls on the right side, it will belong to class w_2. So, the minimum distance classifier is a linear classifier. It classifies every pattern in the dataset regardless of the probability of its actual class. Also, it does not explicitly consider the variance within classes (Fig. 8.3). This classifier works the best when the two classes are more or less equal size (like, they have similar variance).

8.2   Nearest Neighbor Classifier

Among the various methods of supervised statistical pattern classification, the Nearest Neighbor algorithm achieves consistently high performance, without a priori assumptions about the distributions from which the training examples are drawn. The nearest-neighbor method is perhaps the simplest of all algorithms for predicting the class of a test example. The training phase is trivial: simply store every training example, with its label [13]. It can be used even with few examples. To classify a pattern $\mathbf{x}$, it finds its closest neighbor (call it $\mathbf{x}'$) among the training points, and assigns to $\mathbf{x}$ the label of $\mathbf{x}'$. Since it finds only one nearest neighbor, it is often called 1-NN classifier. Consider Fig. 8.4. The red dots are the training examples of a class; whereas the green set is from another class. Here the unknown pattern (blue dot) has the nearest neighbor that belongs to the class of red dots. So, the unknown pattern will be classified to the class of red dots.

8.2.1   Issues Regarding Nearest Neighbor Classifier

The nearest neighbor (NN) technique is very simple, highly efficient and effective in the field of pattern recognition, text categorization, object recognition, etc. The NN rule utilizes only the class label of the nearest neighbor. The remaining test patterns on the feature space are ignored. Surprisingly, it can be shown [9] that, in the large sample case, this simple rule has a probability of error, which is less than twice the Bayes probability of error, and hence, is less than twice the probability of error of any other decision rule. It uses local information that can yield highly adaptive behavior. It can also be lent to parallel implementation. Its simplicity is its main advantage. But the disadvantages can't be ignored even. The memory requirement and computational complexity also matter. In addition, it is highly susceptible to the curse of dimensionality. Moreover, if there are two patterns that lie at the same distance to the unknown pattern, then the classifier will randomly select any of the two, and there is a chance of misclassification. So, a new mechanism was thought to find the k-nearest neighbors instead of just one, where k is a positive integer constant (more than one).

Fig. 8.4 Nearest neighbor classifier

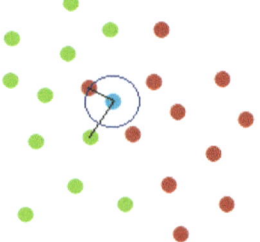

Since the training examples are needed at run time, i.e., they need to be in memory at run time, it is sometimes also called memory-based classification. Now, as induction is delayed to run time, it is considered as a *Lazy Learning Technique*. Also, because classification is based directly on the training examples, it is called *Example-Based Classification* or *Case-Based Classification* [18].

8.2.2   k-Nearest Neighbor Classifier

The k-nearest neighbors are selected based on a distance metric. Then there are a variety of ways in which the k nearest neighbors can be used to determine the class of the unknown pattern. The most straightforward approach is to assign the new pattern to the majority class among the k-nearest neighbors selected. So, the technique is more commonly referred to as k-Nearest Neighbors (k-NN) classification where k-nearest neighbors are used in determining the class. The basic idea is shown in Fig. 8.5, which depicts a 3-NN classifier ($k = 3$) on a two-class problem in a two-dimensional feature space. In this example, the decision for the unknown pattern is straightforward—the unknown pattern has two neighbors of class red dots and one of class green dots, so it is resolved by simple majority voting. As the red dots have a majority among the three nearest neighbors, the unknown pattern will be classified to the class of red dots. So, k-NN classification has two stages: the first is the determination of the nearest neighbors, and the second is the determination of the class label using these neighbors.

Usually for a C class problem, we specify k such that:

- k is greater than C and
- k is not a multiple of C.

8.2.3   Issues Regarding k-Nearest Neighbor Classifier

Inspite of these conditions, the proper choice of k is still a major issue. For example, let us suppose we have a two-class problem and we chose $k = 3$. There may arise a

Fig. 8.5 3-nearest neighbors classifier

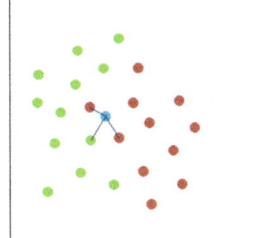

Fig. 8.6 k-nearest neighbors classifier having equal distance from the test data point

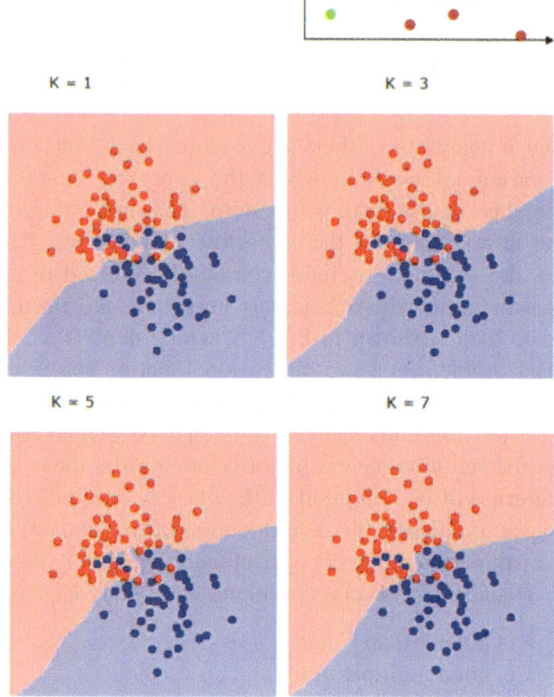

Fig. 8.7 Low values of k create an overfitting complex decision boundary while increasing the value of k simplifies the boundary too much

case (Fig. 8.6) where six points can have equal distances from the sample pattern. $k = 3$ creates a confusion about which three to choose. This might result in an error in classification as all the three would then be chosen randomly.

Smaller values of k make the decision boundary sensitive to noise in the data. Larger values of k reduce the effect of noise on classification, but make boundaries between labeled patterns distinct [31]. The k-NN classifier builds a piecewise linear decision boundary. Higher the value of k, more simple will be the decision boundary (Fig. 8.7).

Using large values of k is detrimental too. It destroys the local nature of estimation. Moreover, the computational cost gets high when dealing with large values of k. Choosing an appropriate k is therefore essential to make the classification more successful. Moreover, some importance should be given to the closer nearest neighbors than the others. The last problem is, however, mitigated by taking into account the weighted k-NN method.

8.2.4 Weighted k-Nearest Neighbor Classifier

It is reasonable to say that one might wish to weigh the existence of neighbors close to an unclassified observation more heavily than the existence of another neighbor, which is at a greater distance from the unclassified observation. Therefore, one would like to have a weighting function w^j, which varies with the distance between the unknown sample and the considered neighbor in such a manner that the value decreases with increasing sample-to-neighbor distance. The value of w^j varies from a maximum of 1 for the nearest neighbor down to a minimum of 0 for the most distant of the k-neighbors. One such weighting function is the "distance-weighted k-nearest neighbour rule" [12].

If the training samples N are larger in number compared to the number of nearest neighbors considered, then it is reasonable to expect that the results obtained through the two rules—the weighted k-nearest-neighbor rule and the simple majority k-nearest-neighbor rule, will be comparable to each other.

8.3 Nearest Feature Line (NFL)

The nearest feature line (NFL) method extends the classification capability of the nearest neighbor (NN) method by taking advantages of multiple (more than one) templates per class. It effectively improves the classification performance especially when the number of templates (sample size) per class is small, a problem frequently encountered in many applications such as face recognition.

Consider a variation in the image space from point z_1 to z_2 and the incurred variation in the feature space from $||x_1$ to $x_2||$. The degree of the change may be measured by $\delta z = ||z_1 - z_2||$ or $\delta x = x_1 - x_2$. When $\delta z \to 0$, $\delta x \to 0$, the locus of x due to the change can be approximated well enough by a straight line segment between x_1 and x_2 when the variation is small enough. Thus, any change between the two can be interpolated by a point on the line. A further small change beyond x_2 can be extrapolated using the linear model. Suppose we have two images, z_1 and z_2, representing faces of the same individual. The corresponding feature vectors are x_1 and x_2. As we move from z_1 to z_2, the corresponding change in the feature space, $x_1 - x_2$, can be approximated by a straight line segment between x_1 and x_2. This approximation becomes more accurate as the change becomes smaller. This concept is fundamental to the NFL method and its effectiveness in classification tasks.

The straight line passing through x_1 and x_2 of the same class, denoted by $\overline{x_1 x_2}$, is called a Feature Line (FL) of that class. The query feature point x is projected onto an FL as a point p (Fig. 8.8).

The nearest feature line (NFL) is given as:

$$d(\mathbf{x}, \overline{\mathbf{x}_{i*}^{w*} \mathbf{x}_{j*}^{w*}}) = \min_{1 \le w \le C} \min_{1 \le i < j \le N_w} d(\mathbf{x}, \overline{\mathbf{x}_i^w \mathbf{x}_j^w}) \tag{8.5}$$

Fig. 8.8 Generalization of
two points $\mathbf{x}_1$ and $\mathbf{x}_2$ to the
feature line $\overline{\mathbf{x}_1\mathbf{x}_2}$. The feature
point $\mathbf{x}$ of a query face is
projected onto the line as
point $\mathbf{p}$.

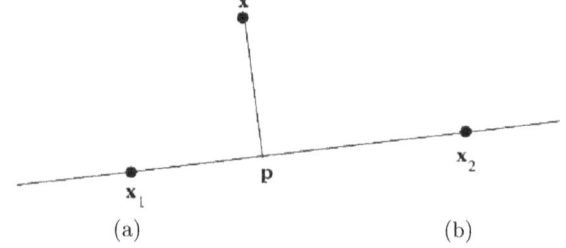

(a)   (b)

Here, $d(\mathbf{x}, \overline{\mathbf{x}_{i*}^{w^*} \mathbf{x}_{j*}^{w^*}})$ represents the distance between a query feature point $\mathbf{x}$ and
a feature line $\mathbf{x}_{i*}^{w^*} \mathbf{x}_{j*}^{w^*}$ of a specific class w^*. This measures how close the query
point is to the feature line. $\mathbf{x}$ is the query feature point that we want to classify.
w represents the class label, ranging from 1 to C, where C is the total number of
classes. In the context of face recognition, this may be different individuals. C is the
total number of classes (e.g., total number of individuals in face recognition). N_w
is the number of templates (or samples) available for class w. $\mathbf{x}_i^w$ refers to the i-th
template (or sample) of class w. $\mathbf{x}_j^w$ refers to the j-th template (or sample) of class
w. $\overline{\mathbf{x}_i^w \mathbf{x}_j^w}$ denotes the Feature Line for class w formed by templates i and j. This is
a straight line in the feature space that best approximates the distribution of features
for class w based on the available templates.

8.4 Bayes' Decision Theory

Bayes' Decision Theory is named after the Reverend Thomas Bayes, an eighteenth-
century statistician, and it forms the basis of various classification and prediction
methods used in diverse fields. At its core, this theory relies on probability to
facilitate informed decision-making. It is particularly valuable when dealing with
uncertain or noisy data. In this section, we are going to discuss Bayesian classifier
built based on the Bayes' Decision Theory.

8.4.1 Bayes' Theorem in Probability

The conditional probability of an event is the probability attained with the additional
information that some other event has already happened. We use $P(B|A)$ for
representing the conditional probability of event B occurring, given that event A
has already occurred. The following formula holds for finding $P(B|A)$ [14]

$$P(B|A) = P(B \cap A)/P(A); \tag{8.6}$$

where $P(B \cap A)$ is the probability of joint occurrence of event A and B.

Adding with the given rule, there is also an intuitive approach for finding a conditional probability. The conditional probability of B given A can be estimated by using the assumption that event A has happened already and, working under that assumption, finding the probability that event B will occur. In this section, the discussion of conditional probability is extended for including applications of Bayes' theorem [41] (or Bayes' rule), which we use for revising a probability value based on additional information that is later obtained. One key to understand the implication of Bayes' theorem is to identify the fact that we are working with sequential events, whereby new extra information is procured for a subsequent event, and that new information is used for revising the probability of the initial event. In this context, the terms prior probability and posterior probability are generally used. A **prior probability** is an initial probability value originally procured prior to any additional information is obtained. A **posterior probability** is a probability value that has been determined by using additional information that is later obtained.

The probability of event A, given that event B has previously occurred, is

$$P(A|B) = \frac{P(A).P(B|A)}{[P(A).P(B|A)] + [P(\bar{A}).P(B|\bar{A})]}. \tag{8.7}$$

The preceding formula for Bayes' theorem and the preceding example use exactly two categories for event A (event happens or does not happen), but the formula can be extended for including multiple categories. When dealing with more than two events of A and $\bar{A}$ ($\bar{A}$ signifies that event A has not occurred), we must keep in mind that the multiple events satisfy two important conditions:

- The events must be disjoint (with no overlapping).
- The events must be exhaustive, which means that they combine to include all possibilities.

Let us consider an example. Suppose there are three jars-A, B, C.

Jar A contains 3 red, 1 blue, and 4 white balls. Jar B contains 1 red, 3 blue, and 2 white balls and Jar C contains 4 red, 2 blue, and 3 white balls.as shown in Fig. 8.9. One jar is chosen at random. We use the following notations:

$J_A \Rightarrow$ The event that Jar A is chosen of the three.
$J_B \Rightarrow$ The event that Jar B is chosen of the three.
$J_C \Rightarrow$ The event that Jar C is chosen of the three.
$R \Rightarrow$ The ball chosen is red.

Fig. 8.9 Sample space representation of events

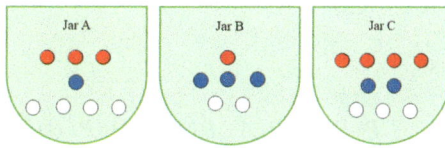

Fig. 8.10 Sample space representation of events

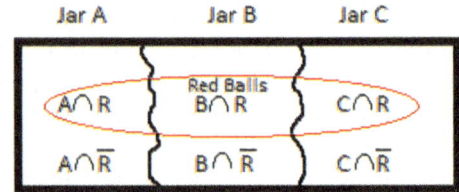

$B \Rightarrow$ The ball chosen is blue.
$W \Rightarrow$ The ball chosen is white.

Now J_A, J_B and J_C are mutually exclusive, i.e., two jars cannot be chosen simultaneously. So, our sample space is divided into three events (Fig. 8.10). As J_A, J_B and J_C are mutually exclusive,

$$P(J_A) = P(J_B) = P(J_C) = \frac{1}{3}. \tag{8.8}$$

So, we can say from the definition of conditional probability (refer Appendix B.3)

$$P(A \cap R) = P(R|A)P(A) = \frac{3}{8} \cdot \frac{1}{3} = \frac{1}{8}. \tag{8.9}$$

Similarly,

$$P(B \cap R) = P(R|B)P(B) = \frac{1}{18}; \tag{8.10}$$

and

$$P(C \cap R) = P(R|C)P(C) = \frac{4}{27}. \tag{8.11}$$

If we want to know, given a selected ball to be red, what is the probability of picking it up from Jar B? This means, we want to find $P(B|R)$. Now,

$$P(B \cap R) = P(B|R)P(R) \tag{8.12}$$

$$\implies P(B|R) = \frac{P(B \cap R)}{P(R)} \tag{8.13}$$

$$\implies P(B|R) = \frac{P(R|B)P(B)}{P(R)}. \tag{8.14}$$

Also,

$$P(R) = P(A).P(R|A) + P(B).P(R|B) + P(C)P(R|C). \tag{8.15}$$

So,

$$P(B|R) = \frac{P(R|B).P(B)}{P(A).P(R|A) + P(B).P(R|B) + P(C)P(R|C)} \qquad (8.16)$$

$$\implies P(B|R) = 0.17. \qquad (8.17)$$

The symbols have their usual meaning.

8.4.2 Bayesian Classifier

Bayesian decision theory is a basic statistical approach with regard to the problem of pattern classification. It is considered to be the ideal scenario where the probability structure underlying the categories is known completely. While this kind of situation infrequently happens in practical scenario, it enables us to determine the optimal classifier against which all other classifiers can be compared. Moreover, in some problems, it allows us to predict the error we will get when we generalize to new patterns.

This approach is based on quantifying the trade-offs between various classification decisions using probability and the costs that come along with such decisions. It makes the assumption that the decision problem is posed in probabilistic terms, and that all the relevant probability values are known.

Let us consider a hypothetical problem of designing a classifier to separate two kinds of chocolates: caramel chocolate and dark chocolates, in a factory. Suppose due to the randomness in the sequence of the types of chocolates, an onlooker looking into the chocolate arrival along the conveyor belt finds it difficult to predict what type will be emerging next. In decision-theoretic terminology, we could say that each chocolate emerges in either of the two possible states: i.e., either it is a caramel chocolate or is a dark chocolate. Let's consider that w represents the state of nature, with $w = w_1$ for caramel chocolate and $w = w_2$ for dark chocolate. The state of nature being so unpredictable, we consider w to be a variable that must be described probabilistically [41].

Considering the assumption that there are no other types of chocolates, $P(w_1) + P(w_2) = 1$. These prior probabilities reflect our prior knowledge of how likely we will be getting caramel chocolate or dark chocolate before the chocolate actually arrives. If we are bounded to take a decision about the type of the chocolate that will be appearing next just by making use of the prior probabilities, we will take the decision of the chocolate to be from class w_1 if $P(w_1) > P(w_2)$; otherwise, decide w_2. This rule might sound good if we need to decide about just one chocolate. But if we need to judge about many chocolates, using this rule repeatedly, we are always going to make the same decision even though we know that both types of chocolates are going to appear. Thus, it is not going to work good enough if we take into consideration only the values of the prior probabilities.

Fig. 8.11 Probability
distribution functions of the
two classes

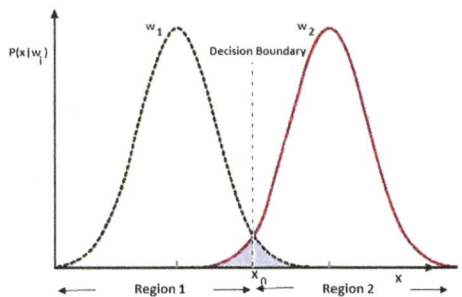

In most circumstances, we are not asked to make decisions using such little information. We might, for example, use another measurement or feature x of the chocolates for improving the classifier. Different chocolates will yield different measurement readings. This kind of variability is expressed by considering x to be a continuous random variable whose distribution is dependent on the state of nature and is expressed as $p(x|w)$. This is the class-conditional probability density function, the probability density function for x given that the state of nature is in w. Then the difference between $p(x|w_1)$ and $p(x|w_2)$ explains the difference in measurements between populations of caramel chocolate and dark chocolate (Fig. 8.11).

Suppose both the prior probabilities $P(w_j)$ and the conditional densities $p(x|w_j)$ are known previously. We make a further measurement of the fat content of chocolate and discover that its value is x. How does this measurement influence our attitude concerning the true state of nature? We note first that the (joint) probability density of finding a pattern that is in category w_j and has feature value x can be expressed as

$$p(w_j, x) = p(w_j|x)p(x) = p(x|w_j)P(w_j). \tag{8.18}$$

Rearrangement of these expressions makes us answer our question, which is known as Bayes' formula:

$$P(w_j|x) = \frac{p(x|w_j)P(w_j)}{p(x)} \tag{8.19}$$

where

$$p(x) = \sum_j p(x|w_j)P(w_j). \tag{8.20}$$

Thus, Bayes' formula can be expressed informally as

$$\text{posterior} = \frac{\text{likelihood} \times \text{prior}}{\text{evidence}}. \tag{8.21}$$

Fig. 8.12 Minimum risk classification

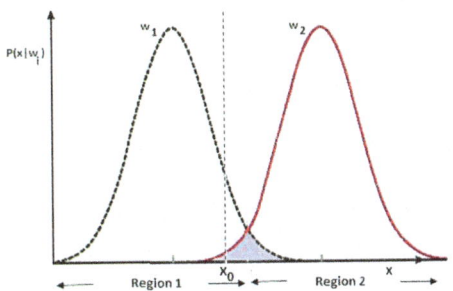

Bayes' formula represents that by observing the value of x the prior probability $P(w_j)$ can be converted to the posterior probability $P(w_j|x)$, i.e., the probability of the state of nature is w_j given that the feature value x has been measured. $p(x|w_j)$ is known as the likelihood of w_j with respect to x, a term chosen to represent that other things being equal, the category w_j, for which $p(x|w_j)$ is large is more "likely" to be the true category. It is the product of the likelihood and the prior probability that is most crucial in determining the posterior probability. The evidence factor $p(x)$ can be viewed as a scale factor that guarantees the fact that the posterior probabilities sum to 1.

If there exists an observation x for which $P(w_1|x) > P(w_2|x)$, we would naturally be inclined to decide that the true state of nature of x is w_1. For an equiprobable case, $(P(w_1) = P(w_2) = \frac{1}{2})$ as shown in Fig. 8.11, there is a minimum error of misclassification shown by the shaded region. If any pattern lies in the shaded part of Region 1 (in Fig. 8.12) though it belongs to w_2, it would be classified to w_1.

The probability of errors is given by

$$P_{\text{error}} = P(w_1) \int_{R_1} P(x|w_2)dx + P(w_2) \int_{R_2} P(x|w_1)dx; \qquad (8.22)$$

i.e.,

$$P_{\text{error}} = \frac{1}{2} \int_{-\infty}^{x_0} P(x|w_2)dx + \frac{1}{2} \int_{x_0}^{\infty} P(x|w_2)dx. \qquad (8.23)$$

This gives the total area under the shaded region. For minimum error criterion, the decision boundary will lie exactly at the middle (passing through $\mathbf{x_0}$ as in Fig. 8.11). Any pattern falling on the left side of the decision boundary will be classified to w_1, and those falling on the right side of the decision boundary will be classified to w_2.

If we want to minimize the average risk for the two classes, we will have to associate a risk factor with the error calculation.

$$P_{\text{error}} = \lambda_{12}P(w_1) \int_{R_1} P(x|w_2)dx + \lambda_{21}P(w_2) \int_{R_2} P(x|w_1)dx. \qquad (8.24)$$

λ_{12} and λ_{21} represent the risk factors. The loss matrix can be therefore written as:

$$L = \begin{bmatrix} 0 & \lambda_{12} \\ \lambda_{12} & 0 \end{bmatrix}. \tag{8.25}$$

$\lambda_{ij} > \lambda_{ii}$ means that correct decisions are penalised much less than wrong ones.

For an example, let us take two diseases—tumor and cancer. There may arise four possible cases during diagnosis.

A tumor diagnosed as tumor: No risk ($\lambda_{11} = 0$)
A cancer diagnosed as cancer: No risk($\lambda_{22} = 0$)
A tumor diagnosed as cancer: Low risk(λ_{12})
A cancer diagnosed as tumor: High risk(λ_{21})

So, as $\lambda_{21} > \lambda_{12}$, for minimum risk, the threshold value moves to the left of middle point as shown in Fig. 8.12. Then, the doctor must decide accordingly as to minimize the risk rather than minimizing error.

8.4.3 Naive Bayes' Classifier

The simplest case is the naive Bayesian classifier, which makes the assumption that *the input features are conditionally independent of each other*. Given three random variables X, Y, and Z, we say X is conditionally independent of Y given Z, if and only if the probability distribution governing X is independent of the value of Y given Z; that is

$$p(X = x_i | Y = y_j, Z = z_k) = p(X = x_i | Z = z_k)(\forall i, j, k). \tag{8.26}$$

As an example, consider three Boolean random variables to describe the current weather: Rain, Thunder, and Lightning. We might reasonably assert that thunder is independent of rain given lightning. Because we know lightning causes thunder, once we know whether or not there is lightning, no additional information about thunder is provided by the value of rain. Of course, there is a clear dependence of thunder on rain in general, but there is no conditional dependence once we know the value of lightning.

The Naive Bayes' algorithm is a classification algorithm, which takes into consideration the assumption of Bayes' rule and a set of conditional independence. Provided the goal of learning $p(w_j | \mathbf{x})$ (probability of a pattern $\mathbf{x}$ belonging to a class w_j) where $\mathbf{x} = \{x_1, x_2, \ldots, x_n\}$ is a feature vector, the Naive Bayes' algorithm takes into consideration the assumption that each feature is conditionally independent of each of the other x_ks given w_j is the class, and also independent of each subset of the other x_ks given class w_j. The value of this assumption is that it dramatically simplifies the representation of $p(\mathbf{x}|w_j)$, and the problem of estimating it from the training data. Considering, the equation obtained from Bayes' rule

$$p(\mathbf{x}|w_j) = p(x_1, x_2, \ldots, x_n|w_j). \tag{8.27}$$

From the general property of probabilities we can write,

$$p(\mathbf{x}|w_j) = p(x_1, x_2, \ldots, x_{n-1}|x_n, w_j)p(x_n|w_j). \tag{8.28}$$

In generic terms, when $\mathbf{x}$ contains n attributes, which satisfies the conditional independence assumption, we have

$$p(\mathbf{x}|w_j) = p(x_1|w_j)p(x_2|w_j) \ldots p(x_n|w_j). \tag{8.29}$$

Once we know this $p(\mathbf{x}|w_i)$, we can find out the class label using the Bayes' formula (Eq. 8.18). The aforementioned model summarizes a Gaussian naive Bayes' classifier, which makes the assumption that the data $X = \mathbf{x}$ is produced by a mixture of class-conditional (i.e., dependent on the value of the class variable w_j) Gaussians. Moreover, the Naive Bayes' assumption introduces the additional constraint that the attribute values x_i are independent of one another within each of these mixture components. In specific problem scenarios where we have more information, we can introduce additional assumptions that will further restrict the number of parameters or the complexity of estimating them.

8.5   One Class Classifier

In machine learning, one-class classification, also known as unary classification, tries to identify objects of a specific class among all objects, by learning from a training set containing only the objects of that class. This is different from and more difficult than the traditional classification problem, which tries to distinguish between two or more classes with the training set containing objects from all the classes.

Consider an example of detecting the changes between two multi-spectral and multi-temporal remotely sensed images using traditional supervised classification. In this case, changes occur only on few small parts of the images. It is very difficult to collect training patterns (objects) from the changed part of the images, and it is also very costly. If it is possible to collect the training patterns (objects) from the changed part of the images, even then the number of patterns (objects) of this class is very limited. In such cases, traditional two-class classification technique may not be a better choice due to unavailability of objects or insufficiency of objects of one (changed) class. On the contrary, one class classification technique is a better choice to do this job.

The problem of classification in pattern recognition can be defined as a problem of assigning an object represented by a vector of feature values to categories of objects (i.e., object class labels). Using a set of objects from the training set, a classifier learns to assign the categories or class labels to previously unseen objects from the test set. Basically, conventional two-class classification algorithms aim to classify an unknown object into one of two predefined object categories. A problem arises when the unknown object does not belong to any of these two predefined categories. This implies that objects of all the categories are not present in the training set or/and generation of objects of a particular category in the training set is very difficult. Sometimes, it also happens that only limited objects of a particular class are collected in the training set, which is unable to represent that class [11, 40]. To solve such classification problems, one-class classification technique was introduced. Basically, one-class classification technique was introduced to distinguish between two classes with the prior knowledge of only one class [3, 21, 38].

In one class classification, the training dataset contains only the observation vectors belonging to a class w_1, while the testing dataset includes observation vectors of both classes w_1 and w_2. Here, the parameters of the discriminant function can be evaluated only for the class w_1. In order to make the classification, an assumption about the distribution of the data in the second class (w_2) can be made; e.g., uniform distribution of $p(x|w_2)$ may be assumed. After that, the calculation of posterior probabilities $p(w_i|x)$ is possible, and the classification is performed using the discriminant function (given in Eq. 7.4). However, in practice, the discriminant function is often compared against a threshold value t:

$$\gamma(X, \Theta) = \begin{cases} w_1, \text{ if } u(X, \Theta) \geq t \\ w_2, \text{ if } u(X, \Theta) < t. \end{cases} \tag{8.30}$$

A number of one-class classification methods have been proposed in the literature [22], varying from modifications of conventional multiclass classification techniques to the methods designed specially for one-class classification problem.

In literature, several algorithms (like density methods, reconstruction methods, and boundary methods) have been developed for doing one-class classification. Density methods are based on the estimation of the probability density functions. Several representative density estimation methods including histograms [1], Markov models [8, 19], Gaussian and mixture of Gaussians models [5, 25], Parzen density estimation [5], and K-nearest-neighbors estimation [5] are used in various applications.

In reconstruction methods, assumptions about underlying data structures are made. The well-known methods belonging to this category include K-means [5, 29], Self-Organizing Maps [23, 42], Principal Component Analysis [10, 35], and autoencoders [20].

On the contrary, in boundary methods, a boundary is built around the training data and the classification is done based on calculated distance between an observation vector and the boundary. K-center [24] and Support Vector Classifier [17, 26] are included in this category.

One-class classification techniques are applied when data from the other classes are extremely hard or impossible to collect. It is applied for text classification [30], written digit recognition [39], information retrieval [30], face recognition [4, 43], medical image analysis [44, 45], bioinformatics [2, 33, 37], spam detection [7], anomaly detection [17, 32], outlier removal [28], machine fault detection [34], novelty detection [6, 16], remote sensing image classification [36], object tracking [15, 27], etc.

References

1. Aboulnaga, A., & Chaudhuri, S. (1999). Self-tuning histograms: Building histograms without looking at data. *ACM SIGMOD Record, 28*(2), 181–192.
2. Bánhalmi, A., Busa-Fekete, R., & Kégl, B. (2009). A one-class classification approach for protein sequences and structures. In *Bioinformatics research and applications* (pp. 310–322).
3. Bhatt, J., & Patel, N. S. (2015). A survey one class classification using ensembles method. *International Journal for Innovative Research in Science and Technology, 1*(7), 19–23.
4. Bicego, M., Grosso, E., & Tistarelli, M. (2005). Face authentication using one-class support vector machines. In *Advances in biometric person authentication* (pp. 15–22).
5. Bishop, C. M. 1995. *Neural networks for pattern recognition.* Oxford University Press.
6. Bowen, R. M. (May 2016). *Online novelty detection system: One-class classification of systemic operation.* PhD thesis, Rochester Institute of Technology.
7. Chaudhary, V., & Sureka, A. (2013). Contextual feature based one-class classifier approach for detecting video response spam on Youtube. In *International conference on Privacy, Security and Trust (PST)* (pp. 195–204).
8. Cho, S., & Park, H. (2003). Efficient anomaly detection by modeling privilege flows using hidden Markov model. *Computers and Security, 22*(1), 45–55.
9. Cover, T., & Hart, P. (1967). Nearest neighbor pattern classification. *IEEE Transactions on Information Theory, 13*(1), 21–27.
10. De Ridder, D., Pekalska, E., & Duin, R. (2002). The economics of classification: Error vs. complexity. In *IEEE International Conference on Pattern Recognition (ICPR) 2* (pp. 244–247).
11. Duda, R. O., Hart, P. E., & Stork, D. G. (2012). *Pattern classification.* Wiley.
12. Dudani, S. A. (1976). The distance-weighted k-nearest neighbor rule. *IEEE Transactions on Systems, Man and Cybernetics, 6*(4), 325–327.
13. Elkan, C. (2011). *Nearest neighbor classification* (vol. 11). Citeseer.
14. Evans, J. S. B. T., Handley, S. J., & Over, D. E. (2003). Conditionals and conditional probability. *Journal of Experimental Psychology: Learning, Memory, and Cognition, 29*(2), 321.
15. Fu, K., Gong, C., Qiao, Y., Yang, J., & Guy, I. (2012). One-class SVM assisted accurate tracking. In *IEEE International Conference on Distributed Smart Cameras (ICDSC)* (pp. 1–6).
16. Ghaoui, L. E., Jordan, M. I., & Lanckriet, G. R. (2002). Robust novelty detection with single-class MPM. In *Advances in neural information processing systems* (pp. 905–912).
17. Heller, K., Svore, K., Keromytis, A., & Stolfo, S. (2003). One class support vector machines for detecting anomalous windows registry accesses. In *Workshop on Data Mining for Computer Security (DMSEC)* (pp. 2–9).

18. Imandoust, S. B., & Bolandraftar, M. (2013). Application of *k*-Nearest Neighbor (KNN) approach for predicting economic events: Theoretical background. *International Journal of Engineering Research and Applications, 3*(5), 605–610.

19. Isaacson, D. L., & Madsen, R. W. (1976). *Markov chains: Theory and applications*. Wiley.

20. Japkowicz, N. (October 1999). *Concept-learning in the absence of counter-examples: An autoassociation-based approach to classification*. PhD thesis, Rutgers University.

21. Khan, S. S., & Madden, M. G. (2009). A survey of recent trends in one class classification. In *Artificial intelligence and cognitive science* (pp. 188–197).

22. Khan, S.S., & Madden, M.G. (2010). A survey of recent trends in one class classification. In L. Coyle, & J. Freyne (Eds.) *Artificial intelligence and cognitive science (AICS 2009)*. Lecture Notes in Computer Science (vol. 6206). Springer. https://doi.org/10.1007/978-3-642-17080-5_21

23. Kohonen, T. (1990). The self-organizing map. *Proceedings of the IEEE, 78*(9), 1464–1480.

24. Lane, T., & Brodley, C. E. (1999). Temporal sequence learning and data reduction for anomaly detection. *ACM Transactions on Information and System Security (TISSEC), 2*(3), 295–331.

25. Lauer, M. (2001). A mixture approach to novelty detection using training data with outliers. In *European Conference on Machine Learning (ECML)* (pp. 300–311).

26. Lazarevic, A., Ertöz, L., Kumar, V., Ozgur, A., & Srivastava, J. (2003). A comparative study of anomaly detection schemes in network intrusion detection. *SDM, 1*, 25–36.

27. Li, L., Han, Z., Ye, Q., & Jiao, J. (2010). Visual object tracking via one-class SVM. In *Asian Conference on Computer Vision (ACCV)* (pp. 216–225).

28. Liu, W., Hua, G., & Smith, J. (2014). Unsupervised one-class learning for automatic outlier removal. In *Proceedings of the IEEE conference on Computer Vision and Pattern Recognition (CVPR)* (pp. 3826–3833).

29. MacQueen, J., et al. (1967). Some methods for classification and analysis of multivariate observations. *Proceedings of the 5th Berkeley Symposium on Mathematical Statistics and Probability, 1*(14), 281–297.

30. Manevitz, L., & Yousef, M. (2002). One-class SVMs for document classification. *Journal of Machine Learning Research, 2*, 139–154.

31. Moraleda, J. G. S., Darrell, T., & Indyk, P. (2008). Nearest-neighbors methods in learning and vision. Theory and practice. *Pattern Analysis and Applications, 11*, 221–222. https://doi.org/10.1007/s10044-007-0076-8.

32. Pauwels, E. J., & Ambekar, O. (2011). One class classification for anomaly detection: Support vector data description revisited. In *Advances in data mining: Applications and theoretical aspects* (pp. 25–39).

33. Reyes, J. A., & Gilbert, D. (2007). Prediction of protein-protein interactions using one-class classification methods and integrating diverse data. *Journal of Integrative Bioinformatics, 4*(3), 1–16.

34. Shin, H., Eom, D., & Kim, S. (2005). One-class support vector machines an application in machine fault detection and classification. *Computers and Industrial Engineering, 48*(2), 395–408.

35. Shyu, M., Chen, S., Sarinnapakorn, K., & Chang, L. W. (2003). A novel anomaly detection scheme based on principal component classifier. Tech. rep., DTIC Document.

36. Song, B., Li, P., Li, J., & Plaza, A. (2016). One-class classification of remote sensing images using Kernel sparse representation. *IEEE Journal of Selected Topics in Applied Earth Observations and Remote Sensing, 9*(4), 1613–1623.

37. Spinosa, E. J., & Carvalho, A. (2005). Support vector machines for novel class detection in bioinformatics. *Genetics and Molecular Research: GMR, 4*(3), 608–615.

38. Tax, D. (May 2001). *One-class classification*. PhD thesis, Delft University of Technology.

39. Tax, D., & Duin, R. (2002). Uniform object generation for optimizing one-class classifiers. *Journal of Machine Learning Research, 2*, 155–173.

40. Theodoridis, S., Pikrakis, A., Koutroumbas, K., & Cavouras, D. (2010). *Introduction to pattern recognition: A matlab approach*. Academic.

41. Triola, M. F., Goodman, W. M., LaBute, G., Law, R., & MacKay, L. (2006). *Elementary statistics*. Pearson/Addison-Wesley.
42. Zanero, S., & Savaresi, S. M. (2004). Unsupervised learning techniques for an intrusion detection system. In *Proceedings of ACM symposium on applied computing* (pp. 412–419).
43. Zeng, Z., Fu, Y., Roisman, G. I., Wen, Z., Hu, Y., & Huang, T. S. (2006). One-class classification for spontaneous facial expression analysis. In *International conference on Automatic Face and Gesture Recognition(AFGR)* (pp. 281–286).
44. Zhang, J., Ma, K., Er, M., & Chong, V. (2004). Tumor segmentation from magnetic resonance imaging by learning via one-class support vector machine. In *International Workshop on Advanced Image Technology (IWAIT'04)* (pp. 207–211).
45. Zhang, Y., Zhang, B., Coenen, F., Xiao, J., & Lu, W. (2014). One-class Kernel subspace ensemble for medical image classification. *EURASIP Journal on Advances in Signal Processing 2014, 1*, 1–13.

Chapter 9
Combination of Classifiers

Classifier combination methods have proven to be an effective tool to increase the performance of pattern recognition systems. The ultimate goal of designing pattern recognition systems is to achieve the best possible classification performance for the task at hand. This objective traditionally led to the development of different classification schemes for any pattern recognition problem to be solved.

It had been observed in classifier design studies that different classifiers yield different results. The sets of patterns misclassified by different classifiers may not overlap. This suggests that different classifiers potentially offer complementary information about the patterns to be classified, which could be harnessed to improve the performance of total classification. The idea is not to rely on a single decision-making scheme. Instead, all the designs, or their subset, are used for decision-making by combining their individual opinions to derive a consensus decision [8] (Fig. 9.1).

In Fig. 9.2, the patterns from two different classes (represented by triangle and circles) cannot be separated using a single line. But using a combination of separation lines, the two classes can be easily separated.

9.1 Classifier Combination Techniques

To introduce a more elaborate description of the classifier combination techniques, we attempt to categorize them in the following subsections.

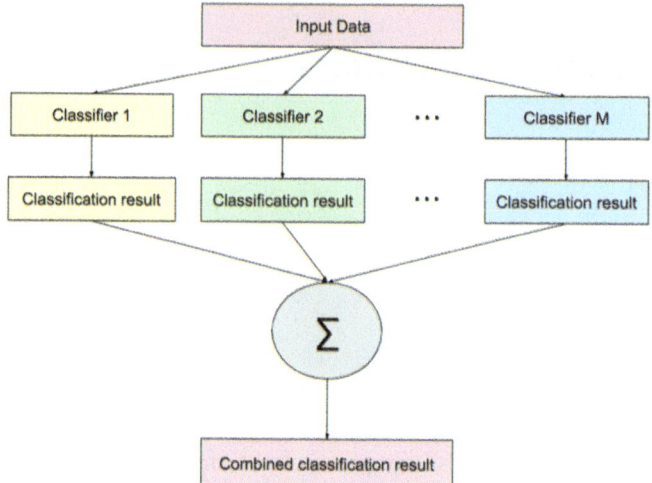

Fig. 9.1 Block diagram of a combination of classifiers

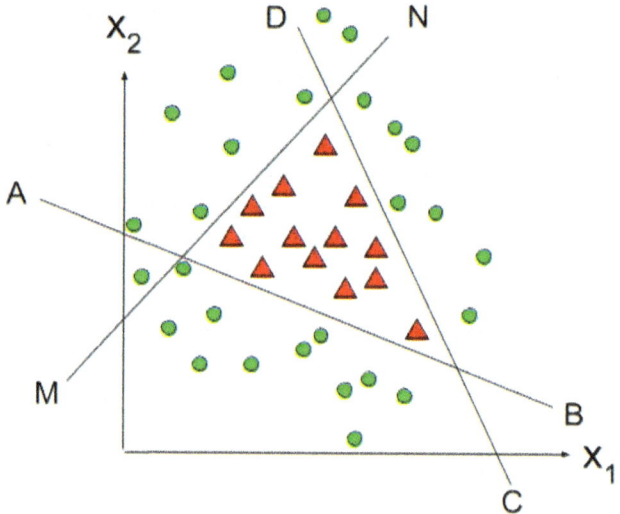

Fig. 9.2 Combination of classifiers improves performance

9.1.1 Score Combination Functions and Combination Decisions

Classifier combination techniques operate on the outputs of individual classifiers and usually fall into one of two categories. In the first approach, the outputs are treated as inputs to a generic classifier, and a combination algorithm is developed by training this "secondary" classifier. For example, a neural network can be used to act on the outputs of the individual base classifiers (say a couple of minimum distance classifiers) and to generate the combined matching score [7]. The benefit of usage from such a generic combinator is that it can learn the combination algorithm and can automatically take into consideration the strengths and score ranges of the individual base classifiers. In the second approach, a function or a rule combines the scores of the base classifiers in a predetermined manner. The final goal of classifier combination is to create a classifier system that operates on the same type of input as the base classifiers and separates the same types of classes. Combination rules are used to make final decision. If we denote the score assigned to class i by the base classifier j as s_i^j, then a typical combination rule is some function f, and the final combined score for class i (from M base classifiers) is

$$S_i = f(\{s_i^j\}_{j=1,...,M}). \tag{9.1}$$

Thus, the combination rules could be observed as a classifier operating on base classifiers' scores, which involves some combination function f. The simplest possible function could be $argmax$. Generic classifiers used for combinations need not have to be necessarily constructed following the afore-described scheme, but in practical scenario, this theme gets generally employed. For instance, in multilayer perceptron classifiers, the last layer has each node containing a final score for one class. These scores are then compared, and the maximum is selected. Similarly, k-nearest neighbor classifier can generate scores for all classes as ratios of the number of representatives of a particular class in a neighborhood to k. The class with the highest ratio is then assigned to a sample.

In summary, combination rules could be considered as a special kind of classifier of particular $argmax\ f$ form. Combination functions f are generally simple functions, such as sum, weighted sum, max, and min. On the other hand, Generic classifiers like neural networks and k-nearest neighbor imply more complicated functions.

For example, a combination of classes is developed for spam detection, and there are three classifiers: Decision Tree, Logistic Regression, and Support Vector Machines. Suppose, for a single email, Decision Tree predicts it as spam, Logistic Regression detects it as not spam, and Support Vector Machine detects as spam, then there are two votes for spam and one vote for non-spam; therefore, the decision for that email will be spam.

9.1.2 Combinations of a Set of Classifiers and Ensembles of Classifiers

One main categorization is based on whether the combination makes use of a fixed (usually less than 10) set of classifiers, as opposed to a large pool of classifiers

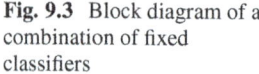

Fig. 9.3 Block diagram of a combination of fixed classifiers

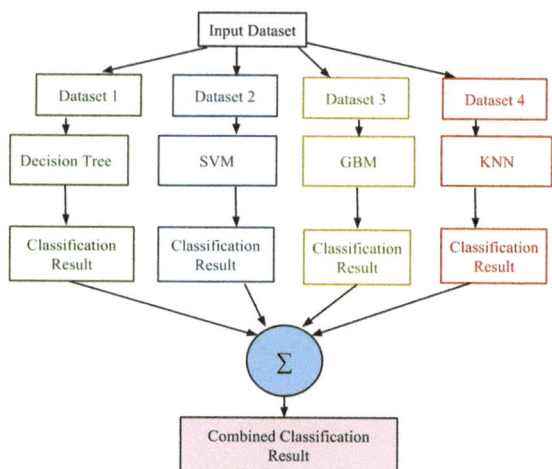

(potentially infinite) from which one selects or produces new classifiers [9]. The **combination of fixed classifiers** makes the assumption that the classifiers are trained on different features or different sensor inputs. The benefit comes from the diversity of the classifiers' strengths on different input patterns. Each classifier might be an expert on certain types of input patterns. Combination of fixed classifiers basically aims to exploit the potential of fixed number of different types (may be one SVM, one KNN, one Decision Tree put together) of classifiers to predict different sample subsets or different feature subsets by applying different input patterns to each of the base classifiers (Fig. 9.3).

In contrast, an **ensemble of classifiers** aims to exploit different ways of combining a large number (potentially infinite) of the same type of classifiers using either different sample subsets or different feature subsets for each of the base classifiers. Figure 9.4 shows an example with the chosen classifier as Decision Tree. The ensemble of classifiers assumes large number of classifiers or ability to generate classifiers. The huge number of classifiers are generally procured by choosing different subsets of training samples from one large training set, or by selecting different subsets of features from the set of all available features, and by training the classifiers with respect to selected training subset or subset of features. Example for this is bagging and boosting of classifiers.

9.2   Output Types of Combined Classifiers

One more technique to categorize classifier combination is by considering the outputs of the base classifiers used in the combination. Three types of outputs are usually considered:

Fig. 9.4 Block diagram of an ensemble of classifiers

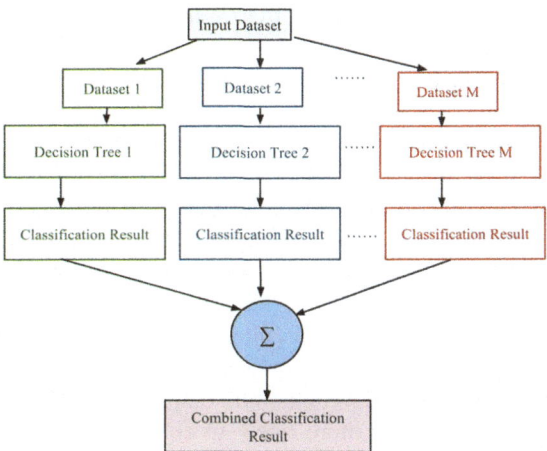

- Type I (abstract level): This is the lowest level because a base classifier provides the least amount of information on this level. Classifier output for a particular data point **x** is merely a single class label or an unordered set of candidate classes (without any ranking associated with it), i.e., none of the labels in the unordered set gets any preference over another label.
- Type II (rank level): In this case, classifier output for a particular data point input $\vec{x}$ is an ordered sequence of labels (with ranking associated with it) as a set of "c-best" possible labels according to the likelihoods that the input samples has those labels. The 1st label is the most likely one and the last label being the most unlikely one. It is to be noted that there are no confidence values linked to the class labels on rank level. Only their position in the c-best list indicates their relative likelihood.
- Type III (measurement level): Adding to the ordered c-best lists of candidate classes on the rank level, classifier output for a particular data point input $\vec{x}$ on the measurement level has confidence values assigned to each entry of the n-best list. The measurement value can be the probability of belonging of $\vec{x}$ to that class or the distance of $\vec{x}$ from the class with that label. These confidences, or scores, could be arbitrary real numbers, which depend on the classification architecture used. Thus, it is the measurement level that has the most information among all three output levels.

Suppose, if we consider a neural network as a classifier then in abstract level, only the best class is provided as output. In rank level, what is the rank of all possible classes provided and in measurement level, what is the confidence associated with those ranks that are also provided. In practice, a combination method can function on any of these levels. The advantage of classifier output on abstract and rank level is that different confidence characteristics have no negative impact on the final outcome, since confidence plays no role in the decision process. Nevertheless, the confidence of a classifier in a particular candidate class generally gives useful information that a simple class ranking is not able to reflect. This makes the suggestion that the use of combination methods that function on the measurement level can exploit the confidence assigned to each candidate class. Currently, most classifiers do give information on measurement level, such that application of combination schemes on the measurement level might be possible for most of the practical applications. On measurement level, however, it has to be taken into consideration that each classifier in a multiple classifier system might generate many different confidence values, with different ranges, scales, means, etc. This may be a minor problem for classifier ensembles generated with bagging and boosting [1, 3, 4] since all classifiers in the ensemble are based on the same classification architecture, only their training sets differ. Each classifier will therefore provide similar output. However, for classifiers based on different classification architectures, this output will in general be different. Since different architectures lead more likely to complementary classifiers, which are especially promising for combination purposes, we need effective methods for making outputs of different classifiers comparable.

9.3   Classifier Ensembles

It is to be assumed that there are only a few classifiers and now making use of a training set some statistical data about these classifiers could be collected. The objective of the combination algorithm is to learn the behavior of these classifiers and generate an efficient combination function. In this section, however, one more technique of combination is discussed that involves methods that not only try to look for the best combination algorithm but also try to find the best set of classifiers for the combination. This kind of combination generally needs a method for producing a large number of classifiers. Few such methods for generating classifiers for these combinations exist. One such technique is based on bootstrapping the training set in order to procure a multitude of subsets and train a classifier on each of these subsets. Another technique randomly selects the subsets of features from one large feature set and trains classifiers on these feature subsets [5]. Another technique applies different training conditions, for instance, chooses random initial weights for neural network training or chooses dimensions for decision trees [10]. The ultimate method for generating classifiers is randomly separating the feature space into the regions related to particular classes. Simple techniques of combination apply some fixed functions (majority voting, bagging) to the outputs of all the base classifiers. The complex methods include techniques like boosting [6], stacked generalization [11], which tries to select only those classifiers that will contribute to the combination.

9.3.1 Bagging

Researchers many a times concentrate on enhancing the performance of single-classifier system mostly because of their lack in sufficient resources for simultaneously developing several different classifiers. A simple method for generating multiple classifiers in those cases is to run several training sessions with the same single-classifier system and different subsets of the training set or slightly modified classifier parameters. Each training session then builds an individual classifier. "Bagging" is the first more systematic approach that became popular. This technique draws the training sets with replacement from the original training set, each set leading to a slightly different classifier after training (Fig. 9.5). The method used for generating the individual training sets is also known as bootstrap technique and looks for decreasing the error of statistical estimators. Let $Q(X, Y|D)$ be a probability distribution that picks a training sample (x_i, y_i) from D uniformly at random. More formally,

$$Q((x_i, y_i)|D) = \frac{1}{n}, \forall (x_i, y_i) \in D \tag{9.2}$$

with $n = |D|$. In practical scenarios, bagging has generated good outcomes. Although the performance gains are generally small when bagging is applied to weak classifiers. In these scenarios, another intensively investigated method for generating multiple classifiers becomes more applicable: boosting.

Fig. 9.5 Block diagram of bagging

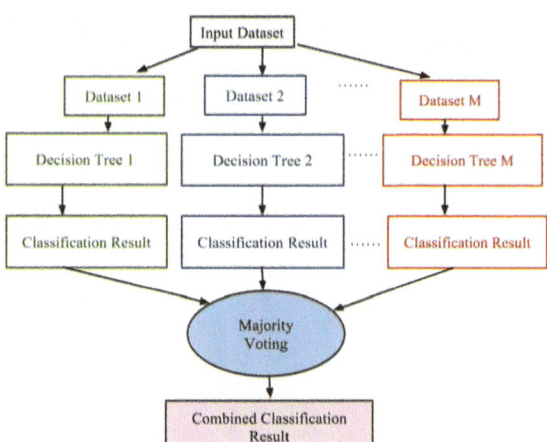

9.3.2 *Boosting*

Boosting has its root in a theoretical framework for studying machine learning and deals with the question whether a weak learner/classifier can be boosted into an arbitrarily accurate learning algorithm. Weak learners are those that are just slightly better classifiers than ones based on completely random prediction, e.g., a decision stump, which is a one-level decision tree using only one feature, so only 1-rule. Boosting attaches a weight to each sample of the training set. The weights are updated after each training cycle according to the performance of the classifier on the corresponding training samples. Initially, all weights are set equally, but on each round, the weights of incorrectly classified samples are increased so that the classifier is forced to focus on the hard examples in the training set. In Fig. 9.6, we see that first the input dataset is fed to a weak learner. Initially, all the data points have equal weightage. The learner finds the misclassified points and computes total error. The dataset is then updated by increasing the weightage of the misclassified points and decreasing the weightage of the correctly classified points at the current stage. Again this updated dataset and weights are fed to the weak learner. The learning continues till it reaches a termination criterion. Create ensemble classifier

$$H_T(\mathbf{x}) = \sum_{t=1}^{T} \alpha_t h_t(\mathbf{x}). \tag{9.3}$$

This ensemble classifier is built in an iterative fashion. In iteration t, we add the classifier $\alpha_t h_t(\mathbf{x})$ to the ensemble. At test time, we evaluate all classifiers and return the weighted sum.

Fig. 9.6 Block diagram of boosting

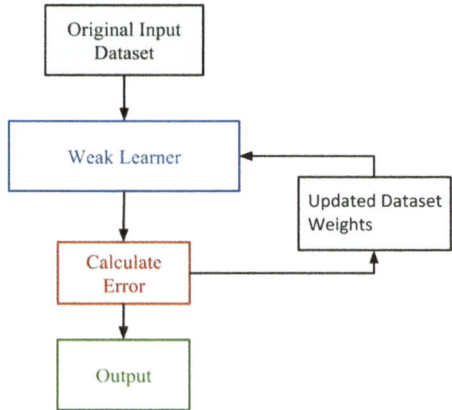

9.3.2.1 Adaboost

A very popular type of boosting is AdaBoost (Adaptive Boosting) [2], which was introduced by Freund and Schapire [3] in 1995 to expand the boosting approach introduced by Schapire. The AdaBoost algorithm generates a set of classifiers and votes them. It changes the weights of the training samples based on classifiers previously built (trials). The goal is to force the final classifier to minimize the expected error over different input distributions. The final classifier is formed using a weighted voting scheme. At each iteration, it gives emphasis to the "hardest" samples. The final classifier is obtained as a weighted average of the previously hierarchically designed classifiers. Adaboost uses M number of classifiers and the classification output for an input data point is a weighted average of the results of the individual base classifiers (the weight vector being a parameter). The parameter vector for each individual base classifier is another parameter to be estimated.

9.3.2.2 Freund and Schapire's model of Adaboost

The input is a training set $(x_1, \omega_1), \cdots, (x_m, \omega_m)$ where each $x_i \in X$ and $\omega \in C$, the labels for the data points x_i. Here we are considering the binary classification, and hence, $C = \{-1, +1\}$ (This algorithm can also be generalized to the multiclass case). In AdaBoost algorithm, a given base classifier that is essentially a weak learner (with hypothesis $h_t : X \to \{-1, +1\}$) repeatedly runs in a series for total T iterations.

A distribution over the training set denotes the set of weights for the training instances. The weight of the training instance on iteration t is denoted by $D_t(i)$. These weights are initialized equally. At each iteration, the misclassified instances are penalized much more than the correctly classified ones in terms of these weights. These changing weights determine which training instances the algorithm needs to focus more on at a particular iteration.

Given further is the error of the base classifier hypothesis:

$$\epsilon_t = Pr_{i \sim D_t}[h_t(x_i) \neq \omega_i] = \sum_{i:h_t(x_i) \neq y_i} D_t(i). \tag{9.4}$$

The value of the error term depends on the distribution D_t.

The algorithm is as follows:

Given: $\{(x_1, \omega_1), \cdots, (x_m, \omega_m)\}$ where $x_i \in X$, $\omega_i \in C = \{-1, +1\}$.

Initialize $D_1(i) = \dfrac{1}{m}$.

Repeat the following steps for $t = 1, \cdots, T$

(i) Train weak learners using distribution D_t.
(ii) Get the hypothesis $h_t : X \rightarrow \{-1, +1\}$ with error ϵ_t as

$$\epsilon_t = Pr_{i \sim D_t}[h_t(x_i) \neq y_i] = \sum_{i:h_t(x_i) \neq y_i} D_t(i). \qquad (9.5)$$

(iii) Compute α_t from error ϵ_t

$$\alpha_t = \frac{1}{2} \ln\left(\frac{1 - \epsilon_t}{\epsilon_t}\right); \qquad (9.6)$$

where α_t is the measure of importance of base classifier with hypothesis h_t.

(iv) Update distribution of data $D_{t+1}(i)$ for the next iteration.

If $h_t(x_i) = \omega_i$,

$$D_{t+1}(i) = \frac{D_t(i)}{Z_t} e^{-\alpha_t}. \qquad (9.7)$$

If $h_t(x_i) \neq \omega_i$,

$$D_{t+1}(i) = \frac{D_t(i)}{Z_t} e^{\alpha_t}; \qquad (9.8)$$

where Z_t is a normalizing factor chosen so as to make D_{t+1} a distribution (the values add up to 1).

The final hypothesis h_t is used to generate the output H

$$H(x) = \text{sign}\left(\sum_{t=1}^{T} \alpha_t h_t(x)\right).$$

Here an important observation is that α_t gets larger as ϵ_t gets smaller (since, we can easily see from Eq. 9.6 that $\alpha_t \geq 0$ if $\epsilon_t \leq \frac{1}{2}$). From Eqs. 9.7 and 9.8, we see that the weights tend to become much more (in terms of positive exponentials) for hard examples and much less for the easy ones (in terms of negative exponentials). The final hypothesis H is a weighted majority vote of the T hypotheses coming from the weak base classifiers where α_t is the weight assigned to h_t.

Adaboost basically penalizes the wrongly classified instances much more (in terms of positive exponentials) heavily than those that are correctly classified (in terms of negative exponentials). Thus, the cost function becomes a complicated exponential one, for which parameter estimation through direct optimization is very complex. Hence, the parameters of the cost function are computed sub-optimally in a stage-by-stage fashion. Boosting has been successfully applied to a wider range of applications. There are some drawbacks of Adaboost, e.g., it is highly sensitive to noise and class imbalance problems. Nevertheless, we will not go more into the details of boosting and other ensemble combinations here. The reason is that the focus of classifier ensemble techniques lies more on the generation of classifiers and less on their actual combination.

References

1. Breiman, L. (1996). Bagging predictors. *Machine Learning, 24*(2), 123–140.
2. Collins, M., Schapire, R. E., & Singer, Y. (2002). Logistic regression, adaboost and bregman distances. *Machine Learning, 48*(1–3), 253–285.
3. Freund, Y., & Schapire, R. E. (1995). A Desicion-theoretic generalization of on-line learning and an application to boosting. *Computational Learning Theory, 55*, 23–37.
4. Freund, Y., Schapire, R. E., et al. (1996). Experiments with a new boosting algorithm. In *International conference on machine learning (ICML)'96* (pp. 148–156).
5. John, G. H., Kohavi, R., Peger, K., et al. (1994). Irrelevant features and the subset selection problem. In *Proceedings of the eleventh international conference on machine learning* (pp. 121–129).
6. Kittler, J., & Roli, F. Multiple classifier systems. In *Proceedings of second international workshop, MCS*.
7. Lee, D. S., & Srihari, S. N. (1995). A theory of classifier combination: The neural network approach. In *Proceedings of the third international conference on document analysis and recognition 1* (pp. 42–45).
8. Murty, M. N., & Devi, V. S. (2011). Combination of classifiers. *Pattern Recognition, 2*, 188–206.
9. Roy, M., Routaray, D., Ghosh, S., & Ghosh, A. (2014). Ensemble of multilayer perceptrons for change detection in remotely sensed images. *IEEE Geoscience and Remote Sensing Letters, 11*(1), 49–53.
10. Webb, A. R. (2003). *Statistical pattern recognition*. Wiley interscience electronic collection. Wiley.
11. Wolpert, D. H. (1992). Stacked generalization. *Neural Networks, 5*(2), 241–259.

Chapter 10
Clustering

Clustering can be used to partition a large set of data into groups, called clusters, so that the data points in a group are similar to each other, while those in different groups are not similar. Clustering is mostly an unsupervised technique. In the clustering process, there are no predefined classes; therefore, it is difficult to find an appropriate metric for measuring if the found clusters are acceptable or not. So, the clustering method is kept simple by assuring low inter-cluster similarity and high intra-cluster similarity. There are numerous machine learning algorithms used for clustering of data whose true class labels are not known. Formally, clustering can be defined as follows:

Considering $\mathbf{X}$ to be a d-dimensional dataset

$$\mathbf{X} = \{\mathbf{x}_1, \mathbf{x}_2, \ldots, \mathbf{x}_N\}.$$

A k-clustering of $\mathbf{X}$ is a partitioning of $\mathbf{X}$ into k subsets (clusters) $c_1, c_2, \ldots, c_k$ in a way that

- Clusters are non-empty, i.e., $c_i \neq \{\}, 1 \leq i \leq k$.
- Clusters cover all points in X, i.e., $\cup_{i=1}^{k} c_i = \mathbf{X}$.
- Clusters do not overlap, i.e., $c_i \cap c_j = \{\}$, if $j \neq i$.
- Points in a cluster are similar.

Clustering basically separates out dense regions leaving sparse regions in between.

10.1 Cluster Analysis

The practice of categorising objects according to their perceived similarities is essential for most of the tasks in scientific studies. Organizing data into sensible groups is one of the most fundamental modes of understanding and learning. Cluster

© The Author(s), under exclusive license to Springer Nature Singapore Pte Ltd. 2025   167
A. Ghosh, *Data Science and Cases in Sustainability*, Mathematics for Sustainable
Developments, https://doi.org/10.1007/978-981-96-8362-8_10

Fig. 10.1 A simple clustering example

analysis is the formal study of algorithms and methods for grouping or classifying objects. An object is described either by a set of measurements or by relationships between the object and other objects. Cluster analysis is the process that groups the data objects using information collected from the data that describes the objects and their relationships. The objective is that the objects in a group be as similar (or related) as possible to each other and as different (or unrelated)as possible from the objects in other groups. For a more distinct clustering, we need to ensure a high intra-cluster similarity and low inter-cluster similarity.

Cluster analysis does not use the category labels that tag objects with prior identifiers. Its objective is to find a convenient and valid organization of the data and not to establish rules for separating future data into categories. Cluster analysis has no mechanism for differentiating between relevant and irrelevant features. Therefore, the choice of features included in a cluster analysis must be underpinned by conceptual considerations. This is very important because the clusters formed can be very dependent on the features specified.

Cluster analysis relates itself with other techniques, which are used for dividing data objects into groups. In Fig. 10.1, a set of unlabeled data points is shown. After clustering, the data are divided into two groups. In clustering, unlike classification, the class labels corresponding to the data points are not explicitly given, rather they are derived from the distribution of data. In contrast, classification model gets a prior training. For this reason, cluster analysis is discussed under the area of unsupervised learning methods. The terms *segmentation* or *partitioning* are sometimes used synonymously for clustering.

10.2   Clustering Techniques

An entire collection of grouping techniques is generally referred to as a *clustering*. Clustering as applied to real-life applications encounters three additional complications:

- Large databases
- Objects with many attributes
- Attributes of different types

These complications tend to impose severe computational requirements that present real challenges to classic clustering algorithms. Clustering can be done in many different ways; some of which are discussed as follows.

10.2.1 Exclusive Clustering

This type of clustering groups the data into k clusters such that each data point belongs to exactly one group. Three types of existing exclusive clusterings are discussed as follows.

10.2.1.1 Partitional Clustering

A partitional clustering is a technique that divides the set of data objects into non-overlapping subsets (clusters) in such a way that each data object becomes part of exactly one subset. Partitional clustering decomposes a dataset into a set of disjoint clusters. Given a dataset of N points, a partitioning method constructs k $(N \geq k)$ partitions of the data, with each partition representing a cluster. We need to know the value of k beforehand (Fig. 10.2).

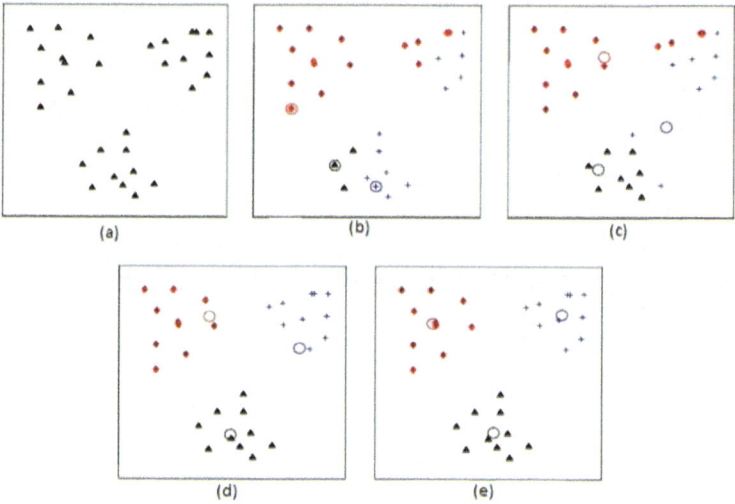

Fig. 10.2 Illustration of partition algorithm: (**a**) Two-dimensional input data, (**b**) initial points are chosen randomly as cluster centers; and initial assignment of the data points is made to clusters, (**c**) and (**d**) intermediate iterations updating clusters and their centers, (**e**) final clustering obtained by the algorithm at convergence

Fig. 10.3 Illustration of agglomerative hierarchical clustering

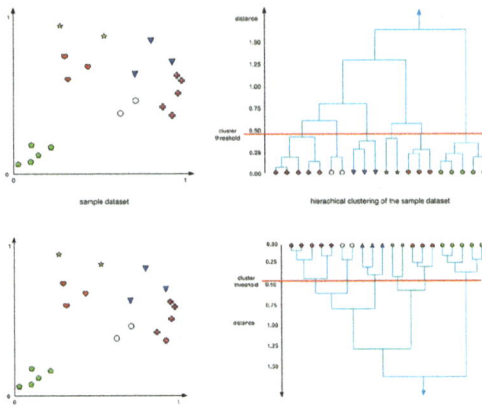

Fig. 10.4 Illustration of divisive hierarchical clustering

10.2.1.2   Hierarchical Clustering

If the clusters are permitted in forming subclusters, then it results into a hierarchical clustering, which is a group of nested clusters that are organized as a tree. Each node (cluster) in the tree (except for leaf nodes) is the union of its children (subclusters), and the root of the tree is the cluster that contains all the objects. A hierarchical clustering can be viewed as a sequence of partitional clusterings, and a partitional clustering can be obtained by cutting the hierarchical tree at a particular level.

- **Agglomerative** methods are those where each data object starts with its own separate cluster. The two "closest" (most similar) clusters are then combined, and this is done repeatedly until all subjects are in a single cluster. At the end, the optimum number of clusters is then chosen out of all cluster solutions.

  Figure 10.3 illustrates the working procedure of agglomerative hierarchical clustering with dendrogram.
- **Divisive** methods are those where all objects start in the same cluster and the aforementioned strategy is applied in reverse order until every object is in a separate cluster.

  Figure 10.4 illustrates the working procedure of divisive hierarchical clustering with dendrogram.

10.2.1.3   Density-based Clustering

In density-based clustering, a data point is declared as a core point if the data density (either deterministically or probabilistically determined) within a certain neighborhood of the considered point is high enough. Each such point is considered as a cluster initially and then grown around it, by shifting the neighborhood region and depending on the density of the points within the newly shifted neighborhood region. The final clusters are obtained when all the given data points are processed.

Fig. 10.5 Illustration of
density-based clustering

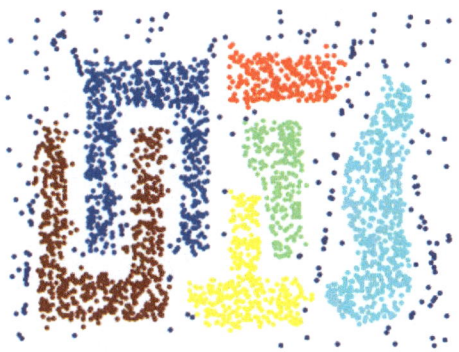

Density-based clustering algorithm is applied when the data are densely placed in
a multidimensional space. Since this kind of clustering algorithms depend on the
density of the data distribution rather than the distance between a data point and a
cluster representative (mean, medoid, etc.), this algorithm imposes no restriction on
the shape of the clusters. It can recover clusters of any shape without having any
prior knowledge of number of clusters. Efficient handling of outliers is something
that comes very naturally with the density-based concept of clustering of the data
points. An isolated point (or a set of points) with considerably less density can
be declared as outlier. Some examples of such kind of algorithms are: DBSCAN,
DBCLASD[12]. Figure 10.5 illustrates the working procedure of density-based
clustering.

10.2.2   Overlapping Clustering

There are various situations where a point could reasonably be kept in more than
one cluster, and these conditions are better taken care by non-exclusive clustering.
In a general way, an overlapping or non-exclusive clustering is used to reflect the
concept that an object can be placed in more than one group. For example, a person
at a university can enrol himself both as a student and as an employee of the
university simultaneously. A non-exclusive clustering is also many a times used
where, for instance, an object is "between" two or more clusters and might as well
be assigned to any of these clusters. Rather than somewhat randomly assigning the
object to a single cluster, it is placed in all the clusters with certain degree; giving
the notion of fuzzy clustering [8]. Figure 10.6 illustrates an example of overlapping
clustering. Fuzzy clustering methods allow the objects to belong to several clusters
simultaneously, with different degrees of membership. In many situations, fuzzy
clustering is more natural than hard clustering, which requires an object to belong
or not to belong to a cluster. For example, objects on the boundaries between several
groups should not be forced to fully belong to one of the groups, but rather are

Fig. 10.6 Illustration of
overlapping clustering

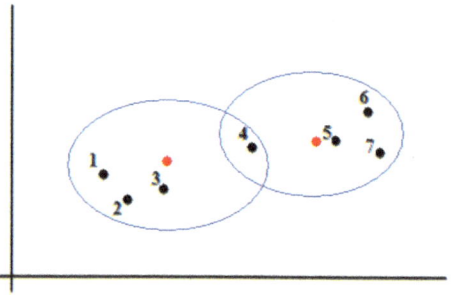

assigned membership values between 0 and 1 indicating their partial belongingness
to different clusters.

Clustering algorithms partition data into a certain number of clusters (subsets,
groups, or categories). There is no universally agreed upon definition of clustering
[9]. Most researchers describe a cluster by considering the internal homogeneity
and the external separation [5], i.e., patterns in the same cluster should be similar
to each other, while patterns in different clusters should not be (Fig. 10.1). Both the
similarity and the dissimilarity should be examinable in a clear and meaningful way.
Here, we give some simple algorithms for clustering, based on the descriptions in
the previous sections.

Major challenges of clustering lie in the cases where the dataset is not naturally
grouped or where we do not know how many groups exist in the data.

10.3   Clustering Tendency

The presence of clusters in a dataset may indicate that the objects belong to different
populations. This suggests that there is a fundamental difference between two or
more groups of samples, e.g., two different products are included in the analysis, or
a shift or drift has occurred in the measurement technique. But, whether clusters
actually exist or not in that data is a matter of concern. Some collections are
better for clustering than others. Tests for clustering tendency attempt to determine
whether worthwhile retrieval performance would be achieved by clustering a
dataset, before investing the computational resources, which clustering the dataset
would entail. One can have an estimate of it by Hopkins statistic test [3].

Hopkins Statistic Test

Let $X = \{x_1, x_2 \cdots , x_N\}$ be a collection of N patterns in a d-dimensional space such
that $x_i = \{x_{i1}, x_{i2}, \ldots, x_{id}\}$. Let $X' \subset X$, where each element $y_j \in X'$ is randomly
chosen from X, such that $|X'| = m << N$. Let there be another set $X_1 \subset X$, each

element of which is randomly chosen from $\mathbf{X}$. For $\mathbf{y_j} \in \mathbf{X}'$, $\mathbf{x_j}$ be its closest data point in $\mathbf{X_1}$. Two distances are defined:

$dist_j$ be the distance between $\mathbf{y_j}$ and $\mathbf{x_j}$ and δ_j be the distance between $\mathbf{x_j}$ and its closest point in $\mathbf{X_1} - \{\mathbf{x_j}\}$.

The Hopkins statistic is defined as,

$$HS = \frac{\sum_{j=1}^{m} dist_j^d}{\sum_{j=1}^{m} dist_j^d + \sum_{j=1}^{m} \delta_j^d}. \tag{10.1}$$

Smaller the value of δ_j, higher is the value of HS, higher is the tendency. Higher the value of δ_j, lower is the value of HS, lower is the tendency.

This statistics compares the nearest-neighbor distribution of randomly selected locations to that for the randomly selected patterns. Under the null hypothesis, (HS_0), the distances from the sampling origins to their nearest patterns should, on the average, be the same as the inter-pattern nearest neighbor distances, implying randomness, and hence, HS should be about 0.5. However, when the patterns are aggregated or clustered, the distance of sampling origins to their nearest neighbors should, on the average, be larger than that of the randomly selected inter-pattern distances. So, HS should be larger than 0.5, approaching 1.0 for very well-defined clustered data.

Figure 10.7 shows an example of three types of data whose Hopkins statistics values (HS_a, HS_b and HS_c) satisfy the relation $HS_a > HS_b > HS_c$. Major aim of testing the clustering tendency is to search for the "correct" number of clusters present in the data.

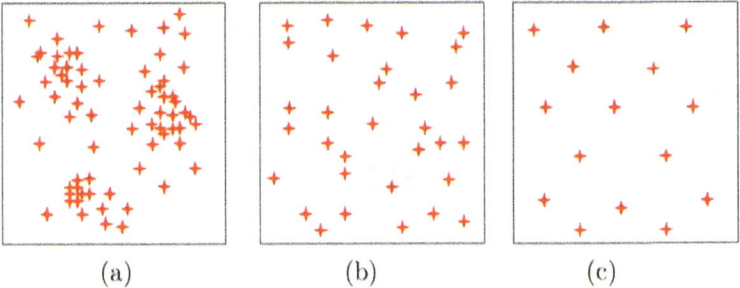

(a)   (b)   (c)

Fig. 10.7 Example of clustering tendency for **(a)** data with cluster structures, **(b)** randomly distributed data, and **(c)** regularly distributed data

10.4   Cluster Validity

For supervised classification, we have a variety of measures like accuracy, precision, recall to evaluate how good a classification model is. However, for cluster analysis, the analogous question is how to evaluate the "goodness" of the resulting clusters? This is a difficult scenario as "clusters lie in the eye of the beholder," i.e., different people may make different clusters with the same dataset. For example, if a class of students is given a set of random numbers and asked to group them, different types of clusters will be seen. Some may form clusters for odd and even numbers, some for prime and non-prime, and so on. The clusters formed may not be wrong in anyway.

In spite of all this, one still wants to evaluate the clusters formed mainly to compare the clustering algorithms, to compare two sets of clusters, and to avoid finding patterns in noise. The basic principle of clustering is to increase inter-cluster variance and decrease intra-cluster variance (Fig. 10.8) and also to increase inter-cluster distance (Fig. 10.9).

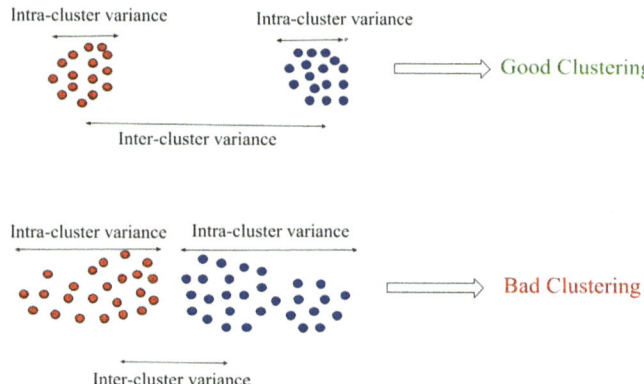

Fig. 10.8 Basic principle of cluster validity depicting variance

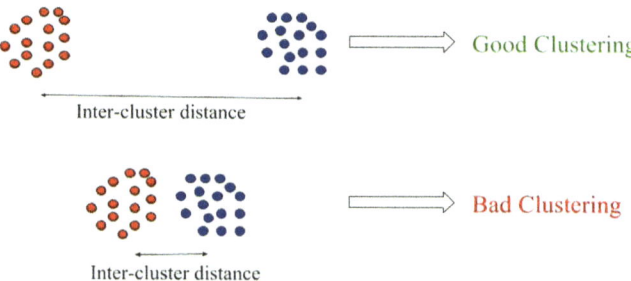

Fig. 10.9 Basic principle of cluster validity depicting distance

The previous statement is nothing but the mathematical explanation of the fact that any two patterns in the same cluster should be increasingly similar and any two patterns from two different clusters should be increasingly apart for achieving a better clustering.

There are several aspects of cluster validation including the following:

- Finding the clustering tendency for a dataset, i.e., distinguish whether non-random structures actually exist in the data.
- Finding the "correct" cluster number.
- Evaluate how well the cluster analysis' outcome fit the data without referencing external information.
- Comparing the results of two different sets of cluster analysis for determining which one performs well.
- Compare the outcome of a cluster analysis to externally known results, e.g., to externally given class labels (if any).

For the last three points, one can further distinguish whether to evaluate the entire clustering or just individual clusters.

10.4.1   Cluster Validity Indices

Numerical measures applied to judge the various aspects of cluster validity are classified into the following three types.

- External index : This is used to measure the extent to which clusters match with externally supplied classes. Example of such a measure is entropy.
- Internal index : This measures the goodness of clusters without using any external information. Sum of Squared Error (SSE) is a typical example.
- Relative index : It compares two different clusterings or clusters. Often an external or internal index is used for this function.

10.4.2   Non-fuzzy Indices

There are some indices that measure the validity of clustering results for hard clustering algorithms, such as, Dunn index [4], DB index [2], silhouette index [10] etc., which are discussed as follows.

Dunn index: Dunn index [4] was introduced in 1974 by J. C. Dunn. The distance between two clusters c_i and c_j is considered to be dist(c_i, c_j), and it is defined as

$$\text{dist}(c_i, c_j) = \min_{\mathbf{x_i} \in c_i, \mathbf{x_j} \in c_j} \text{dist}(\mathbf{x_i}, \mathbf{x_j}). \tag{10.2}$$

Fig. 10.10   Representation of Dunn Index parameters

The diameter of a cluster c_m is

$$\text{diam}(c_m) = \max_{\mathbf{x_i} \in c_m, \mathbf{x_j} \in c_m} \text{dist}(\mathbf{x_i}, \mathbf{x_j}). \tag{10.3}$$

Dunn index D_k for a chosen value of k (number of clusters) is given by

$$D_k = \frac{\min\limits_{i,j;i \neq j} \left(\text{dist}(c_i, c_j) \right)}{\max\limits_{m=1,\cdots,k} \left(\text{diam}(c_m) \right)}. \tag{10.4}$$

Figure 10.10 shows the numerator and denominator of the Dunn Index.

Higher value is suitable for the numerator of the Dunn index, whereas the denominator should have lower value. It is desirable to have clusters with increased inter-cluster distances and small diameters of individual clusters. So, higher value of Dunn index ensures well-separated clusters with small intra-cluster variations. The meaning of $D_k > 1$ is that the clusters are well separated. The number of clusters underlying the data can be computed by plotting the values of D_k versus different possible values of k. The value of k, which maximizes D_k, can be taken as the optimum (!) number of clusters in the dataset. The disadvantage of Dunn index is that it is computationally very expensive and also sensitive to noise.

DB Index: Davies–Bouldin index [2] or DB index was introduced in 1979. Let us consider that the spread of the cluster c_i around the cluster mean is s_i and $\text{dist}(c_i, c_j) \equiv \text{dist}_{ij}$ is the distance between two clusters c_i and c_j (Fig. 10.11). A different similarity index R_{ij} between the ith cluster and the jth cluster is defined using the variables s_i and dist_{ij}, such that the following conditions are satisfied:

$$R_{ij} \geq 0; \; R_{ij} = R_{ji}.$$

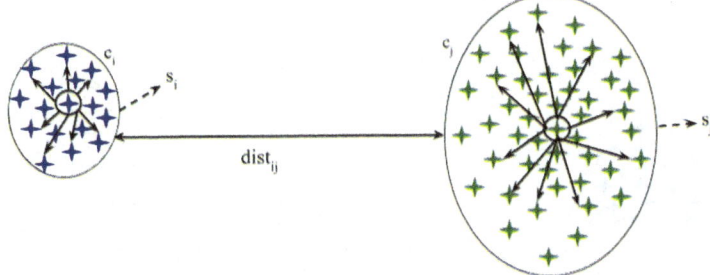

Fig. 10.11 Representation of DB Index parameters

The actual definition of R_{ij}, which satisfies the aforementioned conditions, is given as

$$R_{ij} = \frac{(s_i + s_j)}{\text{dist}_{ij}}. \tag{10.5}$$

Lower value of $s_i + s_j$ and a higher value of dist_{ij} is desirable for a pair of clusters c_i and c_j.

From Eq. 10.5, the following can be derived

If $s_i = 0$ and $s_j = 0$ then $R_{ij} = 0$.
If $s_j > s_m$ and $\text{dist}_{ij} = \text{dist}_{im}$, then $R_{ij} > R_{im}$.
If $s_j = s_m$ and $\text{dist}_{ij} < \text{dist}_{im}$, then $R_{ij} > R_{im}$.

$$R_i = \max_{j=1\cdots k, i\neq j} R_{ij}. \tag{10.6}$$

Finally, DB index for k clusters is defined as

$$DB_k = \frac{1}{k} \sum_{i=1}^{k} R_i. \tag{10.7}$$

DB_k for k clusters is thus the average similarity between each cluster and it's most similar one. As inter-cluster similarity is meant to be minimized for clustering purpose, the minimum value of DB_k is what we have to find out and so also the corresponding value of k.

Silhouette index: Peter J. Rousseeuw first introduced silhouette index [10]. For any data point $x_i \in c_j$, 2 distances (a_i and b_i) are calculated.

$$a_i = \text{dist}_{avg}(x_i, x_l)|x_i \in c_j; x_l \in \{c_j - \{x_i\}\}; \tag{10.8}$$

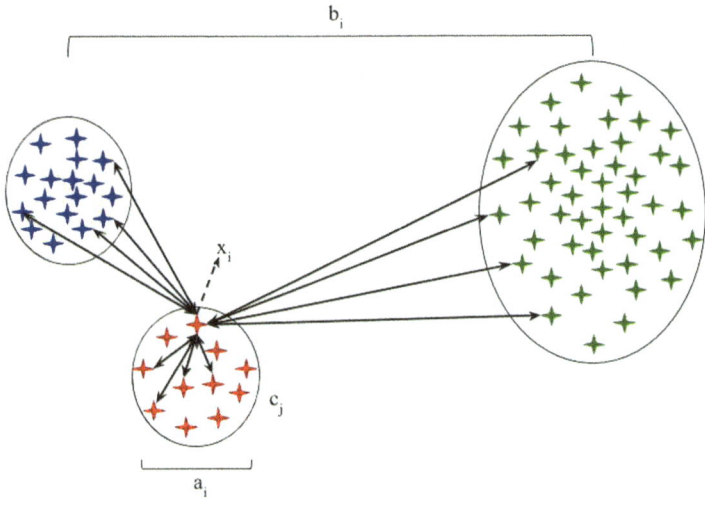

Fig. 10.12 Representation of Silhouette Index parameters

a_i gives the average distance between the data point $\mathbf{x_i}$ and the other data points in the same cluster c_j. The other distance is defined as

$$b_i = \min_{\mathbf{x_i} \in c_j; \mathbf{x_l} \in c_m; j \neq m} \text{dist}_{\text{avg}}(\mathbf{x_i}, \mathbf{x_l}). \tag{10.9}$$

On the other hand, b_i gives the measure of the average distance between the data point $\mathbf{x_i}$ and the data points in the clusters other than c_j. Figure 10.12 depicts a_i and b_i. The silhouette of $\mathbf{x_i}$ is given as

$$s_i' = \frac{b_i - a_i}{\max(b_i, a_i)}. \tag{10.10}$$

The silhouette of cluster c_j is

$$S_j' = \frac{1}{|c_j|} \sum_{i \,|\, \mathbf{x_i} \in c_j} s_i'; \tag{10.11}$$

where $|c_j|$ is the cardinality of cluster c_j. The final silhouette index for k clusters is defined as

$$S_k = \frac{1}{k} \sum_{j=1,\cdots,k} S_j'. \tag{10.12}$$

$S_k \in [-1, +1]$. Higher value of S_k indicates better clustering, hence $b_i > a_i$ is desirable. For the maximum value obtained for S_k, its corresponding k denotes the actual cluster number in the data.

10.4.3 Fuzzy Indices

There are several indices to measure the performance of fuzzy clustering. Among them, partition coefficient (V_{PC}), partition entropy (V_{PE}), and Xie–Beni (V_{XB}) indices are very popular [1, 6, 7].

Partition Coefficient (V_{PC}): Bezdek [1] introduced the Partition Coefficient (V_{PC}) as a validity index to measure the performance of fuzzy clustering techniques (discussed in Chap. 13). It is defined as

$$V_{PC} = \frac{1}{N} \sum_{i=1}^{N} \sum_{j=1}^{k} \mu_{ij}^2; \tag{10.13}$$

where, μ_{ij} is the membership of the pattern $\mathbf{x_i}$ in the jth cluster, $\mathbf{U} = [\mu_{ij}]$. The range of values that V_{PC} can assume is $\left[\frac{1}{k}, 1\right]$. If the value of V_{PC} is closer to 1, then the clustering becomes closer to hard clustering, meaning there is less sharing of the belongingness of a data point between many clusters. On the other hand, if the sharing of the belongingness of a data point between many clusters is more, then the value of V_{PC} becomes closer to $\frac{1}{k}$. $V_{PC} = \frac{1}{k}$ means that either the data do not have any clustering structure or the algorithm cannot find the clustering structure in the data. So, higher value of V_{PC} is desirable to obtain better clustering.

Partition Entropy (V_{PE}): Bezdek [1] has also introduced the Partition Entropy (V_{PE}) measure and defined it as

$$V_{PE} = -\frac{1}{N} \sum_{i=1}^{N} \sum_{j=1}^{k} [\mu_{ij} \log_b \mu_{ij}]. \tag{10.14}$$

It could be deciphered as a scalar measure of the average amount of fuzziness in a given membership matrix. It can be noted that an optimal partition (or an optimal value of k) is obtained by minimizing V_{PE} with respect to $k = 2, 3, \ldots, k_{\max}$ as in the previous case. Lower value of V_{PE} indicates compact clustering or better partitioning. The harder the partitioning is, V_{PE} tends to 0. V_{PE} value tends to $\log_b k$ (b is the base of logarithm), as the clustering becomes fuzzier. $V_{PE} = \log_b k$ means that the either there is no clustering structure in the data or the algorithm is unable to find it. So, lower value of V_{PE} is desirable to obtain better clustering.

Major drawback of both V_{PC} and V_{PE} is that they take into account only the extent of overlapping between the different clusters and does not take into consideration the position of the data points in the d-dimensional space. [1, 6, 7].

Xie–Beni Indices(V_{XB}):   Xie and Beni [11] proposed a validity index (V_{XB}), which considers two properties: compactness and separation. It is defined as

$$V_{XB} = \frac{1}{N} \frac{\sum_{j=1}^{k} \sum_{i=1}^{N} \mu_{ij}^2 \|\mathbf{x_i} - \mathbf{v_j}\|^2}{(\min_{j \neq i} \|\mathbf{v_i} - \mathbf{v_j}\|^2)}. \tag{10.15}$$

$\mathbf{v_j}$ is the representative vector for the jth cluster. Here, the numerator indicates the compactness of the fuzzy partition, while the denominator indicates the strength of separation between clusters. It may be noted that a good partition produces small value of compactness, and that well-separated clusters will have a high value for the denomination. Hence, most desirable partition is obtained by minimizing V_{XB} for $k = 2, 3, \ldots, k_{\max}$. It may be noted that the optimum cluster can be obtained when V_{XB} is very low.

References

1. Bezdek, J. C. (1973). Cluster validity with fuzzy sets. *Journal of Cybernetics, 3*(3), 58–73.
2. Davies, D. L., & Bouldin, D. W. (1979). A cluster separation measure. *IEEE Transactions on Pattern Analysis and Machine Intelligence, 2*, 224–227.
3. Devyver, P. A., & Kittler, J. (1982). *Pattern recognition: A statistical approach*. Prentice-Hall.
4. Dunn, J. C. (1974).   Well-separated clusters and optimal fuzzy partitions.   *Journal of Cybernetics, 4*(1), 95–104.
5. Gower, J. C. (1971). A general coefficient of similarity and some of its properties. *Biometrics, 27*, 857–871.
6. Halkidi, M., & Vazirgiannis, M. (2001). Clustering validity assessment: Finding the optimal partitioning of a data set. *Proceedings of IEEE International Conference on Data Mining* (pp. 187–194).
7. Pal, N. R., & Bezdek, J. C. (1995). On cluster validity for the fuzzy c-means model. *IEEE Transactions on Fuzzy Systems, 3*(3), 370–379.
8. Pal, S. K., & Mitra, S. (1999). *Neuro-fuzzy pattern recognition: Methods in soft computing*. Wiley.
9. Romesburg, C. (2004). *Cluster analysis for researchers*. Lulu Press.
10. Rousseeuw, P. J. (1987). Silhouettes: A graphical aid to the interpretation and validation of cluster analysis. *Journal of Computational and Applied Mathematics, 20*, 53–65.
11. Xie, X. L., & Beni, G. (1991). A validity measure for fuzzy clustering. *IEEE Transactions on Pattern Analysis and Machine Intelligence, 13*(8), 841–847.
12. Xu, X., Ester, M., Kriegel, H. P., & Sander, J. (1998). A nonparametric clustering algorithm for knowledge discovery in large spatial databases. In *Proceedings of the international conference on data engineering (ICDE98)*.

Chapter 11
Clustering Algorithms

Clustering algorithms are mainly categorized into three major types: partitional clustering, hierarchical clustering, and density-based clustering. Some of the popular algorithms would be explored in the following sections.

11.1 Partitional Clustering

Partitional clustering algorithms are those that aim to divide the dataset into several non-overlapping and non-hierarchical clusters so that every data point exclusively belongs to only one cluster. The best such partition is the answer.

11.1.1 k-means Clustering

k-means [5] is one of the simplest clustering algorithms based on iterative partitions. The procedure follows a simple and easy way to group a given dataset through a certain number of clusters (assume k clusters) fixed a priori. The main idea is to determine the centroids, one for each cluster. These centroids should be placed in a cunning way because different locations cause different clustering result. So, a reasonable choice is to place them as much as possible far away from each other. The next step is to take each point of the given dataset, check its distances from each of the assumed centroids, and then associate it to the nearest centroid. When no point of the dataset is pending, the first step is completed and an early grouping is made. At this point, recalculate the k new means as centers of the clusters resulting from the previous step. After we have these k new centroids, a new binding is made between the data points of the set and the nearest new centroids. A loop has thus been generated. As a result of this loop, we may notice that the k centroids change

© The Author(s), under exclusive license to Springer Nature Singapore Pte Ltd. 2025 181
A. Ghosh, *Data Science and Cases in Sustainability*, Mathematics for Sustainable
Developments, https://doi.org/10.1007/978-981-96-8362-8_11

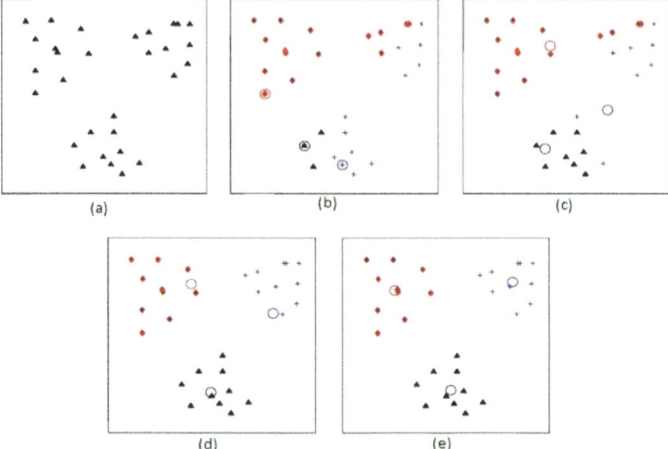

Fig. 11.1 Illustration of k-means algorithm: (**a**) Two-dimensional input data, (**b**) k (three in this case) seed points are chosen randomly as cluster centers; and initial assignment of the data points is made to clusters, (**c**) and (**d**) intermediate iterations updating clusters and their centers, (**e**) final clustering obtained by k-means algorithm at convergence

their locations step by step until no more changes are noticed in any of the means. In other words, centroids do not move any more or the assignment of the patterns becomes stable.

$$J = \sum_{i=1}^{k} \sum_{j=1}^{n_i} ||x_j^{(i)} - \mu_i||; \tag{11.1}$$

The method is described diagrammatically in Fig. 11.1. A stepwise procedure of this algorithm can be described as follows:

   (i) Decide k, and initialize k centers (randomly).
  (ii) Assigning every data point to its nearest center.
 (iii) Replace a center by the mean of the data points assigned to it.
 (iv) After replacing the centers, reassign the objects (data points) to their nearest centers.
  (v) Repeat steps (iii) and (iv) till stabilization; i.e., there is no more change in any of the cluster centers.

The stopping criterion of this algorithm is:

(i) No (or minimum) reassignments of data points to different clusters.
(ii) No (or minimum) change of centroid positions.
(iii) Minimum decrease in the sum of squared error (SSE)

$$SSE = \sum_{j=1}^{k} \sum_{\mathbf{x_i} \in c_j} ||\mathbf{x_i} - \mathbf{v_j}||^2; \tag{11.2}$$

where k is the total number of clusters, $\mathbf{x_i}$ is the ith data point in j^{th} cluster c_j and $\mathbf{v_j}$ is the mean of the data points in the jth cluster c_j.

Finally, this algorithm aims at minimizing an objective function—the average (over all clusters) distance of the members of the same cluster from its center or a squared error clusterings function. Hence, k-means makes circular clusters for Eucledian distance. Given two clusterings, we can choose the one with the smaller error. One easy way to reduce SSE is to increase k, the number of clusters. A good clustering with smaller k can have a lower SSE than a poor clustering with higher k. The time complexity for k-means is O(Nkt) (where t is the number of iterations). k-means algorithm works well (in terms of time complexity) for large datasets.

11.1.2 k-medoids Clustering

In k-means algorithm, the mean of a class may not be an existing point of the data set. In some applications, we want each center to be one of the existing points itself. That is where we need the k-medoids algorithm. For example, k-medoids clustering algorithm is used when the data are categorical in nature. In that case, finding mean is not possible so the cluster representative should be such a point that already belongs to the existing data points and also best represents the underlying clustering structure of the dataset. In this case, instead of choosing the mean as a representative of the cluster, we try to find out the existing points to represent the clusters. Medoid of a set of points is the point in the set that minimizes the average distance (for some given distance function $\delta(x, y)$) to all other points. For a dataset S,

$$\text{medoid}(S) = \arg \min_{x \in S} \sum_{y \in S} \delta(x, y). \tag{11.3}$$

k-medoid is based on centroids (or medoids) calculated by minimizing the absolute distance between the points and the selected centroid, instead of minimizing the square distance. The consequence is that it becomes more robust to noise and outliers than k-means.

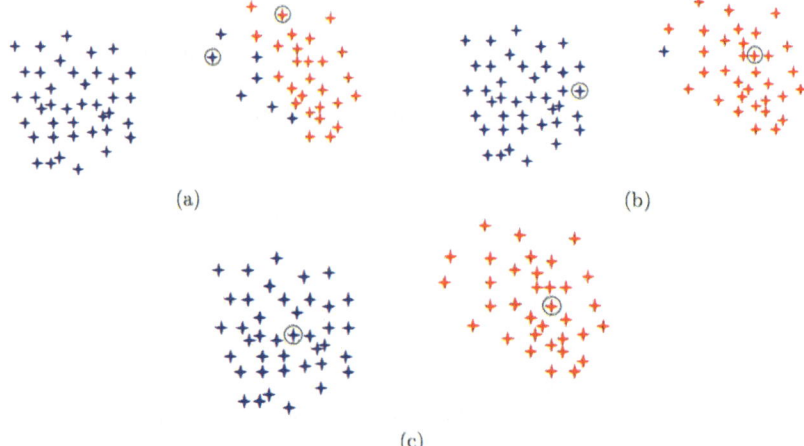

Fig. 11.2 k-medoids clustering: (**a**) Randomly chosen initial cluster centers, (**b**) updated centers, (**c**) final clusters

A stepwise procedure (Fig. 11.2) of this algorithm can be described as:

   (i) Initially guess k, and initialize k centers (randomly).
  (ii) Assign each object to its nearest center.
 (iii) Replace the existing center by the medoid of its group members.
 (iv) After replacing the centers, reassign the objects to the nearest centers.
  (v) Repeat Steps (iii) and (iv) so long as within-cluster variation changes.

The k-medoids algorithm shares the properties of the k-means algorithm. Each iteration decreases the criterion function (total within cluster variation) and the algorithm mostly converges. Different initial seed selection may result in different clustering, and it may not achieve the global minimum. The k-medoids algorithm is computationally harder than k-means. Moreover, k-means is sensitive to extreme points in the data, but k-medoids are not that much sensitive.

11.1.3 Partitioning Around Medoids (PAM)

PAM stands for "partition around medoids." PAM clustering algorithm is intended to find a sequence of the most centrally located objects (called medoids) in a cluster. Objects that are tentatively defined as medoids are placed into a set S of selected objects. If X is the total set of objects, then the set $U = X - S$ is the set of unselected objects. The goal of the algorithm is to minimize the average dissimilarity of objects to their closest selected centroid. Equivalently, we can minimize the sum of the dissimilarities between object and their closest selected object. PAM is developed by Kaufman and Rousseuw in 1987 [4]. The algorithm chooses k-medoids initially and then swaps the medoid object with a non-medoid such that the quality of clusters is improved.

Let us assume that k is the number of clusters, N is the number of objects in the dataset, X is the set of objects to be clustered, S is the current set of selected medoids, and R is the next set of medoids. Two set of medoids S and R are said to be neighbors if $k-1$ elements in S and R are common. R is identical to S except one element s_i that is replaced by any element from U. Therefore, S can have $k*(N-k)$ number of such neighbors.

The following steps are performed sequentially:

(i) Randomly select a set S of k medoids out of X.

(ii) Among the $k*(N-k)$ neighbors of the set S, select the set R, which minimizes ΔJ. Compute the cost function J as

$$J = \sum_{s_j \in S} \sum_{x_i \in c_j} d(x_i, s_j)$$

where s_j is the medoid of the jth cluster and x_i is a data point in j^{th} cluster c_j.

$$\Delta J = J(R) - J(S).$$

- If $\Delta J < 0$; $S = R$.
- If $\Delta J \geq 0$; local minima is reached.

(iii) Repeat step (ii) till the local minima is reached.

Once the set of k medoids that best represents the dataset is reached, each data point is assigned to its nearest medoid. This algorithm treats the small clusters and points far away from the large clusters as outliers. The time complexity for PAM is $O(k(N-k)^2)$. From the time complexity point of view, it works well for small datasets and poorly for large datasets.

11.2   Hierarchical Clustering

Hierarchical clustering is a technique of cluster analysis that aims to develop a hierarchy of clusters. It produces a set of nested clusters organized as a hierarchical tree. It therefore can be visualized as a dendrogram (tree-like diagram that records the sequences of merges or splits) (Fig. 11.3). Number of clusters vary depending on the level of the dendrogram. In general, merging or splitting is performed in a greedy manner.

11.2.1   Similarity Measures

Similarity of two clusters is measured using the two most similar (closest) points or two least similar (most distant) points in different clusters. There are different ways to measure this similarity.

11.2.1.1   Single Linkage

In single-linkage clustering (also called the connectedness or minimum method), we consider the distance between one cluster and another cluster to be equal to the shortest distance from any member belonging to one cluster to any member belonging to the other cluster (Fig. 11.4). If the data consist of similarities, we

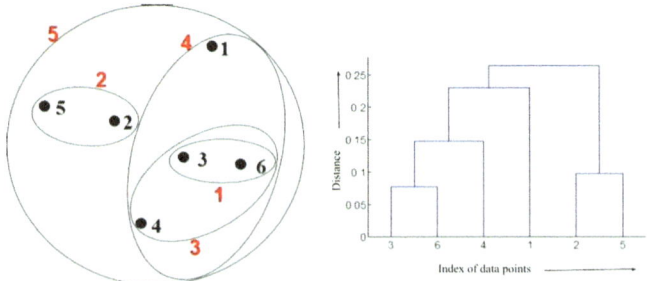

Fig. 11.3 Hierarchical clustering and its corresponding dendrogram

Fig. 11.4 Single linkage

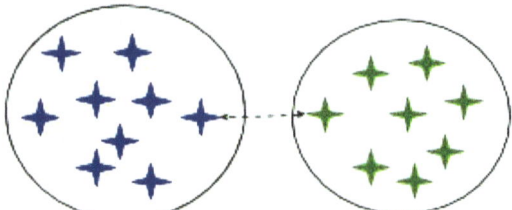

consider the similarity between one cluster and another cluster to be equal to the greatest similarity from any member of one cluster to any member of the other cluster.

$$D(r, s) = \text{Min}\{d(i, j) : \text{where object } i \text{ is in cluster } r \text{ and object } j \text{ is cluster } s\}.$$
(11.4)

11.2.1.2 Complete Linkage

In complete-linkage clustering (also known as the diameter or maximum method), we consider the distance between one cluster and another cluster to be equal to the longest distance from any member of one cluster to any member of the other cluster (Fig. 11.5).

$$D(r, s) = \text{Max}\{d(i, j) : \text{where object } i \text{ is in cluster } r \text{ and object } j \text{ is cluster } s\}.$$
(11.5)

11.2.1.3 Average Linkage

In average-linkage clustering, we consider the distance between one cluster and another cluster to be equal to the average distance from any member of one cluster to any member of the other cluster (Fig. 11.6).

$$D(r, s) = T_{rs}/(N_r * N_s);$$
(11.6)

Fig. 11.5 Complete linkage

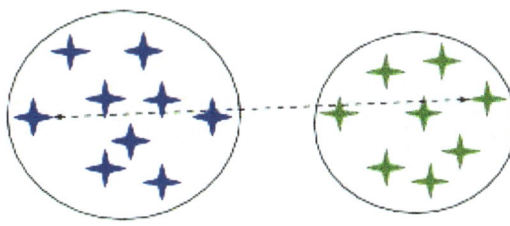

Fig. 11.6 Average linkage

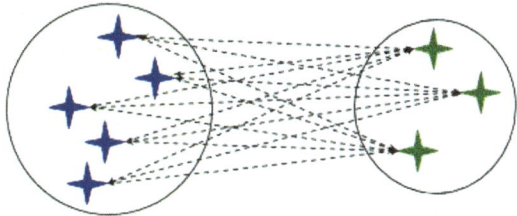

where T_{rs} is the sum of all pairwise distances between data points of cluster r and data points of cluster s.

11.2.2 Adjacency Matrix

The adjacency matrix **A** is also called the connection matrix as it provides information about the connections of the graph that the samples in the data make. The data points or samples act as the nodes in the graph. In graph-theoretic notation, the two data points x_i and x_j are adjacent if there exists an edge between them. In order to compactly represent the structure of a finite graph, the adjacency matrix is used. If a graph has n nodes, the size of the adjacency matrix will be $n \times n$. For any two data points x_i and x_j in a simple graph, the ij^{th} entry of the adjacency matrix will be:

$$A_{ij} = \begin{cases} 1 & \text{if } x_i \text{ and } x_j \text{ are adjacent} \\ 0 & \text{if } x_i \text{ and } x_j \text{ are not adjacent} \end{cases}. \tag{11.7}$$

The diagonal entries A_{ii} are all 0 if there are no loops in the graph.

For an undirected graph, the adjacency matrix is then $A_{ij} = A_{ji}$ for all i, j, which means that the adjacency matrix is symmetric. The adjacency matrix of a directed graph can be nonsymmetric. If the graph is a weighted graph, then the entries A_{ij} represent the weights of the edges between the nodes.

11.2.3 Types of Hierarchical Clustering

Hierarchical clustering could be typically categorized in two ways [6]:

- **Agglomerative**: It is a "bottom up" technique where each observation creates its own cluster in the beginning, and pairs of closest/similar clusters get merged as one goes up the hierarchy.
- **Divisive**: It is a "top down" technique, i.e., all data begin in one cluster, and the clusters are splitted to have more clusters recursively as one moves down the hierarchy depending on some criterion.

Agglomerative clustering methods begin with as many clusters as there are objects where each cluster keeps just one object. The most similar clusters merge with each other to create the next largest cluster. The merging continues till a hierarchy of clusters is created having a single cluster, which contains all the objects at the top of the hierarchy. Figure 11.7 shows the technique when clusters are merged based on nearest neighbors (Fig. 11.8).

Fig. 11.7 Illustration of agglomerative hierarchical clustering

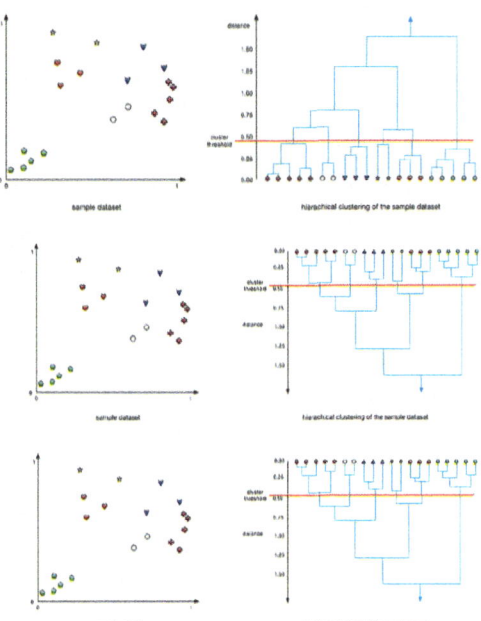

Fig. 11.8 Illustration of divisive hierarchical clustering

Fig. 11.9 Illustration of divisive hierarchical clustering

11.2.3.1 Divisive Clustering

Divisive clustering techniques take the reverse approach from agglomerative technique. This technique starts with all the objects in one cluster and then tries to split that cluster into two smaller clusters. Continue this process until each object forms a single cluster. Figure 11.9 demonstrates the technique.

11.3 Density-Based Clustering

Density-based algorithm is applied to densely placed data in the multidimensional space. The concept of class representatives (mean or medoid, etc.) is not used in density-based clustering. Such algorithms depend on the density of the data distribution. So, density-based algorithms find out clusters of any shape and handle outliers very efficiently.

Mean shift [2] represents a general non-parametric mode finding/clustering procedure. In contrast to the classical k-means clustering approach, there are no embedded assumptions on the shape of the distribution nor the number of modes/clusters. The main idea behind mean shift is to treat the points in the D-dimensional feature space as an empirical probability density function where the dense regions in the feature space correspond to the local maxima or modes of the underlying distribution. For each data point in the feature space, one performs

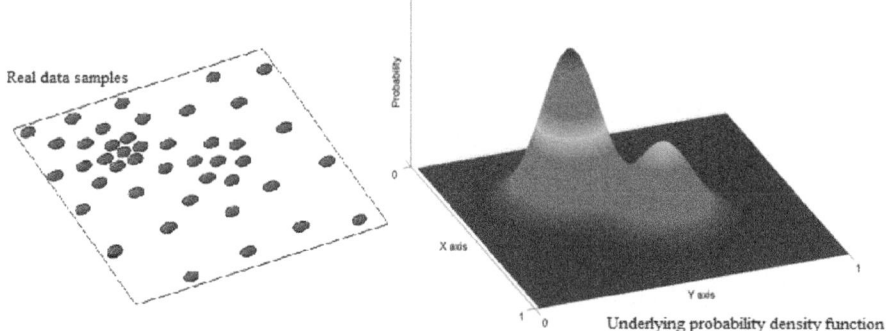

Fig. 11.10 Assumed underlying probability density function from the real data samples

a gradient ascent procedure on the locally estimated density until convergence. The stationary points of this procedure represent the modes of the distribution. Furthermore, the data points associated (at least approximately) with the same stationary point are considered members of the same cluster. It is basically a tool for finding modes in a set of data samples, manifesting an underlying probability density function (*pdf*) in $\mathbb{R}^d$

The density function can be estimated as follows:

Given that there are N data points in a d-dimensional space, i.e., $\mathbf{x_i} \in \mathbb{R}^d$, the data points are sampled from the underlying probability density function as shown in Fig. 11.10. The multivariate kernel density estimate using a radially symmetric kernel (like Gaussian or Epanechnikov [8]), $K(\mathbf{x})$ is given by

$$\hat{f}_K = \frac{1}{Nh^d} \sum_{i=1}^{N} K\left(\frac{\mathbf{x} - \mathbf{x}_i}{h}\right)$$ (11.8)

where h (termed as the *bandwidth* parameter) defines the radius of the kernel. The radially symmetric kernel is defined as

$$K(\mathbf{x}) = c_{k,d} k(||\mathbf{x}||^2)$$ (11.9)

where $c_{k,d}$ represents a normalization constant, which ensures that $K(\mathbf{x})$ integrates to 1. The modes of the density function are located at the zeroes of the gradient function $\nabla f(\mathbf{x}) = 0$. The gradient of the density estimator of Eq. 11.8 is

$$\nabla f(\mathbf{x}) = \frac{2c_{k,d}}{Nh^{d+2}} \sum_{i=1}^{N} (\mathbf{x_i} - \mathbf{x}) g\left(||\frac{\mathbf{x} - \mathbf{x_i}}{h}||^2\right)$$

$$
= \frac{2c_{k,d}}{Nh^{d+2}} \left[\sum_{i=1}^{N} g\left(\left\| \frac{\mathbf{x_i} - \mathbf{x}}{h} \right\|^2 \right) \right] \left[\frac{\sum_{i=1}^{N} \mathbf{x_i} g\left(\left\| \frac{\mathbf{x} - \mathbf{x}_i}{h} \right\|^2 \right)}{\sum_{i=1}^{N} g\left(\left\| \frac{\mathbf{x} - \mathbf{x}_i}{h} \right\|^2 \right)} - \mathbf{x} \right] \qquad (11.10)
$$

where $g(.) = -k'(.)$ denotes the derivative of the selected kernel profile $k(.)$. The first term is proportional to the density estimate at $\mathbf{x}$ (computed with the kernel $G(\mathbf{x}) = c_{g,d} g(\|\mathbf{x}\|^2)$). The second term, called the mean shift vector $\mathbf{m}$, points toward the direction of maximum increase in density and is proportional to the density gradient estimate at point $\mathbf{x}$ obtained with kernel K.

$$
\mathbf{m_{h(x)}} = \left[\frac{\sum_{i=1}^{N} \mathbf{x_i} g(\| \frac{\mathbf{x} - \mathbf{x_i}}{h} \|^2)}{\sum_{i=1}^{N} g(\| \frac{\mathbf{x} - \mathbf{x_i}}{h} \|^2)} - \mathbf{x} \right]. \qquad (11.11)
$$

The mean shift procedure for a given point $\mathbf{x_i}$, at a time t, is as follows [3]:

(i) Compute the mean shift vector $\mathbf{m}(\mathbf{x}_i^{(t)})$.
(ii) Translate density estimation window as $\mathbf{x}_i^{(t+1)} = \mathbf{x}_i^{(t)} + \mathbf{m}(\mathbf{x}_i^{(t)})$.
(iii) Iterate (i) & (ii) until convergence, i.e., $\nabla f(\mathbf{x}) = 0$.

The process is pictorially depicted in Fig. 11.11. Starting at data point $\mathbf{x_i}$, the mean shift procedure is run to find the stationary points of the density function. Superscripts denote the iteration number, the shaded and black dots denote the

Fig. 11.11 Mean shift procedure

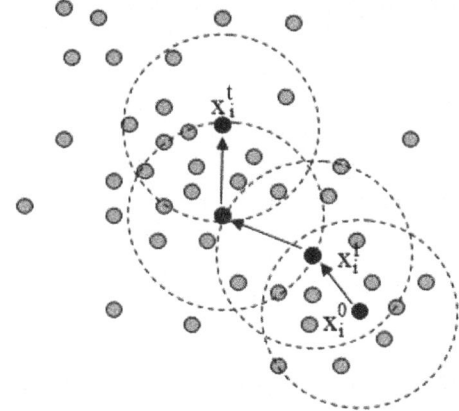

input data points and successive window centers, respectively, and the dotted circles denote the density estimation windows.

The advantage of the algorithm is that no assumption about anything regarding the shape and distribution of different clusters is made. It also does not assume anything about the feature space. The only assumption it makes is the bandwidth or window size or radius of the kernel h. Otherwise, it is like a nonparametric clustering algorithm. A limitation of the standard mean shift procedure is that the value of the bandwidth parameter is unspecified. Choosing the value of h beforehand may be difficult in some cases where adaptive bandwidth selection may be a suitable option.

The most computationally expensive component of the mean shift procedure corresponds to identifying the neighbors of a point in space (as defined by the kernel and its bandwidth). This problem is known as multidimensional range searching in the computational geometry [7] literature. This computation becomes too disorganized to function efficiently for high-dimensional feature spaces.

11.4   Combination of Clustering Algorithms

The task of clustering requires some kind of exploratory knowledge about the underlying clusters. Any particular clustering algorithm assumes some kind of clustering structure of the data without any ground truth information, so the results may be erroneous also. Moreover, in distributed environment, where the data may come from several different sources with different distributions, a combination of clustering algorithms may better reveal the structure of the data than a single clustering algorithm. A combination of clustering algorithms reduces the variance of expectation of error rate.

A better alternative result of cluster analysis is shown by Cluster ensemble method. A set of clusters is produced from the same dataset, and they are combined to form the final cluster. The aim of this combination technique is to enhance the quality of individual data clusterings. The promising results and good number of applications of the new methods make it essential to do critical analysis of the existing techniques and future projections. This section gives the overview of the clustering ensemble methods that would prove to be essential for the community of clustering researchers.

When clustering algorithm is applied to an object's set, it imposes an organization to the data following an internal criterion, the characteristics of the used similarity or dissimilarity function, and the dataset. Thus, if two different clustering algorithms are present and which are applied to the same dataset, it is possible to attain very different results. There is no validity index that impartially gives the outcome of any clustering algorithm. Hence, it can be said that different results produced by different clustering algorithms can be equally plausible, if there lies no prior knowledge about the best way to evaluate the results. Also, it can be assured that for any clustering algorithm, there is a validity index, which will generate satisfactorily its results.

Fig. 11.12 Diagram of the general process of cluster ensemble

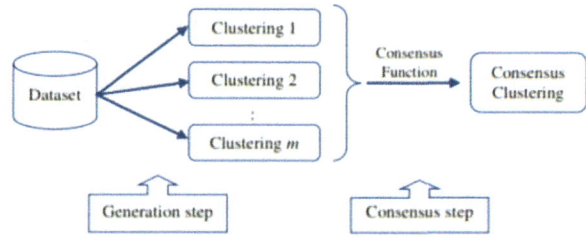

The concept using which various clustering results are combined (cluster ensemble or clustering aggregation) came out as an alternative technique to enhance the quality of outcomes of clustering algorithms. It is developed from the successfulness of the combination of supervised classifiers.

Provided a set of objects, a cluster ensemble technique comprises two principal steps (as shown in Fig. 11.12): generation, which involves creating a set of partitions of these objects, and consensus function, where a new partition, which is the integration of all partitions attained in the generation step, is made. Basically, the goal is to combine "weak" clusterings to a better one.

The idea of ensemble of clustering algorithms is inspired from the success of the ensemble of classifiers. The consensus function is used to combine the results of different clustering algorithms. There are many reasons supporting the fact that designing an ensemble of clustering algorithms is more difficult than designing an ensemble of classifiers. One of the challenges in designing the consensus function is that the number of clusters obtained from a specific dataset produced by different algorithms may be different; but in classification, the number of class labels are known a priori and are constant.

Different clustering algorithms produce different partitions of data and the design of consensus function has different perspectives starting from graph-based, combinatorial/statistical, probabilistic, information theoretical, and even expectation maximizing methods. The consensus function combines results of weak clustering algorithms. Following are two examples of weak clustering algorithms:

(i) Clustering of the data projected (along some random vector) into a one dimensional subspace.

(ii) Clustering by dividing the data into separate groups/clusters using random hyperplanes.

For example, in the Fig. 11.13a, a weak clustering algorithm that divides the dataset using k-means algorithm on the one-dimensional projection of the data points along a random vector is run four times with two different number of clusters ($k = 2, 3$) as parameters. In the Fig. 11.13b, the four different clustering partitions are combined using a consensus function like cluster-based similarity partitioning algorithm (CSPA). In CSPA, the similarity of two data points is the number of partitions/clusterings of the ensemble in which they are clustered together. CSPA is very simple algorithm with very high storage computational complexity. To reduce

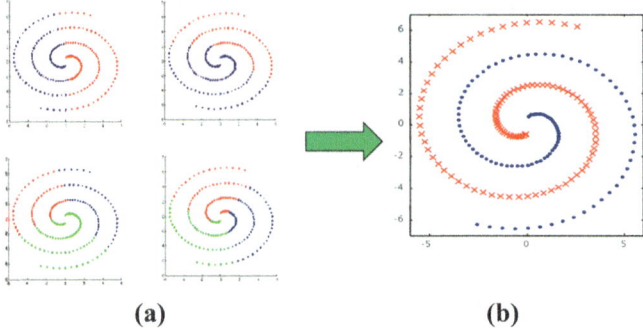

Fig. 11.13 Combination shows good clustering. (**a**) Clustering using k-means algorithm. (**b**) Combination of results

the storage and computational complexity, other algorithms such as meta clustering could be used.

Meta clustering [1] generates many clustering results. It makes use of non-determinism or local minima/maxima, different clustering algorithms, and different parameter settings. Each cluster is represented by a hyperedge. The idea in MCLA is to group and collapse related hyperedges and assign each object to the collapsed hyperedge in which it participates most strongly. The hyperedges that are considered related to the purpose of collapsing are determined by a graph-based clustering (discussed in Chap. 10.1) of hyperedges. We refer to each cluster of hyperedges as a meta-cluster $\mathcal{C}^{(M)}$. Collapsing reduces the number of hyperedges from $\sum_{q=1}^{r} k^{(q)}$ to k; where r is the number of partitions in the meta graph, and k is the original number of clusters. The following section discusses the details of MCLA.

Construct meta-graph: Let us view all the $\sum_{q=1}^{r} k^{(q)}$ indicator vectors $\mathbf{h}$ (the hyperedges of $\mathbf{H}$) as vertices of another regular undirected graph, the meta-graph. The edge weights are proportional to the similarity between vertices. A suitable similarity measure here is the binary Jaccard measure, since it is the ratio of the intersection to the union of the sets of objects corresponding to the two hyperedges. Formally, the edge weight $w_{a,b}$ between two vertices h_a and h_b as defined by the binary Jaccard measure of the corresponding indicator vectors $\mathbf{h}_a$ and $\mathbf{h}_b$ is: $w_{a,b} = \frac{\mathbf{h}_a^{\dagger}\mathbf{h}_b}{\|\mathbf{h}_a\|_2^2 + \|\mathbf{h}_b\|_2^2 - \mathbf{h}_a^{\dagger}\mathbf{h}_b}$. Since the clusters are non-overlapping (e.g., hard), there are no edges among vertices of the same clustering $\mathbf{H}^{(q)}$ and, thus, the meta-graph is r-partite. The concatenated block matrix $\mathbf{H} = \mathbf{H}^{(1,\ldots,r)} = (\mathbf{H}^{(1)} \ldots \mathbf{H}^{(r)})$ defines the adjacency matrix of a hypergraph with n vertices and $\sum_{q=1}^{r} k^{(q)}$ hyperedges. Each column vector $\mathbf{h}_a$ specifies a hyperedge h_a, where 1 shows that the vertex corresponding to the row is a part of that hyperedge and 0 indicates that it is not. Thus, we have mapped each cluster to a hyperedge and the set of clusterings to a hypergraph (Fig. 11.14).

	$\lambda^{(1)}$	$\lambda^{(2)}$	$\lambda^{(3)}$	$\lambda^{(4)}$
x_1	1	2	1	1
x_2	1	2	1	2
x_3	1	2	2	?
x_4	2	3	2	1
x_5	2	3	3	2
x_6	3	1	3	?
x_7	3	1	3	?

$\Leftrightarrow$

	$\mathbf{H}^{(1)}$			$\mathbf{H}^{(2)}$			$\mathbf{H}^{(3)}$			$\mathbf{H}^{(4)}$	
	$\mathbf{h_1}$	$\mathbf{h_2}$	$\mathbf{h_3}$	$\mathbf{h_4}$	$\mathbf{h_5}$	$\mathbf{h_6}$	$\mathbf{h_7}$	$\mathbf{h_8}$	$\mathbf{h_9}$	$\mathbf{h_{10}}$	$\mathbf{h_{11}}$
v_1	1	0	0	0	1	0	1	0	0	1	0
v_2	1	0	0	0	1	0	1	0	0	0	1
v_3	1	0	0	0	1	0	0	1	0	0	0
v_4	0	1	0	0	0	1	0	1	0	1	0
v_5	0	1	0	0	0	1	0	0	1	0	1
v_6	0	0	1	1	0	0	0	0	1	0	0
v_7	0	0	1	1	0	0	0	0	1	0	0

Fig. 11.14 Illustrative cluster ensemble problem with $r = 4$, $k^{(1,\ldots,3)} = 3$, and $k^{(4)} = 2$: Original label vectors (left) and equivalent hypergraph representation with 11 hyperedges (right). Each cluster is transformed into a hyperedge

Cluster hyperedges: Find matching labels by partitioning the meta-graph into k balanced meta-clusters. Each vertex is weighted in proportion to the size of the corresponding cluster. Balancing ensures that the sum of vertex weights is approximately the same in each meta-cluster. This results in a clustering of the **h** vectors. Since each vertex in the meta-graph represents a distinct cluster label, a meta-cluster represents a group of corresponding labels.

Collapse meta-clusters: For each of the k meta-clusters, we collapse the hyperedges into a single meta-hyperedge. Each meta-hyperedge has an association vector, which contains an entry for each object describing its level of association with the corresponding meta-cluster. The level is computed by averaging all indicator vectors **h** of a particular meta-cluster. An entry of 0 or 1 indicates the weakest or strongest association, respectively.

Compete for objects: In this step, each object is assigned to its most associated meta-cluster: Specifically, an object is assigned to the meta-cluster with the highest entry in the association vector. Ties are broken randomly. The confidence of an assignment is reflected by the winner's share of association (ratio of the winner's association to the sum of all other associations). Note that not every meta-cluster can be guaranteed to win at least one object. Thus, there are at most k labels in the final combined clustering λ. Figure 11.15 illustrates meta-clustering as an example where $r = 4$, $k = 3$, $k^{(1,\ldots,3)} = 3$, and $k^{(4)} = 2$. It shows the original 4-partite meta-graph. Edge darkness increases with edge weight. The three meta-clusters are shown in red/$\circ$, blue/$\times$, and green/$+$. Consider the first meta-cluster, $C_1^{(M)} = \{h_3, h_4, h_9\}$ (the red/$\circ$ markers in Fig. 11.15). Collapsing the hyperedges yields the object-weighted meta-hyperedge $h_1^{(M)} = \{v_5, v_6, v_7\}$ with association vector $(0, 0, 0, 0, 1/3, 1, 1)^{\dagger}$. Subsequently, meta-cluster $C_1^{(M)}$ will win the competition for vertices/objects v_6 and v_7, and thus represent the cluster $C_1 = \{x_6, x_7\}$ in the resulting integrated clustering. Our proposed meta-clustering algorithm robustly outputs $(2, 2, 2, 3, 3, 1, 1)^{\dagger}$, one

Fig. 11.15 Illustration of
Meta-Clustering Algorithm
(MCLA) for a cluster
ensemble

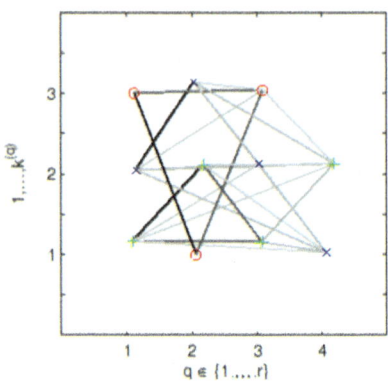

of the six optimal clusterings, which is equivalent to clusterings $\lambda^{(1)}$ and $\lambda^{(2)}$. The uncertainty about some objects is reflected in the confidences 3/4, 1, 2/3, 1, 1/2, 1, and 1 for objects 1 through 7, respectively.

References

1. Caruana, R., Elhawary, M., Nguyen, N., & Smith, C. (2006).   Meta clustering.   In *Sixth international conference on data mining, ICDM'06* (107–118).
2. Cheng, Y. (1995).   Mean shift, mode seeking, and clustering. *IEEE transactions on pattern analysis and machine intelligence, 17*(8), 790–799.
3. Comaniciu, D., & Meer, P. (2002). Mean shift: A robust approach toward feature space analysis. *IEEE Transactions on Pattern Analysis and Machine Intelligence, 24*(5), 603–619.
4. Kaufman, L., & Rousseeuw, P. (1987). *Clustering by means of medoids.* North-Holland.
5. MacQueen, J. (1967). Some methods for classification and analysis of multivariate observations. *Proceedings of the Fifth Berkeley Symposium on Mathematical Statistics and Probability, 1*(14), 281–297.
6. Maimon, O., & Rokach, L. (2005). *Data mining and knowledge discovery handbook*, vol. 2. Springer.
7. Matoušek, J. (1994). Geometric range searching. *ACM Computing Surveys (CSUR), 26*(4), 422–461.
8. Worton, B. J. (1989). Kernel methods for estimating the utilization distribution in home-range studies. *Ecology, 70*(1), 164–168.

Chapter 12
Outliers

Let us assume that there are multiple groups each having some distinct features and the normal objects are "clustered" into these multiple groups. It might arise that certain patterns are far away from all these groups of normal patterns, like the red data points shown in Fig. 12.1. An *outlier* can be defined as an observation that tends to deviate drastically from all other observations. An outlier may arise because of variability in the measurement, or it may indicate experimental error; the latter is sometimes excluded from the dataset. Outliers are different from noise. Noise is defined as a random error or variation in a measured variable, but outliers violate the mechanism that generate normal data. Example of outliers can be given as a credit card transaction where a suspiciously high money transaction than the usual transactions occurs. Outliers can be detected using statistical methods, proximity-based methods, or clustering-based methods [4] (Fig. 12.2).

Statistical approach assumes that a generative model generates the objects in a dataset. The objects lying in the lower probability regions of the model are identified as outliers. Proximity-based approaches have the intuition that objects far away from others are outliers. There are basically two types of proximity-based outlier detection methods—distance-based and density-based [2]. Distance-based methods detect the points that stay far away from most of the data points as outliers. Figure 12.3 depicts an outlier detection method based on distance from the neighborhood. As shown in the example, the outliers does not belong to any of the clusters.

Density-based techniques detect points as outliers, which have a low neighborhood density, i.e., there are very few or no points in a close neighborhood of the outlier (Fig. 12.4). As shown in the example, outliers belong to a comparatively sparse region than other dense data points.

Clustering-based approaches [6] identify an object as an outlier if it does not belong to any cluster (Fig. 12.5). As shown in the example, outliers does not have the compactness like any one of the clusters.

© The Author(s), under exclusive license to Springer Nature Singapore Pte Ltd. 2025   197
A. Ghosh, *Data Science and Cases in Sustainability*, Mathematics for Sustainable
Developments, https://doi.org/10.1007/978-981-96-8362-8_12

Fig. 12.1 Supervised outlier
detection

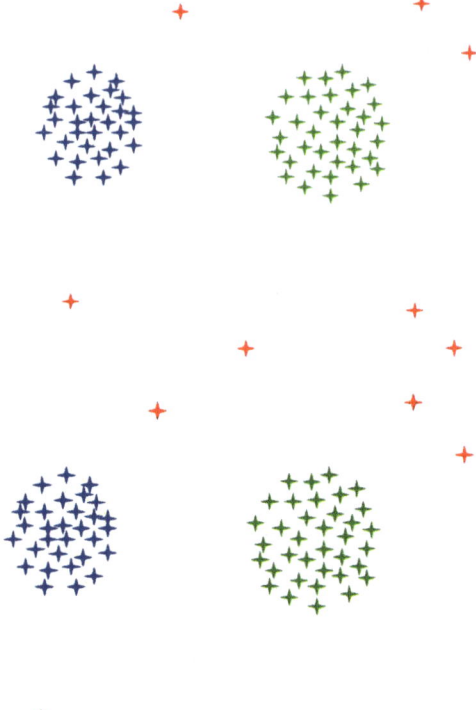

Fig. 12.2 An example of
outliers

The distance between the object and its closest cluster is large or the object
belongs to a small or sparse cluster. Outlier detection has important applications
in the field of data mining, such as fraud detection, customer behavior analysis,
intrusion detection, abnormality detection in medical sciences.

12.1 Importance of Outlier Detection

There are multiple reasons that make the outlier detection ability in various application domains a highly desirable characteristic. Some of the popular applications of
outlier detection are discussed in this section. For each of the application domain,
we discuss the significance of outliers, the challenges that are unique to that domain,
and the popular techniques that have been adopted for outlier detection in that
domain.

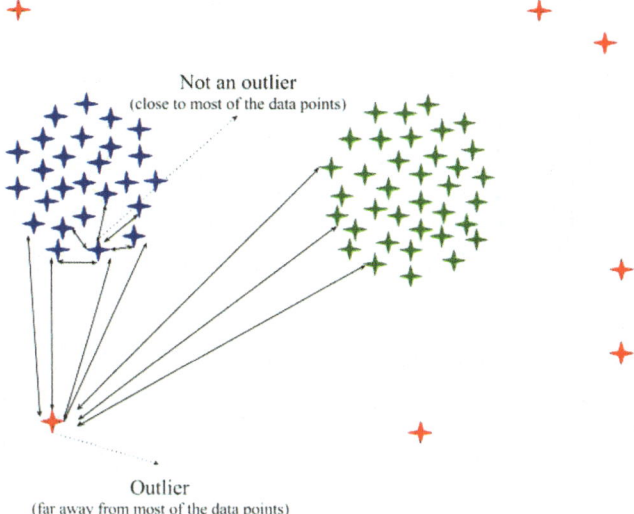

Fig. 12.3 An example of distance-based outlier detection

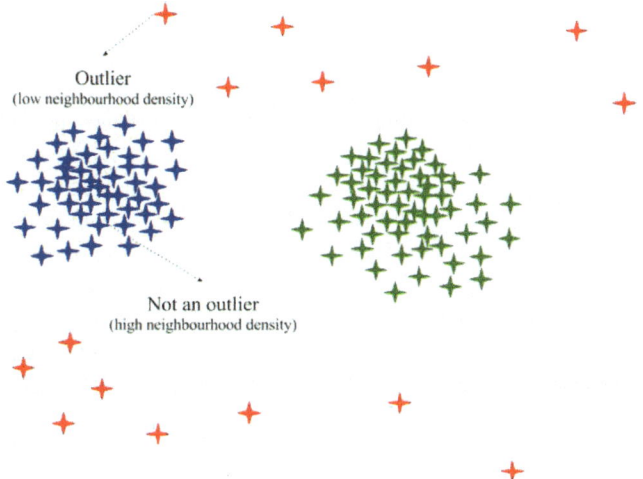

Fig. 12.4 An example of density-based outlier detection

12.1.1 Intrusion Detection

Intrusion detection can be defined as identification of malign activities (like break-ins, penetrations, and other types of computer abuse) in a computer-related system [8]. From a computer security perspective, these malicious activities or intrusions are very interesting. An intrusion differs from the normal behavior that is shown by

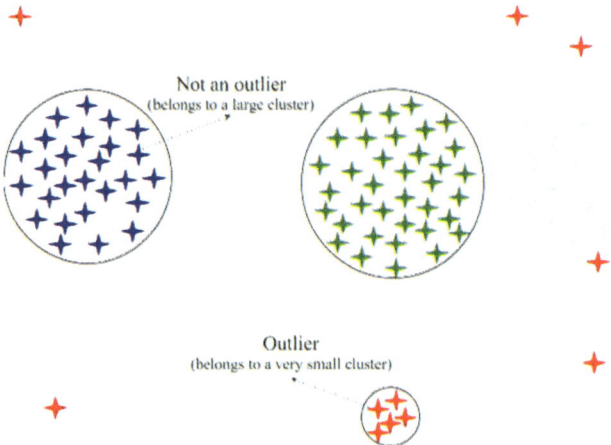

Fig. 12.5 An example of clustering-based outlier detection

the system. This property allows the problem to be directly formulated as an outlier detection problem. Outlier detection techniques are widely applied for intrusion detection.

In this domain, the main challenge faced by outlier detection is the high volume of data. There comes the need to make the outlier detection schemes computationally efficient for handling these large-sized inputs within a very short period of time. Also, since the data generally appear to be a streamed data, so online analysis becomes a requirement. Another problem is the false alarm rate, which also arises because of the large size of the input. Now, because the data lead into millions of data objects, a small percentage of false alarms might make analysis massive for an analyst. Typically, labeled data that correspond to normal behavior are easily available, while labeled data for outliers that are required for training purpose are not readily available. Thus, in this domain, semi-supervised and unsupervised outlier detection techniques have been more favored over supervised methods.

12.1.2 Fraud Detection

Fraud detection can be defined as identification of criminal acts that occur in commercial places like banks, insurance agencies, credit card companies, stock market, and cell phone companies. The spiteful users can be the ones who fake themselves to be customers (also known as identity theft), or they might be the actual customers of the affected organization. The fraud generally happens when these users use the resources related to the organization in an illegal way [5]. To prevent the economic losses, the affected organizations generally want to immediately identify such frauds. Some of the specific areas/applications of fraud detection are insurance claim fraud detection, credit card fraud detection, mobile/cellular fraud detection, etc.

12.1.3 Medical and Public Health Data

Outlier detection in the medical and public health domains typically is related to patient records [7]. The data can have outliers due to various reasons like instrumentation errors or recording errors or some abnormal patient conditions. Thus, the outlier detection is a very critical problem for this domain, and also high degree of accuracy becomes a major requirement for this domain.

12.1.4 Fault Detection in Mechanical Units

In this domain, the outlier detection methods track the performance of industrial components like turbines, motors, oil flow in pipelines, or other mechanical components, and defects detection, which may have occurred due to wear and tear of components or some other abrupt incidents [9].

12.1.5 Structural Defect Detection

Cracks in beams and strains in airframes are the few examples of structural defects and damages, and the detection of these structural defects and damages is basically related to the industrial fault detection [3]. Owing to the continued usage and the normal wear and tear, industrial units get damaged. Such damages require early detection for prevention of further escalation and fatal losses. The normal data and the models trained using these data typically become static over time. The data may have spatial correlations.

Some popular outlier detection techniques are discussed in the following sections, and a few more are discussed in Chapter 31.

12.2 Principle of Outlier Detection

An outlier is defined as an observation that tends to deviate strikingly from other observations present in the sample. An outlier may imply to be bad data. For instance, it may have been the incorrect coding of data or an experiment that did not run correctly. If this can be determined, that an outlying point is actually erroneous, then that outlying value should be removed from the data analysis (or corrected if possible). But this cannot be inferred for all the cases, like in some cases, it may not be possible to determine if an outlying point is erroneous or belongs to a bad data. It is possible that outliers may indicate something scientifically interesting, or outliers may be caused due to some random variation. For any event, we typically do not desire to simply remove the outlying observation from the analysis. But in situations like when the data contain significant outliers, we may need to consider using some of the robust statistical techniques. The underlying distribution of the data guides in identifying an observation as an outlier. Hence, the outlier detection techniques can be broadly classified as supervised, semi-supervised, and unsupervised techniques.

12.2.1   Supervised Outlier Detection Techniques

This technique assumes that the training data are readily available and are composed of labeled instances for the normal classes as well as the outlier classes. The general technique in this scenario is to develop predictive models designed for both normal and outlier classes. Next, the unseen data instance gets compared with the two trained models for determining which class the unseen data belong to. Accurate models can be built using supervised outlier detection methods since they have an explicit notion of the normal and outlier behavior. One drawback for this technique is that it might be prohibitively expensive to get correctly labeled training dataset. A lot of effort is required to obtain the labeled training dataset since data labeling is generally manually done by a human annotator. Certain techniques artificially inject outliers in a normal dataset (which has no outliers) to obtain a fully labeled training dataset and then apply the supervised outlier detection techniques in the test data to detect outliers [1].

12.2.1.1   Supervised Outlier (Anomaly) Detection

In supervised outlier detection, we have labels of data indicating which one is an outlier (anomaly, $y = 1$) and which one is not an outlier (not anomaly, $y = 0$). This way of outlier detection is a more standard one.

Supervised outlier detection should not be confused with classification. Supervised outlier detection should be used when there is a large number of negative ($y = 0$) examples and the positive class is really very much skewed. It is hard to know how the outlier would be like. A totally new kind of anomaly can occur suddenly. Rather, it is easier to know how the positive ones are like. So, in supervised outlier detection, the negative examples are preferred over the rest for learning purpose. In case of classification, both the positive and negative examples are likely to follow the trend of the labeled training data. So, in classification, every class is given importance for the learning purpose.

Sometimes, unsupervised outlier detection can fail to detect an outlier that has a label ($y = 1$) but is having a high probability value, calculated using the existing features. Such examples may indicate a necessary inclusion of a new feature or a generated feature, which would help to distinguish between the normal and anomalous examples, in terms of the probability density value. Since the correlation between different features is not considered automatically by the algorithm, in this way we can capture them manually.

12.2.2 Semi-Supervised Outlier Detection Techniques

This technique assumes that a few labeled instances are available for only one class. Now often, it gets difficult to collect labels for rest of the other classes. For instance, in the case of space craft fault detection, an accident would signify as an outlier scenario, which is not very easy to model. A typical technique to handle this situation using this technique is to develop model using only the available class and to treat any test instance, which does not fit in this model to represent the other class.

12.2.3 Unsupervised Outlier Detection Techniques

This third category is the most widely applicable technique since it does not make any assumption about the labeled training data availability. Although this technique makes some assumptions about the data like a pattern that occurs frequently is generally considered as a normal data while rarely occurring data are typically considered as an outlier. These unsupervised outlier detection techniques generally suffer from the problem of higher false alarm rate, because many a times the underlying assumptions do not hold true. Clustering-based methods consider a cluster of small sizes, including the size of one observation, as clustered outliers. Some examples for such methods are the partitioning around medoids (PAM) (discussed in Chap. 11).

12.2.3.1 Unsupervised Outlier (Anomaly) Detection using Density Estimation

Let N be the number of available examples and d the number of features.

Available set: $\{\mathbf{x_1}, \mathbf{x_2}, \cdots, \mathbf{x_N}\}$.

Each of the components (x_{if}) of the feature vector$(\mathbf{x_i})$ for i^{th} training example is assumed to be following a specific Gaussian distribution with specific parameters $(\bar{x}_f, \sigma_f^2)$. These parameters are estimated from the given data, the estimations are necessarily the corresponding maximum likelihood estimations (Fig. 12.6).

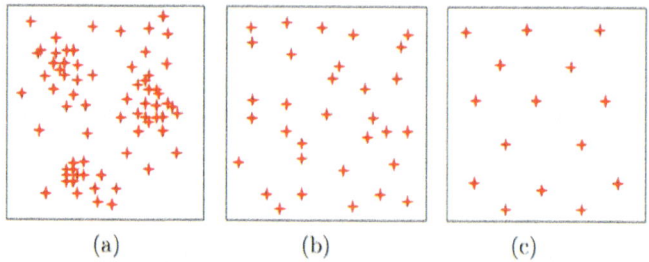

(a)   (b)   (c)

Fig. 12.6 Example of unsupervised outlier detection based on density

$\bar{x}_f, \sigma_f^2$ are estimated for f^{th} feature using the following formulae:

$$\bar{x}_f = \frac{1}{N} \sum_{i=1}^{N} x_{if};$$ (12.1)

$$\sigma_f^2 = \frac{1}{N} \sum_{i=1}^{N} (x_{if} - \bar{x}_f)^2; \mathbf{x} \in \mathbb{R}^d.$$ (12.2)

Probability of random vector $\mathbf{x}$ is estimated as the product of the probabilities of the individual features as shown as follows:

$$p(\mathbf{x}) = p(x_1; \bar{x}_1, \sigma_1^2) p(x_2, \bar{x}_2, \sigma_2^2) \cdots p(x_d; \bar{x}_d, \sigma_d^2) = \prod_{f=1}^{d} p(x_f, \bar{x}_f, \sigma_f^2).$$ (12.3)

Different features are assumed to be independent while estimating the density.

For instance, let us have $d = 3$, i.e., the feature vector has three features. Let $\mathbf{x}$ be a new datapoint, $\mathbf{x} = \{x_1, x_2, x_3\}$. Each of the features is generated from the three different Gaussian distributions as shown in Fig. 12.7. Let us assume a new data point with the feature vector $\mathbf{x} = \{a, b, c\}$. So from the figure, we get the probability $p(\mathbf{x}) = p(x_1 = a) * p(x_2 = b) * p(x_3 = c)$.

The steps of this algorithm are described as follows.

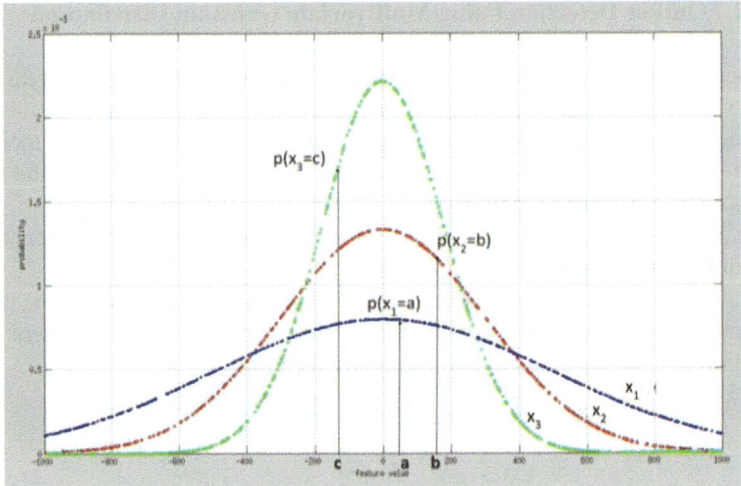

Fig. 12.7 Separate Gaussian distributions of three features x_1, x_2, x_3

(i) Choosing the examples $\mathbf{x_i}$ that might be showing some anomalous behavior (e.g., some user in a system with anomalous behavior).

(ii) Find parameters $(\bar{x}_1, \bar{x}_2, \cdots, \bar{x}_d)$ and $(\sigma_1, \sigma_2, \cdots, \sigma_d)$ corresponding to every feature from the available data.

$$\bar{x}_f = \frac{1}{N} \sum_{i=1}^{N} x_{if}.$$

$$\sigma_f^2 = \frac{1}{N} \sum_{i=1}^{N} (x_{if} - \bar{x}_f)^2.$$

(iii) Given a new example $\mathbf{x}$, calculate

$$p(\mathbf{x}) = \prod_{j=1}^{d} p(x_f, \bar{x}_f, \sigma_f^2) = \prod_{f=1}^{d} \frac{1}{\sqrt{2\pi}\sigma_f} exp\left(-\frac{(x_f - \bar{x}_f)^2}{2\sigma_f^2}\right).$$

1. The example $\mathbf{x}$ is detected as outlier if $p(\mathbf{x}) < \varepsilon$; ε is a pre-assigned positive constant.

12.2.3.2   Outlier Detection Using Multivariate Gaussian Distribution

The independence assumption of the different features in the previously discussed unsupervised outlier detection algorithm can pose some constraints on the shape of the region that contains the normal data points. In order to take the dependence among features into consideration, multivariate Gaussian distribution of the feature vector is taken for anomaly detection.

Here, $\bar{\mathbf{x}}$ is the mean vector. $\boldsymbol{\Sigma}$ is the covariance matrix (capturing the dependence among different features).

Now, the probability density of an example is no more considered as the product of the probability densities of individual features; rather, the parameters of the distribution are calculated using the following formula:

$$\bar{\mathbf{x}} = \frac{1}{N} \sum_{i=1}^{N} \mathbf{x_i}; \tag{12.4}$$

$$\Sigma_{fj} = \frac{1}{N} \sum_{i=1}^{N} (x_{if} - \bar{x}_f)(x_{ij} - \bar{x}_j); \tag{12.5}$$

where, $\mathbf{x} \in \mathbb{R}^d$, $\bar{\mathbf{x}} \in \mathbb{R}^d$, $\boldsymbol{\Sigma} \in \mathbb{R}^{d \times d}$, $N > d$, and $\boldsymbol{\Sigma}^{-1}$ should exist. $\boldsymbol{\Sigma}$ being non-invertible is an indication that some of the features are either duplicated or are linearly dependent on each other. The probability density function looks like

$$p(\mathbf{x}) = \frac{1}{(2\pi)^{\frac{d}{2}} |\boldsymbol{\Sigma}|^{\frac{1}{2}}} exp(\mathbf{x} - \bar{\mathbf{x}})^T \boldsymbol{\Sigma}^{-1}(\mathbf{x} - \bar{\mathbf{x}}). \tag{12.6}$$

The corresponding algorithm is as follows (Fig. 12.8).

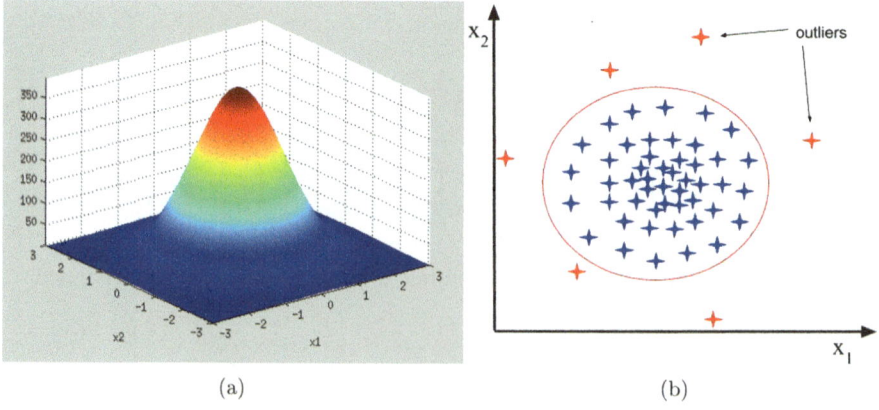

(a)   (b)

Fig. 12.8 (**a**) Gaussian distribution of features $\mathbf{x} = [x_1, x_2]$, (**b**) Data points with probability density value less than the threshold ε are detected as outliers

  (i) Find parameters ($\bar{\mathbf{x}}$ and $\mathbf{\Sigma}$) using Eqs. 12.4 and 12.5.
 (ii) Given a new example $\mathbf{x}$, calculate $p(\mathbf{x})$ using Eq. 12.6.
(iii) Given example is detected as outlier if $p(\mathbf{x}) < \varepsilon$.

This procedure is computationally costly; but, unlike the algorithms discussed in the previous sections, it automatically captures the correlations between different features.

12.3   Appropriate Features for Outlier Detection

The features that have Gaussian distribution or that can be transformed into Gaussian using some kind of transformation are chosen for outlier detection. Some examples of such transformation of a feature x_f may be like the following:

$$x_f = \log(x_f).$$

$$x_f = \log(x_f + c).$$

$$x_f = \sqrt{x_f}.$$

$$x_f = x_f^{\frac{1}{3}}.$$

Any transformation of a feature can be designed so as to make it fit into a Gaussian distribution.

References

1. Abe, N., Zadrozny, B., & Langford, J. (2006). Outlier detection by active learning. *Proceedings of the 12th ACM SIGKDD international conference on knowledge discovery and data mining* (pp. 504–509).
2. Breunig, M. M., Kriegel, H.-P., Ng, R. T., & Sander, J. (2000). LOF: Identifying density-based local outliers. *ACM SIGMOD Record, 29*(2), 93–104.
3. Chan, C., & Pang, G. K. (2000). Fabric defect detection by fourier analysis. *IEEE Transactions on Industry Applications, 36*(5), 1267–1276.
4. Duan, L., Xu, L., Liu, Y., & Lee, J. (2009). Cluster-based outlier detection. *Annals of Operations Research, 168*(1), 151–168.
5. Fawcett, T., & Provost, F. (1997). Adaptive fraud detection. *Data Mining and Knowledge Discovery, 1*(3), 291–316.
6. He, Z., Xu, X., & Deng, S. (2003). Discovering cluster-based local outliers. *Pattern Recognition Letters, 24*(9), 1641–1650.

7. Leyland, A. H., & Goldstein, H. (2001). *Multilevel modelling of health statistics*. Wiley.
8. Phoha, V. V. (2007). *Internet security dictionary*. Springer.
9. Thomson, W. T., & Fenger, M. (2001). Current signature analysis to detect induction motor faults. *IEEE Industry Applications Magazine, 7*(4), 26–34.

Chapter 13
Fuzzy Set Theoretic Approach to Pattern Recognition

Fuzzy sets were basically introduced by Lotfi Zadeh in 1965 to represent/manipulate data and information possessing under non-statistical uncertainties [15]. Fuzzy logic provides an inference morphology, which allows the approximate human reasoning capabilities to make it apply to knowledge-based systems. The concept of fuzzy logic gives a mathematical strength that makes it easier to capture the uncertainties associated with the human cognitive processes, such as thinking and reasoning. Few of the vital characteristics of fuzzy logic are mentioned as follows:

(i) In fuzzy logic, exact reasoning is treated as a limiting case of approximate reasoning.
(ii) Knowledge is interpreted as a collection of elastic or, equivalently, fuzzy constraint on a collection of variables.
(iii) Inference is viewed as a process of propagation of elastic constraints.
(iv) Fuzzy systems are suitable for uncertain or approximate reasoning, especially for a system with a mathematical model that is difficult to derive.
(v) Fuzzy logic allows decision-making with estimated values under uncertain or incomplete information

Fuzzy logic includes 0 and 1 as extreme cases of truth (or "fact") but also includes various states of truth in between these extreme cases of 0 and 1; thus, for instance, the result of a comparison between two things could not be represented as "tall" or "short" but could be represented as "0.38 of tallness."

13.1 Fuzzy Sets

Crisp sets are sets those we use in most of our real-life decision-making. In a crisp set, an element is either a member of the set or not a member of the set. For instance, suppose a jelly bean belongs to the class of food known as candy while potatoes do not belong to this class of food called candy. On the other hand, fuzzy sets allow the elements to get partially allocated to a set. Each element is given a degree of membership in a set. This membership value can range from 0 (not an element/member of the set) to 1 (a full member of the set). It is clear that the extreme membership values of 0 and 1 would be represented as crisp sets. A membership function is a function mapping the element values and their membership degree to a set. For example, suppose we have a set of tall men. If we say that the people having height equal to or more than 6 feet are tall, then it becomes a crisp set. This set can be represented graphically as shown in Fig. 13.1.

On the contrary, the fuzzy set theoretical approach to the set of tall men provides a much better representation of the tallness of a person. The set, shown in Fig. 13.2, is defined by a continuous function.

So, for the two people shown in Fig. 13.2, the first person has a membership of 0.3 and so is not very tall. The second person has a membership value of 0.95 and thus is definitely tall. However, he does not belong to the set of tall men, although he definitely has a high degree of membership in the fuzzy set of tall men. Thus, this representation is different from the way the bivalent (crisp) sets work (Fig. 13.3).

Fig. 13.1 Graphical illustration of a crisp set

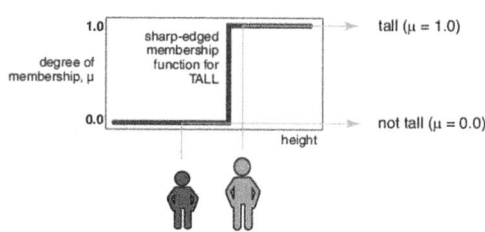

Fig. 13.2 Complement of a fuzzy set

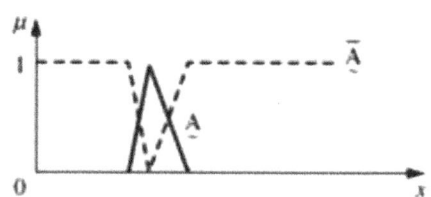

Fig. 13.3 Graphical illustration of a fuzzy set

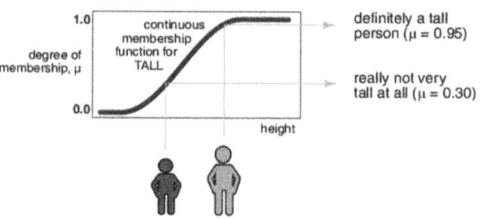

Fuzzy logic is a superset of conventional (Boolean) logic that has been extended for handling the concept of partial truth—i.e., the truth values between "completely true" and "completely false." Note that a fuzzy set tries to model a vaguely defined concept having no clear boundary. Here some basic concepts and tools of fuzzy set theory are presented. It deals with some basic operations on fuzzy sets.

13.2   Theory of Fuzzy Sets

A fuzzy subset A of the universe $\tilde{U}$ is defined as a collection of ordered pairs

$$A = \{(a_i, \mu_A(a_i)), \forall a_i \in \tilde{U}\} \tag{13.1}$$

where $\mu_A(a_i)$, $(0 < \mu_A(a_i) \leq 1)$ denotes the degree of belonging of the element a_i to the fuzzy set A.

Because fuzzy sets are a generalization of the classical sets, the embedding of conventional models into a larger setting endows fuzzy models with greater flexibility to capture various aspects of incompleteness or imperfection (i.e., deficiencies) in whatever information and data are available about real processes. In other words, the membership functions possess *elasticity*; thus, the higher the value of membership of an element to a set, the less the imprecisely defined concept of the fuzzy set must be stretched to accommodate the element. Hard membership functions, of course, are *inelastic*.

13.3   Definitions

In this section, we provide a few terminologies associated with fuzzy logic.

Universe of Discourse
The Universe of Discourse is the range of all possible values for an input to a fuzzy system.

Membership Functions
The only condition that must be satisfied by a membership function is that it must vary only between the values 0 and 1. The function itself could be an arbitrary curve whose shape can be defined as a function that suits us from the point of views of convenience,simplicity, efficiency, and speed.

A classical set may be written as

$$A = \{a_i | a_i > 6\}. \tag{13.2}$$

Fig. 13.4 Support of a fuzzy
set

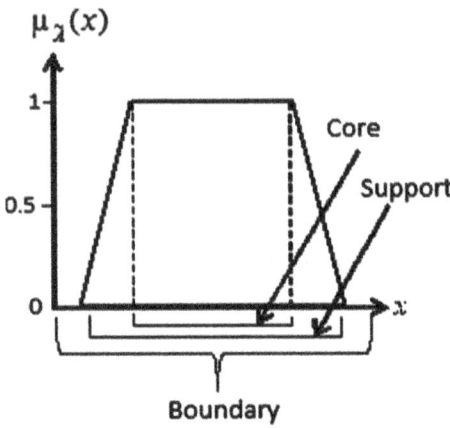

A fuzzy set is an extension of a classical set. If $\tilde{U}$ is the universe of discourse and its elements are denoted by a_i, then a fuzzy set A in $\tilde{U}$ is defined as a set of ordered pairs.

$$A = \{a_i, \mu_A(a_i)|a_i \in \tilde{U}\};\qquad\qquad (13.3)$$

$\mu_A(a_i)$ is called the membership function (or MF) of a_i in A. The membership function maps each element of $\tilde{U}$ to a membership value in (0, 1].

Support
The Support of a fuzzy set A is the crisp set of all points in the Universe of Discourse $\tilde{U}$ such that the membership function of those points to belong to the fuzzy set A is non-zero (Fig. 13.4).

Crossover Point
A Crossover Point of a fuzzy set is the element in $\tilde{U}$ for which the membership value is 0.5.

Fuzzy Singleton
A fuzzy singleton is a fuzzy set whose support is a single point in $\tilde{U}$ with a membership value equal to 1.

Fuzzy Numbers
A fuzzy number is a quantity whose value is imprecise, instead of being exact as is the case with "ordinary" (single-valued) numbers. Any fuzzy number can be thought of as a function whose domain is a specified set (usually the set of real numbers, and whose range is the span of non-negative real numbers between, and including, 0 and 1000). Each numerical value in the domain is assigned a specific "grade of membership" where 0 represents the smallest possible grade, and 1000 being the largest possible grade.

13.4   Basic Operations on Fuzzy Sets

Let A and B be two fuzzy subsets of $\tilde{U}$. Some of the basic operations that can be performed on A and B are as follows:

Union

The membership function of the Union of two fuzzy sets A and B with membership functions μ_A and μ_B respectively is defined as the maximum of the two individual membership values. This is called the maximum criterion.

$$\mu_{A \cup B} = \max(\mu_A, \mu_B). \tag{13.4}$$

The Union operation in fuzzy set theory is equivalent to the **OR** operation in Boolean algebra (Fig. 13.5).

Intersection

The membership function of the Intersection of two fuzzy sets A and B with membership functions μ_A and μ_B respectively is defined as the minimum of the two individual membership values. This is called the minimum criterion.

$$\mu_{A \cap B} = \min(\mu_A, \mu_B). \tag{13.5}$$

The Intersection operation in fuzzy set theory is equivalent to the **AND** operation in Boolean algebra (Fig. 13.6).

Complement

The membership function of the Complement of a fuzzy set A with membership function μ_A is defined as the negation of the specified membership function. This is called the negation criterion.

$$\mu_{\overline{A}} = 1 - \mu_A. \tag{13.6}$$

Fig. 13.5 Union of two fuzzy sets

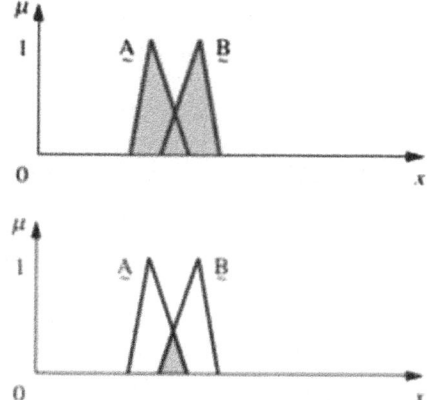

Fig. 13.6 Intersection of two fuzzy sets

The Complement operation in fuzzy set theory is equivalent to the **NOT** operation in Boolean algebra.

13.5   Fuzzy Set Theory Rules

The following rules, which are common in classical set theory, can also be applied to Fuzzy set theory.

De Morgan's law

$$\overline{(A \cap B)} = \overline{A} \cup \overline{B} \tag{13.7}$$

and

$$\overline{(A \cup B)} = \overline{A} \cap \overline{B}. \tag{13.8}$$

Associativity

$$(A \cap B) \cap C = A \cap (B \cap C) \tag{13.9}$$

and

$$(A \cup B) \cup C = A \cup (B \cup C). \tag{13.10}$$

Commutativity

$$A \cap B = B \cap A \tag{13.11}$$

and

$$A \cup B = B \cup A. \tag{13.12}$$

Distributivity

$$A \cap (B \cup C) = (A \cap B) \cup (A \cap C) \tag{13.13}$$

and

$$A \cup (B \cap C) = (A \cup B) \cap (A \cup C). \tag{13.14}$$

Let us now discuss some measures that give the degree of fuzziness in a fuzzy set A. The degree of fuzziness expresses the average amount of ambiguity in making a decision, i.e., whether an element belongs to the fuzzy set (or possesses the concept or property represented by the fuzzy set) or not. Such a measure (I, say) should have the following properties.

(i) $I(A) = $ minimum   if $\mu_A(a_i) = 0$ or 1, $\forall a_i$, i.e., for a crisp set, the measure of fuzziness is minimum.

(ii) $I(A) = $ maximum   if $\mu_A(a_i) = 0.5$, $\forall a_i$.

(iii) $I(A) \geq I(A^*)$,   where A^* is a sharpened or crisp version of A.

(iv) $I(A) = I(A^c)$ where A^c is the complement of A.

One such measure of fuzziness is entropy defined as follows.

Entropy : The entropy as a measure of fuzziness of a fuzzy set $A = \{(a_i, \mu(a_i)), i = 1, 2, \cdot, N\}$ is defined as:

$$I(A) = H(A) + \overline{H}(A); \tag{13.15}$$

where

$$H(A) = - \sum_{i=1}^{N} \mu_A(a_i) ln(\mu_A(a_i)). \tag{13.16}$$

Here, μ_A represents the membership function of the fuzzy set A, and N represents the number of elements in support of A.

The aforementioned measure lies in [0, 1] and satisfies properties (i) through (iv).

Index of Fuzziness

Index of fuzziness gives the measure of average vagueness present in a fuzzy set $X = \{x_1, x_2, \cdots, x_N\}$. The index of a crisp(non-fuzzy) set should be 0. A fuzzy set X^* is considered to be the crisp set of X, if:

$$\mu_{X^*}(x) = \begin{cases} 0, \text{ if } \mu_X(x) < 1 \\ 1, \text{ otherwise.} \end{cases}$$

In this setting, the index of fuzziness is calculated by basically finding the distance between X and X^* by making use of the following formula.

$$\psi_k(X) = \frac{2}{N^{\frac{1}{k}}} [\sum_{i=1}^{N} |\mu_X(x_i) - \mu_X^*(x_i)|^k]^{\frac{1}{k}}. \tag{13.17}$$

If we want to consider different distance norms we can put different values of k. At $k = 1$, we get L_1 distance between X and X^* as the index of fuzziness. Similarly, at $k = 2$, we get L_2 distance between X and X^* as the index of fuzziness.

13.6 Probability versus Fuzzy Membership

Both fuzzy logic and probability theory are aimed to model uncertain situations, and their measuring value lies in the interval $[0, 1]$, the key difference is in their way of modelling. Probability is associated with events and not facts, and those events will either occur or won't occur. There is no fuzziness associated with it. It depends only on the frequency of occurrence of these events. In fuzzy logic, we basically try to capture the essential concept of vagueness. Fuzzy logic is all about degree of truth. Probability theory cannot deal with the things that are not entirely true or entirely false. In a nutshell, we can say that fuzzy logic captures the concept of partial truth, whereas probability theory captures the concept of partial knowledge.

To understand this, let us take an example. Consider a busy street in a city where some accidents occur. Probability will tell you about the frequency of accidents that can occur or whether the accident will occur or not (deals with only the occurrence of the event). Fuzziness will tell you about the severity of the accident if it occurs. Fuzziness doesn't deal with the frequency of occurring of the event but with extent to which it occurs. In this example, it can be fatal accidents or light accidents.

13.7 Fuzzy Systems

Any system that makes use of the concept of fuzzy mathematics can be looked into as fuzzy system. A block schematic diagram of fuzzy system is represented in Fig. 13.7. Various elements of a fuzzy system are:

Input vector: $\mathbf{x} = \{x_1, x_2, \ldots, x_d\}^T$, where d is number of features, crisp values, which are transformed into fuzzy sets by the fuzzification block.

Output vector: $\mathbf{y} = \{y_1, y_2, \ldots, y_c\}^T$, where c is number of classes, comes out of the defuzzification block, which transforms an output fuzzy set back to a crisp value.

Fig. 13.7 Elements of fuzzy a system

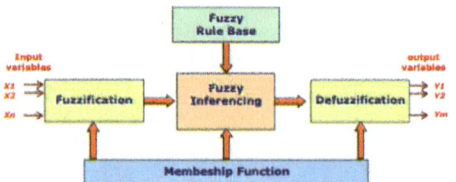

Fuzzification: It is a technique to transform crisp values into grades of membership in linguistic terms of fuzzy sets.

Fuzzy rule base: It is a collection of propositions that contains linguistic variables. The rules are expressed by the form:

$$\textbf{IF } (x \text{ is A}) \textbf{ AND } (y \text{ is B}) \textbf{ THEN } (z \text{ is C})$$

where x, y, and z represent variables (e.g. distance, size) and A, B, and C are linguistic variables (e.g. "far," "near," "small"). For example, one could make up a rule that says:

$$\textbf{IF } \text{temperature is high } \textbf{AND} \text{ humidity is high } \textbf{THEN} \text{ room is hot.}$$

The **if** part is called the antecedent, and the **then** part is called the consequent.

Membership function: This gives a measure of the degree of similarity of elements in the universe of discourse to the fuzzy set.

Fuzzy inferencing: It combines the facts procured from the fuzzification with the rule base and does the fuzzy reasoning process.

Defuzzification: It translates back the result to the real-world values.

13.8 Fuzzy Classification

The fuzzy classification models classify patterns using an informal knowledge about problem domain for classification [10]. The goal of fuzzy classification is to create fuzzy "category membership" functions by converting objectively measurable parameters to "category memberships." These category memberships do not refer to the final class but to overlapping ranges of feature values. For example, skin color of different people may be divided into black, brown, dusky, medium, and fair. These are called *soft labels*. This dismisses the standard assumption in pattern recognition that the classes are mutually exclusive. For example, a non-fuzzy classifier will assign either of the crisp labels "rain" or "no rain," based on whether it rains on a particular day or not. This is called crisp assignment (hard labels) where any pattern fully belongs to one and only one class. A fuzzy classifier can allocate degrees of membership (soft labels) to all four classes {"rain," "cloud," "wind," "sunshine"}, which accounts for the possibility of winds and cloudy weather throughout the day. This is called soft assignment where a pattern belongs to many classes with different degrees of memberships for the corresponding classes.

The simplest fuzzy rule-based classifier is a fuzzy if–then system, similar to that used in fuzzy control. Consider a 2D example with three classes. A fuzzy classifier could be developed by assigning classification rules, e.g.,

$$\textbf{IF } x_{i1} \text{ is medium } \textbf{AND } x_{i2} \text{ is small } \textbf{THEN} \text{ class is } \omega_1,$$

$$\textbf{IF } x_{i1} \text{ is medium } \textbf{AND } x_{i2} \text{ is large } \textbf{THEN} \text{ class is } \omega_2,$$

IF x_{i1} is large **AND** x_{i2} is small **THEN** class is ω_2,

IF x_{i1} is small **AND** x_{i2} is large **THEN** class is ω_3.

Let each pattern be denoted as $\mathbf{x_i} = (x_{i1}, x_{i2})$. The two features x_{ij} and x_{ik} (of the pattern x_i) are numerical, but the rules use linguistic values. Given M possible linguistic values for each feature, and d features are present in the problem, the number of possible different if–then rules of this conjunction type (AND) would be M^d. If the fuzzy classifier comprises all such rules, then it will turn into a simple look-up table. However, unlike look-up tables, fuzzy classifiers could give results for combinations of linguistic values that are not included as one of the rules. Each linguistic value is represented by a membership function.

$$\tau_1(\mathbf{x_i}) = \mu_{\text{medium}}(x_{ij}) AND \mu_{\text{small}}(x_{ik}). \tag{13.18}$$

The **AND** operation is typically implemented as minimum or the union of two fuzzy sets. The rule "votes" for the class of the consequent part. The weight of this vote is $\tau_1(\mathbf{x})$.

To find the outcome of the classifier, the votes of all rules are aggregated. For any given pattern $\mathbf{x_i} = (x_{i1}, x_{i2})$, the degree of satisfaction of the antecedent part of the rule is used for determining the firing strength of the rule.

The soft class label for pattern $\mathbf{x}$ comprises membership values $g_k(x) \in [0, 1]$, $k = 1, 2, \ldots, C$, where C denotes the number of classes. Let $i \rightarrow k$ denote that rule i votes for class k. Then,

$$g_k(\mathbf{x}) = \max_{i \rightarrow k} \tau_i(\mathbf{x}). \tag{13.19}$$

The three classes presented in the example can be looked into as the RGB (Red Green Blue) colors of an image as represented by Fig. 13.8a. Class 1 is encoded as red, class 2 as green, and class 3 as blue. The dark regions belong to those points in the feature space that have very low membership values among all of the three classes. Further analysis is required to label such points. If a crisp label is required, $\mathbf{x}$ is assigned to the class with the largest $g_k(\mathbf{x})$. Figure 13.8b shows the crisp classification regions. The critical questions that arise now are: how fuzzy classifiers gets trained, from where the membership functions for the linguistic variables arise,

Fig. 13.8 (**a**) The three-class output of the fuzzy if–then classifier encoded as RGB, (**b**) the classification regions formed from the crisp three-class output of the fuzzy if–then classifier

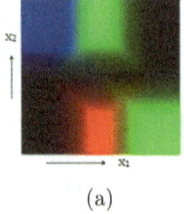
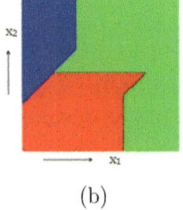

(a)   (b)

how are the rules developed, and how are the consequents determined? There are variety of techniques to train and fine-tune like fuzzy neural networks [9], genetic algorithms [5, 13], also various heuristic techniques for rule extraction based on the data geometry. Online training of fuzzy classifiers has also been explored [1].

Some fuzzy rule-based classifiers suffer from combinatorial explosion of the number of rules [7]. The main reason for that is the classifier is trained by doing partition of the data space along each feature [14]. A careful pay-off between accuracy and transparency must be made [11].

Fuzzy k-NN algorithm assigns memberships to all classes for a data point rather than assigning the data point to a particular class only [6]. Consider $X = \{x_1, x_2, \ldots, x_N\}$ to be the set of N labeled data points. Let μ_{ij} be the membership value of ith labeled data point to the jth class, and $\mu_j(x)$ be the assigned membership of unknown data point x to the jth class to be computed. For labeled data points, complete membership in their own class and nonmemberships in all other classes are considered. The fuzzy k-NN algorithm works as follows:

(i) Input: x, an unknown data point.
(ii) Set k, $1 \leq k \leq N$.
(iii) Initialize $i = 1$.
(iv) Compute the Euclidean distance from the unknown data point x to labeled data point x_i.
(v) If ($i \leq k$), include x_i in the set of k-nearest neighbors of x.
(vi) Else If x_i is closer to x than any previous k-nearest neighbors, delete the farthest of the k-nearest neighbors and include x_i in the set of k-nearest neighbors of x.
(vii) Increment i.
(viii) Repeat Steps 4–7, until i is equal to N.
(ix) Initialize $j = 1$.
(x) Compute $\mu_j(x)$ using Eq. 13.20.
(xi) Increment j.
(xii) Repeat Steps 10–11, until j is equal to number of classes.

The assigned membership of an unknown data point x to the jth class is given as follows:

$$\mu_j(x) = \frac{\sum_{i=1}^{k} \mu_{ij}(1/\|x - x_i\|^{2/(m-1)})}{\sum_{i=1}^{k}(1/\|x - x_i\|^{2/(m-1)})}. \tag{13.20}$$

According to Eq. 13.20, the assigned memberships of x are influenced by the inverse of the distances from the nearest neighbors and their class memberships μ_{ij}. The variable m determines how heavily the distance is weighted when calculating each neighbor's contribution to the membership. As m increases, the neighbors are more evenly weighted, and their relative distances from the data point being classified to have less effect.

13.9   Fuzzy Clustering

Hard clustering methods are based on the concept of classical set theory, and it requires that an object either does belong to a cluster or does not belong to a cluster. Hard clustering is the practice of partitioning the given data into a specified number of mutually exclusive subsets. In fuzzy clustering, every object gets assigned to every cluster with a membership weight, which ranges between 0 (absolutely doesn't belong to the cluster) and 1 (completely belongs to the cluster). In other words, clusters are represented as fuzzy sets [3, 8]. The membership weights actually indicate the strength of the association between the data element and the cluster. Fuzzy clustering helps to find natural vague boundaries in data. The fuzzy k-means [12] (FKM) method is one of the most popular clustering methods.

In Fuzzy clustering methods, the objects are allowed to belong to several clusters simultaneously, with different degrees of membership. In many of the situations, fuzzy clustering methods are way more natural than hard clustering methods. Objects that are present on the boundaries between several groups are not forced to fully belong to one of the groups/classes, but rather those objects are assigned membership degrees between 0 and 1 indicating their partial membership.

13.9.1   Fuzzy K-Means Clustering

FKM is a clustering technique where a data point is allowed to belong to two or more clusters [2]. In this algorithm, based on the distance between the cluster center and the data point, each of the data points corresponding to each cluster center is assigned membership degree. The more the data are near to the cluster center, the more is its membership toward that cluster.

The algorithm begins with an initial guess for the cluster centers, which are intended to mark the mean location of each of the cluster. The initial guess for these cluster centers is most likely incorrect. Next, every data point is assigned a membership grade for each cluster. Iteratively, the cluster centers and the membership grades for each data point are updated, and the cluster centers are moved toward the proper location within a dataset. These iterations for the updation process are based on minimizing an objective function that represents the distance from any given data point to a cluster center weighted by that data point's membership grade.

In Fig. 13.9, few steps of fuzzy k-means algorithm are shown. In Fig. 13.9a, there are initially six points to be clustered. $k = 2$ for this example, therefore two random centers (points 2 and 6) are chosen in Fig. 13.9b. The fuzzy membership values to each of the clusters are computed for all the data points. Figures Fig. 13.9c, d,

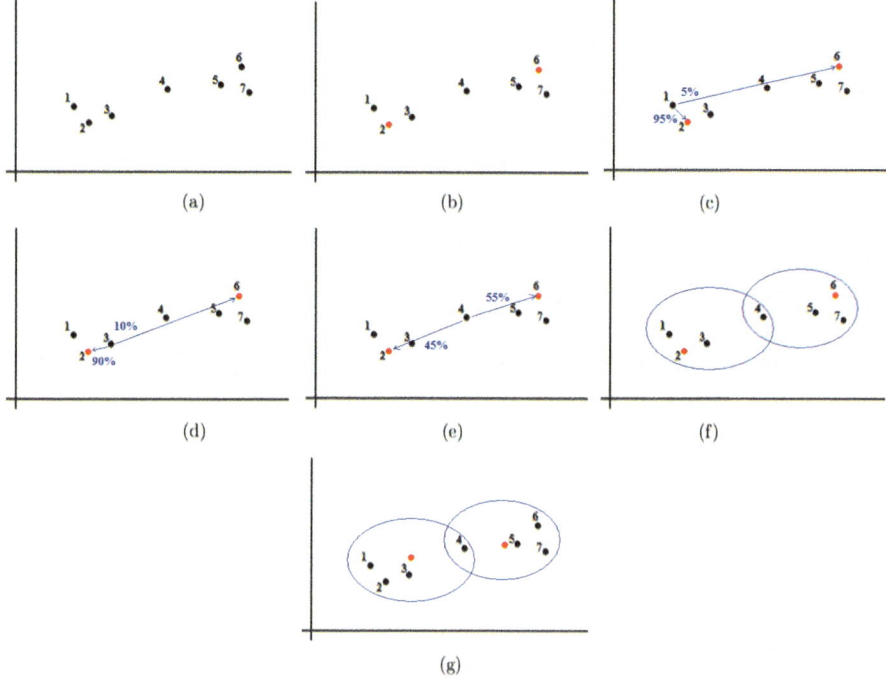

Fig. 13.9 Diagrammatic illustration of fuzzy k-means clustering

and e show the distance of points 1, 3, and 4 respectively from the centroids, and the corresponding membership percentage. Clusters are formed with the given data points and their memberships (as shown in Fig. 13.9f). The new cluster centers are computed (Fig. 13.9g). These steps are repeated till convergence.

The algorithm composes the following steps:

(i) Randomly select cluster centres.
(ii) Initialize $U = [\mu_{ij}]$ matrix, $U^{(0)}$. Calculate μ_{ij} using

$$\mu_{ij} = \frac{1}{\sum_{l=1}^{k} \left(\frac{||\mathbf{x_i} - \mathbf{v_j}||}{||\mathbf{x_i} - \mathbf{v_l}||} \right)^{\frac{2}{m-1}}};$$

(13.21)

where U is the membership matrix, μ_{ij} denote the membership of the ith pattern to the jth class, and m is the fuzzifier that determines the level of fuzziness ($m(> 1) \in \mathbb{R}$).

(continued)

(iii) At t^{th} step: calculate the centers $[v_j]^{(t)}$ using $U^{(t)}$ matrix as

$$\mathbf{v_j} = \frac{\sum\limits_{i=1}^{N} \mu_{ij}^m \mathbf{x_i}}{\sum\limits_{i=1}^{N} \mu_{ij}^m}. \tag{13.22}$$

(iv) Update $U^{(t)}$ to $U^{(t+1)}$ using Eq. 13.21.
 (v) If $||U^{(t+1)} - U^{(t)}|| < \varepsilon$
       OR the minimum J is achieved, where

$$J_m = \sum_{i=1}^{N} \sum_{j=1}^{k} \mu_{ij}^m ||\mathbf{x_i} - \mathbf{v_j}||^2 \tag{13.23}$$

then STOP; otherwise return to Step 2.

where N denotes the number of data points, k represents the number of clusters, μ_{ij} is the degree of membership of $\mathbf{x_i}$ in cluster j, $\mathbf{x_i}$ is the ith data, $\mathbf{v_j}$ is the center of the cluster, and $|| * ||$ is any norm expressing the similarity between any data and the center.

Before using the fuzzy k-means algorithm, the following parameters must be specified:

• The number of clusters, k;
• The fuzziness exponent, m;
• The termination tolerance, ε.

This algorithm provides good results for the overlapped datasets and gives comparatively better result than k-means algorithm. Unlike the k-means algorithm, where data point must exclusively belong to one cluster, here data point is assigned membership to each of the clusters. As a result, a data point may belong to more than one cluster.

However, the algorithm has some pitfalls too. A priori specification of the number of clusters is necessary. If we lower the value of ε, we can get better results, but it comes at the expense of more number of iterations. Moreover, Euclidean distance measures can unequally weight underlying factors.

13.9.2 Fuzzy C-Shell Clustering

In objective function-based clustering algorithms, each cluster is usually represented by a prototype, and the sum of distances from the data points to the prototypes is used as the objective function. This method has been traditionally used to detect "compact" or "filled" clusters in feature spaces, whose prototypes are typically represented by cluster centers and cluster covariance matrices. Lately, this approach has been extended to the case of hollow or shell-like clusters by using shells (manifolds) for prototypes and measuring the distances to the shells rather than to the cluster centers.

In most of the fuzzy clustering algorithms, each cluster is represented by a cluster representative. Since the clusters can assume different kinds of shapes and different level of compactness, this becomes a driving factor behind choosing the type of cluster representative to achieve efficient clustering results. So, each fuzzy clustering algorithm uses a specific cluster representative and hence a specific distance metric (for calculating the distance between data point and cluster representative) to deal with clusters of a certain kind of shape. For example, if the clusters are very compact, then a point representative (with d dimensions) can be used. If the clusters are not very compact, then specific type of hypersurfaces can represent the clusters more suitably. Cluster representatives take the form of following quadratic hypersurfaces like hypersphere, hyperellipse. The fuzzy C-shell (FCS) clustering algorithms use hypersurface as cluster representative.

In fuzzy C-shell clustering, the cluster representative is a hyperellipse given the following equation:

$$(\mathbf{x} - \mathbf{v_j})^T \Lambda_j (\mathbf{x} - \mathbf{v_j}) = 1. \tag{13.24}$$

The Λ_j contains the major, minor axes lengths and orientation information about the hyperellipse (E_j) representing the jth cluster and $\mathbf{v_j}$ the center of E_j. The distance of a data point from hyperellipse (E_j) is measured in terms of normalized radial distance $dist_{Rij}$ (as an approximation of perpendicular distance). $E_j = (\mathbf{v_j}, \Lambda_j)$.

$$\text{dist}^2(\mathbf{x_i}, E_j) = \text{dist}^2_{Rij} = ||\mathbf{x_i} - \mathbf{z}||^2; \tag{13.25}$$

where $\mathbf{z}$ is the point of intersection E_j and line through $\mathbf{x_i}$ and $\mathbf{v_j}$. $dist_{Rij}$ is normalized using the following equation:

$$\text{dist}^2_{Rij} = \frac{(\sqrt{(\mathbf{x} - \mathbf{v_j})^T \Lambda_j (\mathbf{x} - \mathbf{v_j})} - 1)||\mathbf{x_i} - \mathbf{v_j}||^2}{(\mathbf{x} - \mathbf{v_j})^T \Lambda_j (\mathbf{x} - \mathbf{v_j})}. \tag{13.26}$$

The objective function becomes

$$J = \sum_{i=1}^{N} \sum_{j=1}^{k} \mu_{ij}^{q} \mathrm{dist}_{Rij}^{2}(\mathbf{x_i}, E_j).$$

(13.27)

It follows from Eq. 13.26 that the objective function becomes a function of $\mathbf{v_j}$ and $\mathbf{\Lambda_j}$, $\forall j = 1, 2, \cdots, k$. So, $\mathbf{v_j}$ and $\mathbf{\Lambda_j}$, $\forall j = 1, 2, \cdots, k$ are the parameters to be updated iteratively so that it leads to the minimum value of cost function J. The resulting clusters are obtained by optimizing J in Eq. 13.27 w.r.t $\mathbf{v_j}$ and $\mathbf{\Lambda_j}$.

The FCS clustering algorithms are widely applied in the fields of pattern recognition and computer vision. However, the available FCS algorithms are valid only for detecting the hyperspherical shell or hyperellipsoidal shell-type clusters (as shown in Fig. 13.10), which limits their applications.

13.9.3 Gustafson–Kessel Clustering (GKC)

It is already stated that since FCM employs Euclidean norm to measure the dissimilarity between patterns and cluster centers, only spherical clusters can be detected properly using it. Gustafson and Kessel introduced [4] adaptive distance norm to measure the distance between clusters using fuzzy covariance matrix (a fuzzy equivalent of the classical covariance)—a representation of cluster centers along with data points. The cluster centers are found using the following equation:

$$\mathbf{v}_j = \frac{\sum_{i=1}^{N} (\mu_{ij})^m . x_i}{\sum_{i=1}^{N} (\mu_{ij})^m}.$$

(13.28)

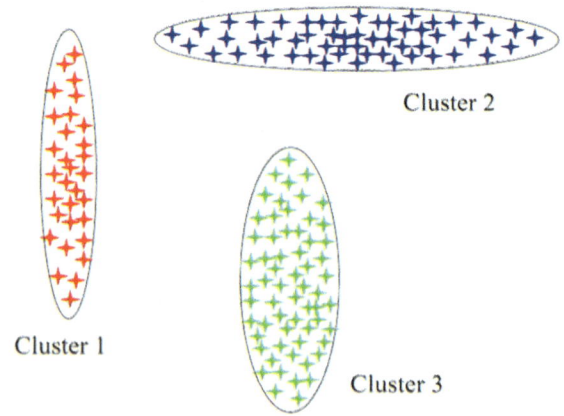

Fig. 13.10 Clusters formed by fuzzy C-shell clustering

Cluster 2

Cluster 1

Cluster 3

Fig. 13.11 Clusters formed by GK clustering

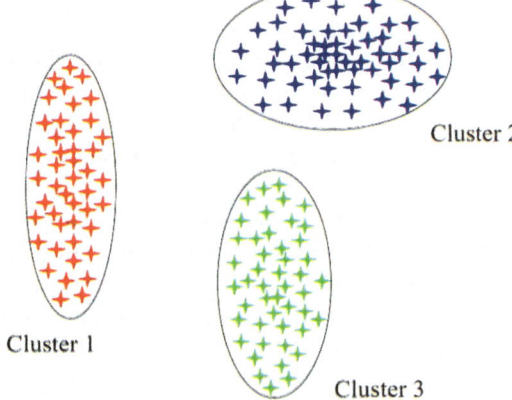

Cluster 2

Cluster 1

Cluster 3

Using GKC [4] ellipsoidal clusters (Fig. 13.11) can be detected. Each cluster has its own norm-inducing matrix A_i, a positive definite symmetric one, for automatically adapting its shape. The fuzzy covariance matrix F_i of the ith cluster is expressed as

$$F_i = \frac{\sum_{k=1}^{N}(\mu_{ik})^m (\mathbf{x}_k - \mathbf{v}_i)(\mathbf{x}_k - \mathbf{v}_i)^T}{\sum_{k=1}^{n}(\mu_{ik})^m}. \tag{13.29}$$

The distance d_{ikA_i} is computed as

$$d_{ikA_i} = \sqrt{(\mathbf{x}_k - \mathbf{v}_i)^T A_i (\mathbf{x}_k - \mathbf{v}_i)}, \tag{13.30}$$

where the norm-inducing matrix $A_i = [\rho_i det(F_i)]^{1/D} F_i^{-1}$, D is the dimension of input patterns. ρ_i is a predefined constant, which controls the shape of the ith cluster. Thus,

$$\mu_{ik} = \frac{1}{\sum_{j=1}^{K}(\frac{d_{ikA_i}}{d_{jkA_i}})^{\frac{2}{(m-1)}}}. \tag{13.31}$$

The objective function J_m will now be of the form

$$J_m(\mathbf{X}; U, \mathbf{V}, A) = \sum_{i=1}^{K}\sum_{k=1}^{N}(\mu_{ik})^m D_{ikA_i}, \tag{13.32}$$

where $D_{ikA_i} = d_{ikA_i}^2$.

It can be noted that in the first step of the algorithm, though U is initialized randomly, ρ_i has to be set properly to detect the proper shapes of the clusters. The role of the parameter m is the same as that used in FKM. The algorithm works as follows:

Input : Unlabeled data set consisting of N patterns.
Output : Prototypes $\mathbf{V}$ and fuzzy partition matrix U.
Step 1: Set fuzzifier m, initialize U randomly and ρ_j reasonably.
Step 2: Compute each cluster center using Eq. 13.28.
Step 3: Compute all the distances d_{ikA_i} using Eq. 13.30; i=1,2,...K; k=1,2,...n.
Step 4: Update fuzzy partition matrix using Eq. 13.31.
Step 5: Compute $\Delta = ||(U_t - U_{t-1})||$; t denotes t^{th} iteration .
Step 6: Check if $\Delta < \epsilon$; ϵ is a predefined small positive constant.
Step 7: If the aforementioned condition is not true, go to Step 2.
Step 8: Stop.

The Gustafson–Kessel clustering algorithm is used to detect ellipsoidal clusters with approximately the same size but different shapes. The algorithm can detect planar clusters with j^{th} cluster c_j having center $\mathbf{v_j}$ and covariance $\mathbf{\Sigma_j}$. This algorithm considers distance between the ith data point and the jth cluster calculated using the following equation:

$$dist^2_{GK}(\mathbf{x_i}, c_j) = |\Sigma_j|^{\frac{1}{d}} (\mathbf{x_i} - \mathbf{v_j})^T \Sigma_j^{-1} (\mathbf{x_i} - \mathbf{v_j}). \tag{13.33}$$

The objective function or cost function becomes:

$$J = \sum_{i=1}^{N} \sum_{j=1}^{k} \mu_{ij} dist^2_{GK}(\mathbf{x_i}, c_j). \tag{13.34}$$

It follows from Eq. 13.33 that the objective function becomes a function of $\mathbf{v_j}$ and $\mathbf{\Sigma_j}$, $\forall j = 1, 2, \cdots, k$. So, $\mathbf{v_j}$ and $\mathbf{\Sigma_j}$, $\forall j = 1, 2, \cdots, k$ are the parameters to be updated iteratively so that it leads to the minimum value of cost function J. The resulting clusters are obtained by optimizing J in Eq. 13.34 w.r.t $\mathbf{v_j}$ and $\mathbf{\Sigma_j}$, $\forall j = 1, 2, \cdots, k$.

References

1. Angelov, P. P., & Zhou, X. (2008). Evolving fuzzy rule-based classifiers from data streams. *IEEE Transactions on Fuzzy Systems, 16*(6), 1462–1475.
2. Dunn, J. C. (1973). A fuzzy relative of the ISODATA process and its use in detecting compact well-separated clusters. *Journal of Cybernetics, 3*(3), 32–57.
3. Ghosh, A., Mishra, N. S., & Ghosh, S. (2011). Fuzzy clustering algorithms for unsupervised change detection in remote sensing images. *Information Sciences, 181*(4), 699–715.
4. Gustafson, D., & Kessel, W. (1978). Fuzzy clustering with a fuzzy covariance matrix. In *Proceedings of the IEEE conference on decision and control 2* (pp. 761–766).

5. Ishibuchi, H., Nozaki, K., Yamamoto, N., & Tanaka, H. (1995). Selecting fuzzy if-then rules for classification problems using genetic algorithms. *IEEE Transactions on Fuzzy Systems, 3*(3), 260–270.
6. Keller, J. M., Gray, M. R., & Givens, J. A. (1985). A fuzzy k-nearest neighbor algorithm. *IEEE Transactions on Systems, Man, and Cybernetics, 15*(4), 580–585.
7. Kuncheva, L. I. (2000). *Fuzzy classifier design*, vol. 49. Springer.
8. Łęski, J. (2003). Towards a robust fuzzy clustering. *Fuzzy Sets and Systems, 137*(2), 215–233.
9. Nauck, D., Klawonn, F., & Kruse, R. (1997). *Foundations of neuro-fuzzy systems*. Wiley.
10. Pal, S. K. (1992). Fuzzy sets in image processing and recognition. *IEEE international conference on fuzzy systems* (pp. 119–126).
11. Pal, S. K., & Mitra, S. (1999). *Neuro-fuzzy pattern recognition: Methods in soft computing*. Wiley.
12. Pal, N. R., Pal, K., Keller, J. M., & Bezdek, J. C. (2005). A possibilistic fuzzy c-means clustering algorithm. *IEEE Transactions on Fuzzy Systems, 13*(4), 517–530.
13. Roubos, J. A., Setnes, M., & Abonyi, J. (2003). Learning fuzzy classification rules from labeled data. *Information Sciences, 150*(1), 77–93.
14. Setnes, M., Babuška, R., Kaymak, U., & van Nauta Lemke, H. R. (1998). Similarity measures in fuzzy rule base simplification. *IEEE Transactions on systems, man, and cybernetics, part B: Cybernetics, 28*(3), 376–386.
15. Zadeh, L. A. (1965). Fuzzy sets. *Information and control, 8*(3), 338–353.

Part IV
Machine Learning

Chapter 14
Rules of Thumb

For designing machine learning algorithms, basic intuition could be taken from some "rules of thumb" described as follows.

14.1 Occam's Razor

Occam's razor (also known as Ockham's razor) is a concept that says that having unnecessary information discarded is the quickest way to understand the truths or the best descriptions when trying to understand anything. An English theologian and philosopher William of Ockham (1287–1347) spent most of his life establishing a theory that reconciled religious belief with demonstrative and largely true beliefs, primarily by distinguishing between them. Occam deemed that religious belief is unworthy of certain justifications while earlier philosophers were attempting to justify God's existence through rational pieces of evidence. He opposed the theories of the independent existence of attributes such as truth, endurance, and reliability, which were inherited from the classical era. He argued that such principles have meaning only when they are used as explanations for certain entities and were really typical of human cognition.

Occam is called out for his emphasis on language as a method for thought and also focused on observation for checking the truth. The foundation for current scientific and mathematical inquiry is perceived to be his radical ideas and writings.

Occam's emphasis on the use of parsimony, something we could call a minimalist attitude, contributed to the subsequent development of the terminology, Occam's razor. Some of his arguments are "Plurality is not to be assumed without necessity" and "What can be done with fewer (assumptions) is done in vain with more." The implication of this approach is the idea that among multiple conflicting theories, the easiest or most apparent interpretation should be preferred unless proven wrong.

© The Author(s), under exclusive license to Springer Nature Singapore Pte Ltd. 2025   231
A. Ghosh, *Data Science and Cases in Sustainability*, Mathematics for Sustainable
Developments, https://doi.org/10.1007/978-981-96-8362-8_14

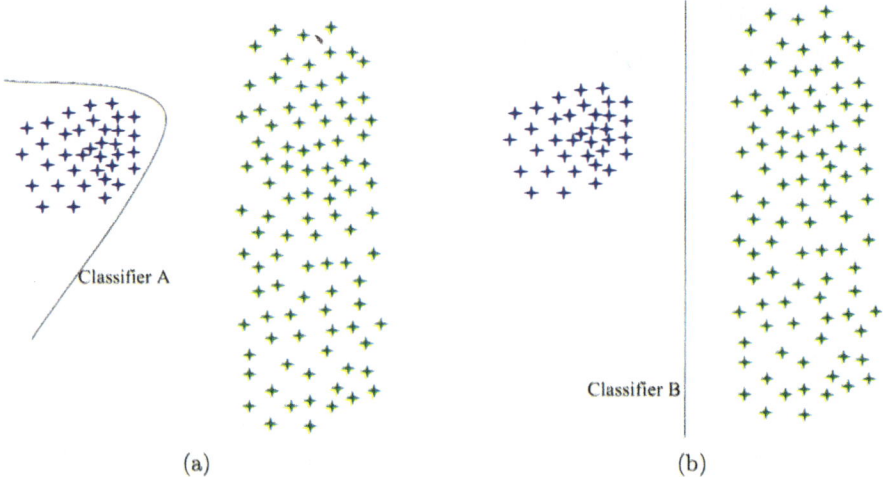

Fig. 14.1 Two classifiers with same training error: (**a**) Complex classifier, (**b**) Simple classifier

Occam's Razor suggests that the most uncomplicated classification is the most optimal. Therefore, it is important to choose the simplest classifier that produces similar training errors. In Fig. 14.1, the data from two different classes come from two different distributions and both classifiers A (Fig. 14.1a) and B (Fig. 14.1b) give the same training error (in this case error is 0, since the data are separable); but classifier B is simpler, so this one should be used.

14.2   No Free Lunch Theorem

The *No Free Lunch Theorem* was first applied [3] to the paradigm of supervised machine learning. The hypothesis indicates that there exist different universes (or tasks) for which one algorithm performs better than all the others. There are no universally applicable reasons, independent of context or use, to choose one classifier system over another if the aim is to achieve efficient generalization quality. If in a specific case, one algorithm seems like it outperforms another, it is because it suits the particular problem in a better way and not because the algorithm is generally superior.

For example, if two training inputs (0, 0) and (1, 1) are presented to a classifier, both being labeled as "correct," two simple hypotheses exist that fit the data. Firstly, it can be that every input possible will lead to the "correct" classification. Secondly, every combination, except the two given sample entries, will lead to the "false" classification and that these two samples from the training sets are the only examples of the "correct class." In summary, for a given training set, at least two generalization hypotheses are always possible, which would be the complete opposite of each other. This suggests that any training algorithm needs a kind of "bias" for these possible conclusions to be differentiated.

Several algorithms might have such powerful biases that only certain functions are learned by them. For reference, a basic form of the neural network, the perceptron, is biased or to say is inclined to learn only linear functions (a line in two dimensions, a plane in three dimensions, and a hyperplane in higher dimensions). For situations where the data are not necessarily linearly separable, this may be a limitation. If the data have the property of linear separability, its identification by a perceptron will be effective. In that case, however, there will be issues with complex and sophisticated nonlinear methods.

For example, if our data look like $((0, 0), FALSE), ((1, 0), TRUE), (0, 1), TRUE)$, and $((1, 1), TRUE)$, which is basically a Boolean OR function, we can expect our perceptron to learn this kind of linear function very easily. So, if the actual function is Boolean OR, then the perceptron could have been fairly extended its interpretation of the aforementioned four training samples to a dataset with more samples also. However, a more complex and flexible algorithm may not have generalized its bias to such a kind of linear function.

Bad compatibility of a problem with the learning system's assumptions may result in a degradation of classification performance. For example, if the provided data resemble two overlapped Gaussians, the decision boundary formed by graph-based models will go straight through the densest area and produce poor results. In the scenario of structured data, on the other hand, as the discovery of traditional DNA subsequences, graph-based models will yield good results since a graph can be used to represent the molecular structure of DNA very efficiently. Therefore, a conservation law is portrayed in *No Free Lunch* theorem. It says if the quality of an algorithm is greater than average for a particular tasks, it must be less than average for others.

14.3 Simpson's Paradox

The hypothesis of Simpson's Paradox reveals the sort of problems that result from merging data from different groups. Suppose we consider many groups and for each group, we form a relationship or correlation. Simpson's paradox suggests that the correlation that we found before can modify when we merge all groups and view the data in composite form. This is most often because of the unknown variables not taken into account, while it is sometimes because of the numeric aspects of the data.

Let us have a look at the following example to get a little more comprehension of Simpson's paradox. Two surgeons are available in a certain hospital. Surgeon A operated on 100 clients, which includes 95 survivors. Surgeon B operated on 80 patients out of which 72 survived. We consider that both the surgeons did the surgery in the same hospital and with same equipment, and the necessary criterion to evaluate is the survival rate through the procedure. So, we are looking for the best of both surgeons.

We analyze the data and compare the survival rate for surgeon A's patients and surgeon B's patients.

Ninety five out of 100 patients who were operated by surgeon A survived. So, surgeon A has 95% survival rate.

Seventy two out of 80 patients who were operated by surgeon B survived. So, surgeon B has 90% survival rate.

Going by this analysis, when it comes to our treatment, which surgeon must we opt for? Surgeon A appears to be a better choice. Is this valid, though? What if we researched for more the information, and figured out that the hospital had actually taken two different types of surgeries but then brought together all the details to comment on each of its surgeons.

Some were deemed to be high-risk emergency procedures, and others were quite normal and had been planned in advance. So, not all operations were equivalent.

Of the 100 patients who received operation from surgeon A, 50 were critical cases out of which three did not survive. This means for a critical procedure, the death rate of surgeon B is $3/50 = 6\%$. The remaining 50 had routine surgeries and two of them died. This implies that a patient handled by surgeon A for a routine operation has a death rate of $2/50 = 4\%$.

We are now looking at the surgeon B records more closely and discover that out of 80 patients, 40 were critical patients, out of which seven are dead. This means for a critical procedure, the death rate of surgeon B is $7/40 = 17.5\%$. The remaining 40 had routine procedures, with only one dead. This indicates that with surgeon B doing a routine procedure, the death rate is $1/40 = 2.5\%$.

This brings a serious question now. If your procedure is to be a regular operation, surgeon B is the best option. However, surgeon A is best if we look at the critical surgeries done by the surgeons. That becomes very contradictory to the earlier fact. The latent parameter of the type of surgery, in this case, affects the consolidated surgical results.

Simpson's paradox was named for its coiner Edward Simpson. In the 1951 paper "The Interpretation of Interaction in Contingency Tables" from the Journal of the Royal Statistical Society [2], he first identified this phenomenon. Pearson [1], and Yule [4] discovered the same phenomenon half a century ago than Simpson. So, this paradox is also often referred to as the Simpson–Yule effect. For fields as varied as sport stats and unemployment information, there are many specific uses for the phenomenon. Whenever the details are aggregated, we must watch out for this paradox to arise.

For example, using the following scatterplot, which is based on performance monitoring data, the impact on speed is checked. There are three servers (Server A, Server B, and Server C).

Only by looking at Fig. 14.2a, it may seem that all the servers are performing equally. Figure 14.2b reveals that data coming from different servers are aggregated into a single scatterplot. It also reveals that only server C has steady response time and the other two servers do not. The paradox is getting more attention nowadays with the progress of big data analysis. This is because the Big Data in many places are generated by integrating data from different sources.

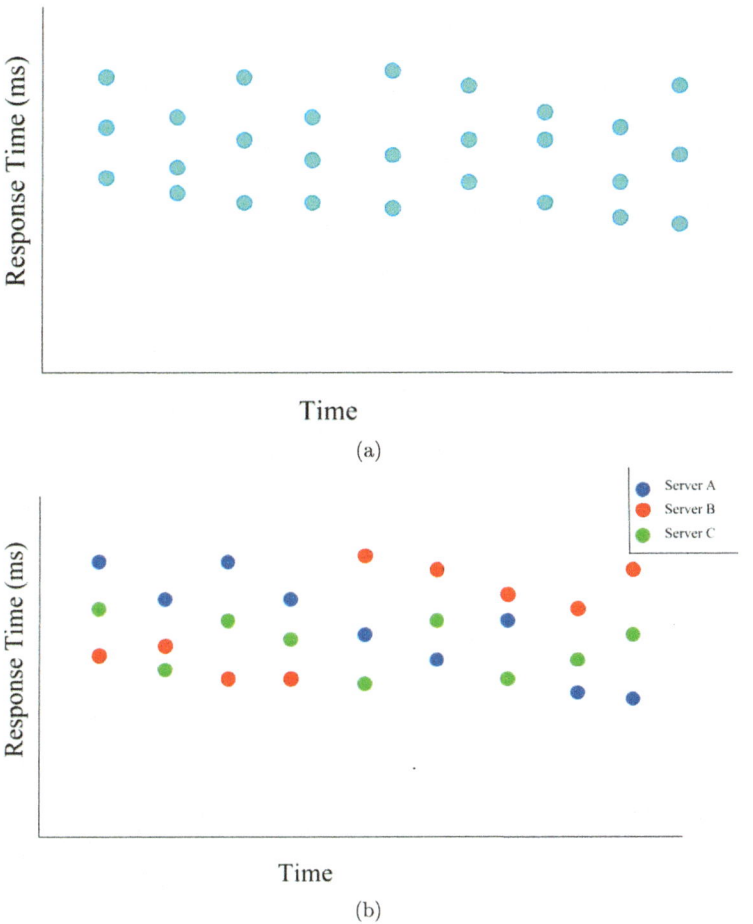

Fig. 14.2 Example of Simpson's Paradox for the performance of three servers: (**a**) aggregated scatterplot, (**b**) separate scatter plots

References

1. Pearson, K., Lee, A., & Bramley-Moore, L. (1899). Mathematical contributions to the theory of evolution. VI. Genetic reproductive selection: Inheritance of fertility in man, and of fecundity in thoroughbred racehorses. *Philosophical transactions of the Royal Society of London. Series A, containing papers of a mathematical or physical character 192* (pp. 257–330).
2. Simpson, E. H. (1951). The interpretation of interaction in contingency tables. *Journal of the Royal Statistical Society. Series B (Methodological), 13*, 238–241.
3. Wolpert, D. H., & Macready, W. G.(1997). No free lunch theorems for optimization. *IEEE Transactions on Evolutionary Computation, 1*(1), 67–82.
4. Yule, G. U. (1903). Notes on the theory of association of attributes in statistics. *Biometrika, 2*(2), 121–134.

Chapter 15
Artificial Neural Networks

The Artificial Neural Network (ANN) is a pattern recognition model motivated by the manner in which the human nervous system (for instance, the brain) interprets any information. A vast number of highly integrated neural components (neurons), working together for specific problems, make up the artificial neural system. Like humans, ANNs learn from instances. An ANN may be optimized through a learning process for a particular application, for instance, pattern recognition or object classification. Biological systems learn through the modifications of the synaptic interactions between the neurons. This also extends to ANNs.

The first artificial neuron was introduced in 1943 by the neurophysiologist Warren McCulloch and the logician Walter Pitts [3]. Drawing inspiration from neurobiology, they introduced a computational model tailored by the cognitive capacities of biological neurons and investigated the deduction of basic concepts from individual observations. However, with the technology they had at the time could not do much. Several significant developments have been made in this field with the introduction of powerful and cheap computer emulations.

Neural networks are distinguished from traditional systems in their approach toward problem-solving. The generic conventional machine uses an algorithmic method to solve a problem. This means that the system follows a fixed and rigid set of steps and guidelines to do any task. Without the information of the precise steps to be taken by the machine, the machine cannot solve that problem. It limits the ability of traditional machines to solve problems, which we already recognize and know the solution to. But if machines could do tasks that are not necessarily clearly defined and not that easy to comprehend (for example, choosing a ripe fruit among many), they would be much more useful.

Biological neural networks interpret data using a vast number of highly linked computing units (neurons) to address a specific problem by working in coordination with each other. They cannot be hardwired to accomplish a particular task. We have to choose the examples for learning; otherwise, the network may function improperly. As the network will automatically figure out how the problem can be solved, the only drawback is that it can be uncertain.

Conventional machines, on the other hand, use an intuitive approach to addressing problems; it needs to be known and specified in unambiguous simple directions how the question should be solved. Such commands are then translated to a high-level language system and to the computer-readable machine code. The system's result is completely predictable, and it is down to a software or hardware error that anything goes wrong.

There is no competition in the field of neural networks and conventional algorithmic computers. They, rather, complement each other. There are certain tasks that are better suited for a programmatic approach such as arithmetical operations, and there are certain other tasks that are more suitable for neural networks. Therefore, most tasks involve systems that use a combination of both approaches (usually a traditional machine is used for keeping track of a neural network).

15.1 Neural Computing

Artificial neural networks are a means of modeling the ability of the human nervous system to solve complex problems. First of all, from an information processing viewpoint, therefore, we need to take into account the fundamental properties of biological or normal neural networks.

15.1.1 Biological Neural Networks

The brain and the remainder of the nervous system contain several different cell varieties, but the core functioning component is a cell defined as the neuron. Each perception, gesture, feeling, memory, and emotion are the product of neuronal signals.

There are three sections of a neuron (Fig. 15.1). The body of the cell comprises the nucleus, which creates the bulk of the chemicals that the neuron needs to survive and work. Dendrites spread like the tree roots from the cell body and receive signals from other neighboring nerve cells. Signals are then transferred through one cell to another cell body through the help of dendrites. The space between dendrites of different cells is filled with some synaptic fluids, which determine how much information will be passed to the next cell. This is how a signal may pass from the cell body to a particular neuron, muscle cell, or cells in another organ. Typically, several support cells surround the neuron. Many cell types are bundled into forming an insulated layer around the axon. This sheath may contain a fatty molecule known as myelin, which ensures axon protection and makes neuronal impulses travel around quickly. Axons can be very small, for instance, the axons that carry impulses from one neuron in the cortex to another neuron are less than the thickness of a hair.

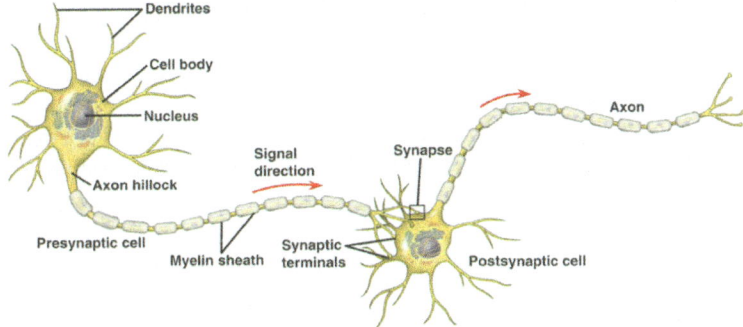

Fig. 15.1 Parts of a human neuron

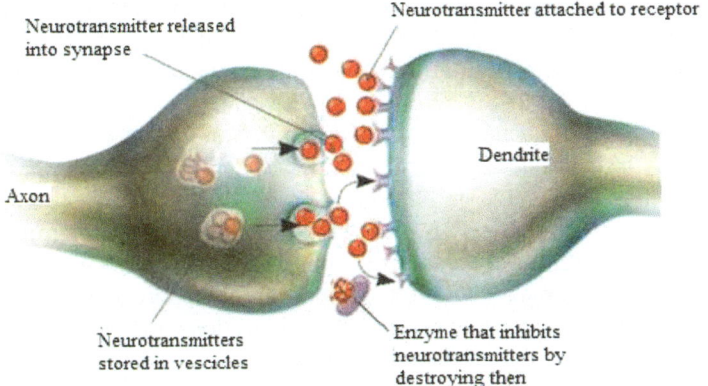

Fig. 15.2 Transmission of information through Synapse

Also, axons like those that carry signals from the brain throughout the backbone can be very long.

Through observing the synapse—the space between the dendrites of one neuron and the dendrites of another neuron, researchers have learned a lot about the processing capabilities of our brain. A single neuron receives stimuli in the human brain through a variety of delicate structures known as dendrites. The neuron sends timed electric spikes via a long, thin fiber known as an axon, which is fragmented into millions of branch-like structures. Synapses convert activation of the axon at the end of each branch into electric signals that suppress or activate the signals in the connecting neurons (Fig. 15.2). When the signals make it to the end of the axon, the release of some small sacs is stimulated. The sacs discharge fluids called neuro-transmitters into the synapse. The synaptic fluid may also contain chemicals that raise (excite) the electrical potential or decrease (inhibit) it. These neurotransmitters pass through the synapse and attach themselves to the cell receptors. Such receptors may alter the recipient cell's characteristics. The signal will proceed to transfer to the next cell if the recipient cell is also a neuron. These neurotransmitters cross the synapse and bind to the receptors of the cells. These receptors can change the characteristics of the recipient neuron. If the receiving cell is also a neuron, the message is passed to the next node. When a neuron is given an excitatory input, which is high enough relative to its inhibitory input, i.e., the input potential crosses a threshold value, the axon passes down an electrical spike [2]. Training takes place by adjusting the influences of the synaptic cleft so that one neuron modifies its effect on another.

15.1.2 Artificial Neural Networks

The effort to imitate the human neural networks may be known as artificial neural networks. The understanding of important components of biological neurons lays the groundwork for the design of architectures of artificial neural networks. The analogy between biological and artificial neurons can be understood as depicted in Table 15.1.

The basic idea behind neural networks is that they are composed of a number of nodes and may be connected to each other (anyway) via connections mimicking the neurons in the brains (Fig. 15.3).

The network shown in Fig. 15.3 is not so beneficial for processing information. In ANN, neurons are usually organized into several layers to get a systematic processing of the information. Also, the connections are directed as to specify in which direction information is flowing. The connections are assigned weights to specify their importance over the other in making a decision at the node (Fig. 15.4). To get a neural network functioning, we need to trigger a node at first with some input and that node triggers the other nodes. In a typical ANN, input units store the inputs, hidden units transform the inputs into an internal numeric vector, and an output unit transforms the hidden values into prediction.

Now, let us take a single neuron for simplicity. The nodes of the network contain the primitive functions and the composition rules are carried implicitly in the interconnection pattern of the nodes. Figure 15.5 demonstrates what the structure of a general artificial neuron containing n inputs [5]. Any given input channel i can pass on only one real value x_i. In the artificial neuron's body, the basic function f

Table 15.1 Analogy between biological and artificial neural networks

Biological Neuron	Artificial Neuron
Cell Body	Node
Dendrite	Input channel
Axon	Output channel
Synapse	Weights
Potential	Weighted sum
Threshold	Bias weight
Signal	Activation

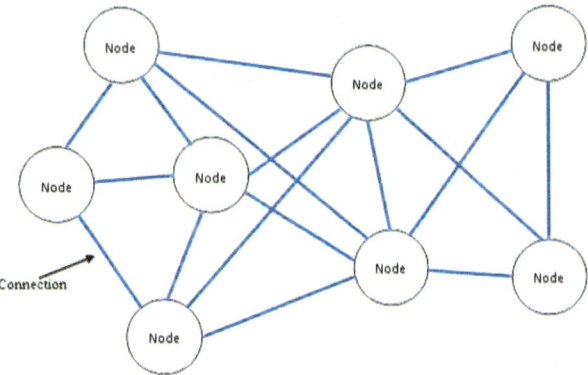

Fig. 15.3 Artificial neural network connections

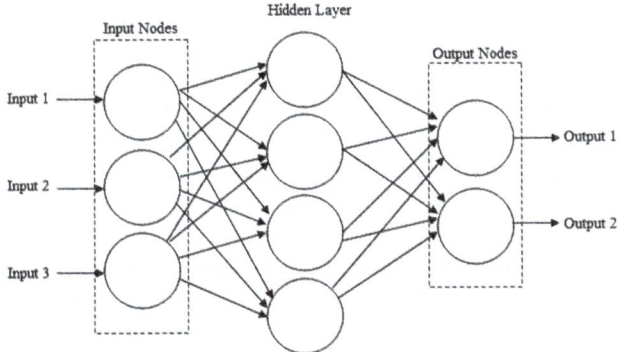

Fig. 15.4 A representative diagram of a neural network

Fig. 15.5 An abstract neuron (artificial)

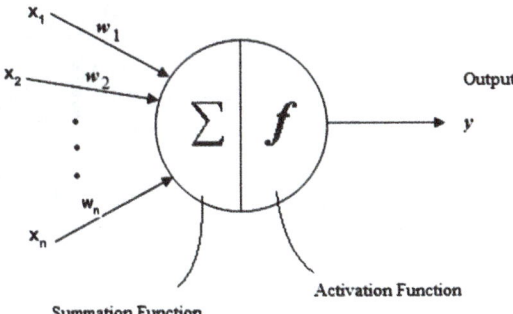

can be randomly chosen. The input channels typically have a corresponding weight. The received signals x_i are multiplied by such weights w_i. In a particular neuron, the received data are combined (normally by simply adding different signals), and the basic function is validated [1].

If we presume that each node in the artificial neural network will convert its input X to a particular output o as per the basic function, then it is only a network of the basic function. The presumption on basic functions generally differs across different types of Artificial Neural Networks. The types of ANNs also differ on the basis of the structure of the interconnections and the time required for the transmission of signals.

$$o = \sum_{i=1}^{n} x_i w_i. \tag{15.1}$$

15.2   Components of Neural Networks

An artificial neural network's mathematical system comprises basic units of computation (neurons) and weighted paths between neurons. In this case, the weightage of the interaction between two neurons i and j (or the weight of link) is labeled w_{ij}. Information that gets conveyed through the connection may either be excitative (positive) or inhibitory (negative) governed by positive or negative weights between neurons, respectively. The activation function controls the activation of a neuron that depends on the network input and threshold value (Fig. 15.6). When the network input exceeds their threshold value, the neurons get activated. The learning strategy is basically an algorithm that can be used to change and thus train the neural

Fig. 15.6 Data processing within a neuron

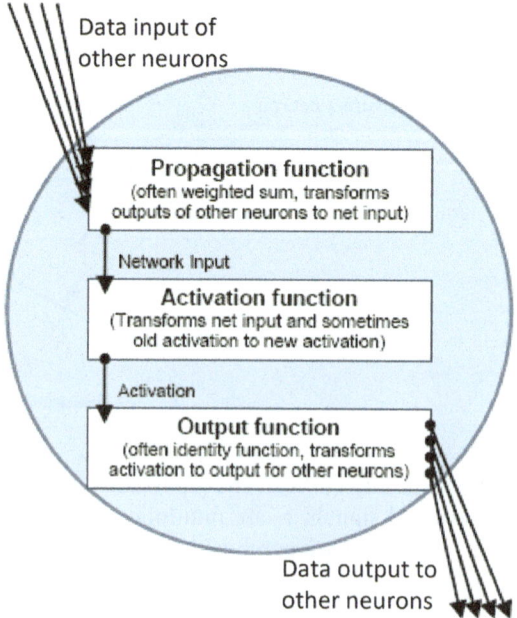

Data input of other neurons

Propagation function
(often weighted sum, transforms outputs of other neurons to net input)

Network Input

Activation function
(Transforms net input and sometimes old activation to new activation)

Activation

Output function
(often identity function, transforms activation to output for other neurons)

Data output to other neurons

network, so that for a given input, the neural network can now produce a desired output.

15.3 Properties of Neural Networks

The computational power and the capacity of neural networks to learn and generalize arise from a large distributed architecture in parallel. In spite of all this, we have a hard road till we can build a network design that imitates a human brain. The common features are:

- A single real vector (called its state) reflects its performance and normally defines each node/ neuron/ processor.
- The nodes are connected densely. Each node is influenced by the weight or power of the connection according to the cumulative outputs of other preceding nodes.
- The new state of a neuron is given as a nonlinear function of the potential that other neurons connected to the node will produce.
- The network input is provided by initially setting a subset of node states to certain values.
- The transmission process takes place according to the nature of a specific network by the transformation of the states of all the nodes and continues to stabilize them.
- The network training (learning) is a mechanism by which the connection weight values are modified to achieve the desired efficiency.

The use of neural networks offers the following useful properties:

- **Nonlinearity:** An artificial neuron can be linear or nonlinear. It is basically made of an interconnection of nonlinear neurons and therefore is inherently nonlinear.
- **Mapping capabilities:** The neural networks exhibit mapping capabilities, that is, they can map input patterns to their associated output patterns.
- **Adaptivity:** Neural networks can adapt the free parameters to changes in the surrounding environment. Free parameters are the synaptic connections strength (neuron structure).
- **Ability to generalize:** The neural networks have the capability to generalize. Thus, they can predict new outcomes from past trends.
- **Robust:** They are fault-tolerant robust systems . Therefore, from incomplete, partial, or noisy patterns, they can recall full patterns .
- **Parallel and distributed processing:** They can process information at high speed, in parallel, and in a distributed manner.

15.4   Neural Network Architectures

It is understood that neural networks are universal function approximators. There are various designs to approximate every linear or nonlinear function. Various architectures require functions of different complexity and capacity to be built. However, there are three basic elements in an ANN model:

- **Set of synapses or connecting links:** Each connecting link carries its own weight or strength. The synaptic weight of an artificial neuron can fall in the range that includes positive as well as negative real values.
- **An adder:** This component sums up the input signals, weighted according to the weights carried by the corresponding synapses of the neuron.
- **Activation functions:** Activation functions are used for capping the output amplitude of the neuron.
- The model also includes a **bias** applied externally. The bias has the ability to increase or decrease the total input value of the activation function.

In this context, we describe some neural network architectures based on how the information is processed.

15.4.1   Feed-forward Neural Networks (FFNNs)

Feed-forward networks are defined as networks without cycles (feedback loops) . A directed graph is used to represent a collection of neurons connected together in a network (Fig. 15.7).

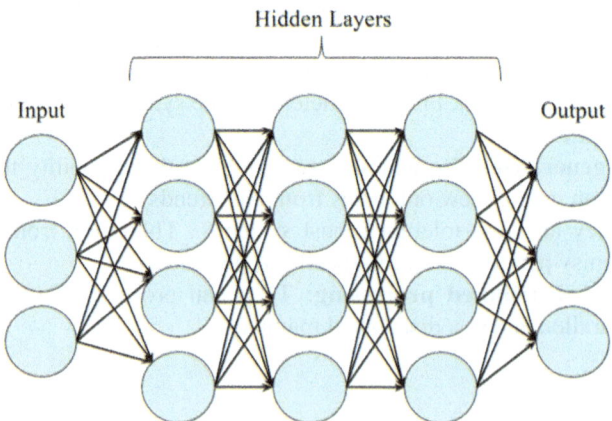

Fig. 15.7 An example of a feed-forward neural network architecture

Input layer: Neurons needed in this layer are as many as the number of features being input. It includes passive nodes that do not partake in the direct modification of the signal but only relay the signal to the next level.

Hidden layer: Hidden (between input and output) layers, of any arbitrarily chosen number, with arbitrary neurons can be created. The nodes in this layer modify the signal and are thus called active nodes.

Output layer: Neurons needed in the output layer are as many as the classes the data should be classified into. There are also active nodes in this network. The network provides no feedback, and information flows only in the forward direction. The communication is carried out from layer to layer. In fact, information is fed in the forward. That is why it is called feed-forward network. There can be one or more hidden layers in the network. FFNNs with one hidden layer have been shown to approximate almost all continuous functions [4].

15.4.2   Feedback Neural Networks

The structure of a feedback neural network generally works on incorporating feedback from outputs to adjust the weights. The neuron output is either transferred directly or indirectly through other connected neurons to its input. This is used for specific pattern recognition tasks such as language understanding. Feedback or recurrent architectures for neural networks can take many forms. The generic Multilayer Perceptron (MLP) and an additional feedback loop are common types of such feedback networks. It takes advantage of the MLP's efficient nonlinear mapping capability, as well as memory storage capacity from the feedback loop. Others have more standardized architectures and can probably have a stochastic activation function in any neuron connected to each other (Fig. 15.8).

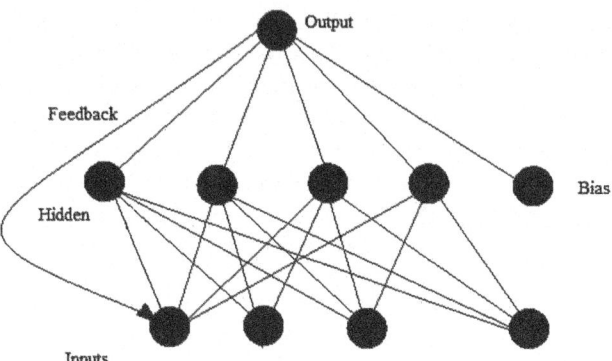

Fig. 15.8 Feedback neural network architecture

Fig. 15.9 Associative neural
network architecture

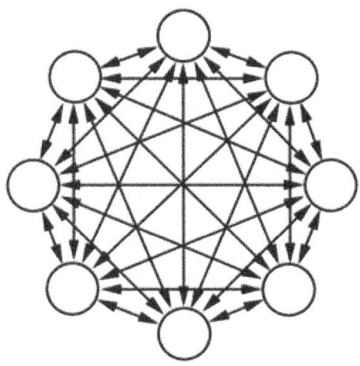

15.4.3 Associative Neural Networks

There is no hierarchical structure for these networks. There are also bidirectional connections. The associative neural network (ASNN) is an ensemble-based system, dependent on neuronal correlations of the brain's activity and structure. The method works by representation of neural network short and long-term memory. The long-term memory is a collection of neural network weights, while the short-term memory is retained as a set of internal neural network representations of the input sequence. The structure permits the ASNN in the short-term memory to include new data cases and provides a high level of generalization (Fig. 15.9).

15.5 Neural Activation Functions

Each node has an activation function in a neural network, which determines the node's output to a given input value. Neurons are "switches" that generate a "1" if activated enough and a "0" if not. This transfer function activates or transfers the input data to the output signals. A feature of an artificial neuron template is defined by the behavior of the activation function. Often it is important to have continuous and differentiable activation functions. Four forms of functions are generally utilized: step (threshold), sigmoid, rectified linear, and Gaussian. The list is not complete, though (Fig. 15.10).

15.5.1 Step

Based on the largest input being bigger than or less than a certain threshold value, the output is set at one of two points. This Heaviside function is used to design standard "All-or-None" behavior. This is also named as "step function." It also looks like a

Fig. 15.10 Step (threshold)
function

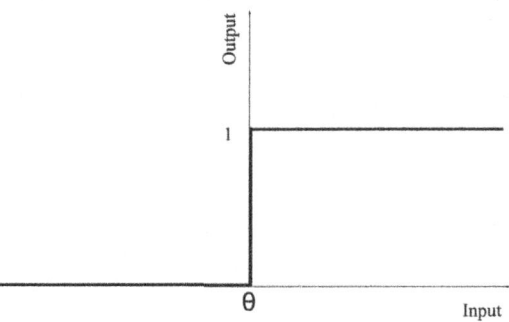

Fig. 15.11 Sigmoid function

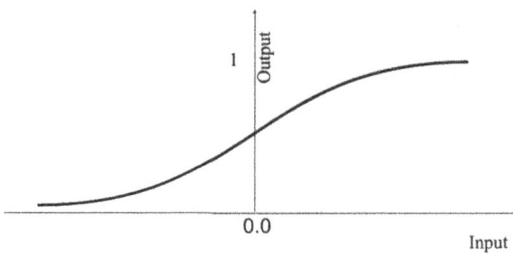

ramp function. However, it changes the output value instantly whenever a threshold
value θ is crossed.

$$f(x) = \begin{cases} 0, \ if \ x < \theta \\ 1, \ if \ x \geq \theta \end{cases}. \tag{15.2}$$

15.5.2 Sigmoid

The use of step function at the output limits the signals to be binary. So, when
neurons are connected in a network, there is no possibility of getting continuously
graded output. This may be overcome by "softening" the step function to a
continuous "squashing" function so that the output depends smoothly on the
activation. One of the convenient forms for such a type of function is sigmoid
function. The sigmoid group consists of two types: logistic and tangential functions.
The codomain of logistic function is set as being from 0 and 1 and -1 to $+1$ for
tangential function (Fig. 15.11).

$$f(x) = \frac{1}{1 + e^{-\beta x}}. \tag{15.3}$$

This sigmoid function can be configured with a parameter β that gives a measure
for the steepness. Given that β is larger than zero, the function is monotonically
increasing, continuous, and differentiable on the whole domain (Fig. 15.12).

Fig. 15.12 Piecewise linear
function

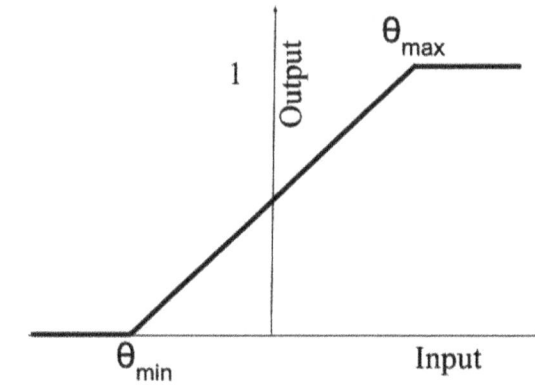

Fig. 15.13 Gaussian
function

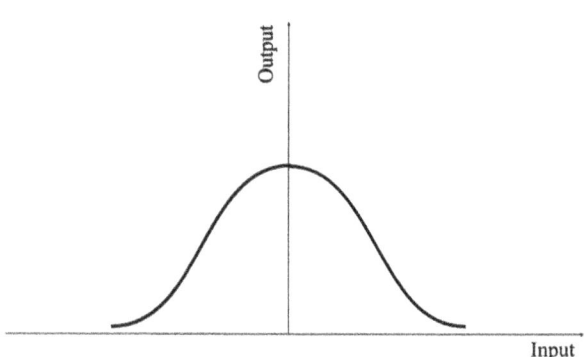

15.5.3   Piecewise Linear

The output is proportional to the total weighted input.

$$f(x) = \begin{cases} 0, & if \ x \leq \theta_{min} \\ mx + b, & if \ \theta_{max} > x > \theta_{min} \\ 1, & if \ x \geq \theta_{max} \end{cases} \tag{15.4}$$

While the trigger is smaller than the θ_{min} limit, the neuron shows $y = 0$ for output. When the trigger reaches the θ_{max} threshold value, the effect is $y = 1$. A linear interpolation of stimulation specifies the neuron's activation output in the interval between the two threshold values $\theta_{max} > x > \theta_{min}$ (Fig. 15.13).

15.5.4   Gaussian

Gaussian functions are continuous bell curves. Totally dependent on the closeness of net input to the average value selected, the node's output value (big/small) is represented by assigning a class (1/0).

$$f(x) = \frac{1}{\sqrt{2\pi}\sigma} e^{-\frac{(x-\mu)^2}{2\sigma^2}}.$$ (15.5)

The maximal functional value of a Gaussian function is found for zero activation. This function is even, i.e., $f(-x) = f(x)$. For increased absolute activation value, the functional value decreases. Such a decrease can be handled by the σ function. Greater values of σ lead to a reduction in functional values with a greater variance from the maximum.

15.6 The McCulloch–Pitts Model of Neuron

The McCulloch–Pitts neural model is also called linear threshold gate. The neuron works on the following basic assumptions:

- The output is either 0 or 1. These are binary units.
- The threshold of every neuron is θ.
- The neuron accepts signals, all of which are identical weights, from excitatory synapses. Nevertheless, several inputs can be obtained from the same origin, which is why the excitatory weights have positive values.
- Inhibitory inputs provide an absolute denial of all excitatory inputs.
- The neurons are synchronously modified in each stage by adding up the weighted excitatory inputs and setting the output to 1 if the sum is higher or equal to the threshold, and the neuron does not obtain an inhibitory intake; otherwise, it is 0.

Thus, it can be described as a neuron with a collection of input values I_1, I_2, I_3, ..., I_n and one output value y. The linear threshold gate simply classifies the set of inputs into two different classes. Thus, the output y can only hold two values. Figure 15.14 shows a symbolic representation of the linear threshold gate.

$$\text{sum} = \sum_{i=1}^{n} x_i w_i$$ (15.6)

and

$$y = f(\text{sum}).$$ (15.7)

We can summarize these two rules as follows:

$$y = \begin{cases} 1, \text{ if } & (\sum_{i=1}^{n} I_i w_i \geq \theta) \text{ AND (no inhibition)} \\ 0, \text{ otherwise.} \end{cases}$$ (15.8)

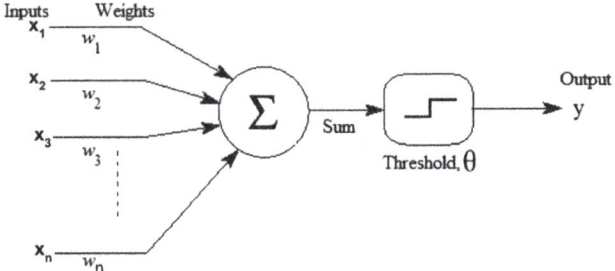

Fig. 15.14 Symbolic illustration of McCulloch–Pitts neuron

Fig. 15.15 A two-input
neuron acting as an AND gate

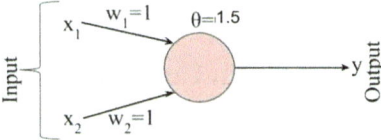

Now, using this simple neuron model, we can create some interesting things. Here
are a few examples:

15.6.1   AND Gate

Figure 15.15 shows a two-input neuron that gives an output of 1 only when both
the inputs are 1. If we set, $w_1 = 1$, $w_2 = 1$ and $\theta = 1.5$, then the following table
reflects the same for the input–output combination for the neuron. This is nothing
but the truth table of an AND gate.

x_1	x_2	y
0	0	0
0	1	0
1	0	0
1	1	1

15.6.2   OR Gate

Figure 15.16 shows a two-input neuron that gives an output of 0 only when all inputs
are zero. If we set $w_1 = 1$, $w_2 = 1$ and $\theta = 0.5$, then the following table depicts the

Fig. 15.16 A two-input
neuron acting as an OR gate

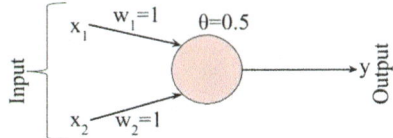

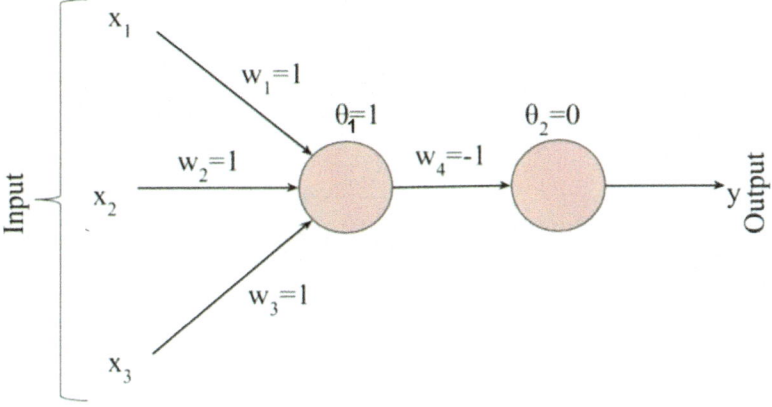

Fig. 15.17 A three-input NOR gate realization using two neurons

input–output combination of the neuron. This is easy to say that this is functioning
like an OR gate.

x_1	x_2	y
0	0	0
0	1	1
1	0	1
1	1	1

15.6.3   NOR Gate

Figure 15.17 shows a three-input neuron that in combination with another neuron
gives an output of 1 only when all inputs are zero. x_1, x_2, and x_3 are the inputs
received from external world to the first neuron. It is an OR gate for these inputs.
The second neuron works upon the output of the first neuron and has no clue what
the initial inputs were. This neuron negates the output of the first neuron. If we set,
$w_1 = 1$, $w_2 = 1$, $w_3 = 1$, $w_4 = -1$, $\theta_1 = 1$, and $\theta_2 = 0$, then the following
input–output table reflects the activities of the two neurons. This is the truth table of
a NOR gate.

x_1	x_2	x_3	y
0	0	0	1
0	0	1	0
0	1	0	0
0	1	1	0
1	0	0	0
1	0	1	0
1	1	0	0
1	1	1	0

Fig. 15.18 Three-input neuron functioning as a NAND gate

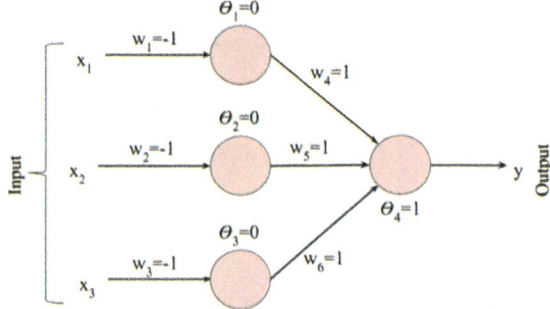

15.6.4 NAND Gate

Figure 15.18 shows a three-input McCulloch–Pitts neurons. A NAND gate gives a zero only when all inputs are 1. This gate needs four neurons. The output of the first three is the input for the fourth neuron. If we set, $w_1 = -1$, $w_2 = -1$, $w_3 = -1$, $w_4 = 1$, $w_5 = 1$, $w_6 = 1$, $\theta_1 = 0$, $\theta_2 = 0$, $\theta_3 = 0$, and $\theta_4 = 1$, then the following truth table reflects the same for the NAND gate.

x_1	x_2	x_3	y
0	0	0	1
0	0	1	1
0	1	0	1
0	1	1	1
1	0	0	1
1	0	1	1
1	1	0	1
1	1	1	0

The results obtained for NAND and NOR gates are interesting, but these networks displayed no learning. Basically, they are logical "hard-wired" units. Weights had to be calculated and the neurons connected to the ideal specification in the proper way. Therefore, over traditional digital logic circuits, there is no major advantage. Weights usually have random initialization and are then modified through the automated use of certain training rules (e.g., error-correction rule), to provide minimum error with respect to the system's intended outputs. These gates are primarily significant since they demonstrated that the basic neuron units had the capability to compute. The linearly inseparable XOR problem could therefore not be handled. Perceptrons too cannot be particularly regarded as a strong model because they are capable of learning only linearly separable functions.

15.7   The Perceptron

In his 1958 paper [6], Frank Rosenblatt introduced the next major advancement called the perceptron. It varied from the McCullouch–Pitts neuron on the following grounds:

- All the weights and thresholds were not identical.
- The value of the weights can be positive or negative.
- There is no absolute inhibitory synapse.
- The neurons have the same two states, but the output function $f(sum)$ goes from $[-1,1]$, not $[0,1]$.
- The perceptron has a learning rule.

No specification for a perceptron is defined, but the concept is used most of the time to define a feed-forward network with shortcut connections. It includes several input neurons and some computing neurons that process information. The input neuron gives an identity mapping. It relays the received data, as it is. It therefore reflects the function of identity. The hidden layer neurons interpret the input data using transformation functions, i.e., they do not reflect a function of identity. A binary neuron combines all input stimuli of the weighted sum as the propagator signal. Then a binary threshold function decides the final activation from the neuron. This gives us a complete description of the single-layer perceptron as shown in Fig. 15.19.

One can identify the perceptron's learning mechanism as an example of a combination of reinforcement and supervised learning. There are other variants too that combine applications of supervised learning and error correction (corrective learning). The perceptron, however, can be used only to classify **linearly separable** classes. Let us consider a two-class classification problem as shown in Fig. 15.20.

Fig. 15.19 A single-layer
perceptron having two
neurons in the input layer and
one in the output layer

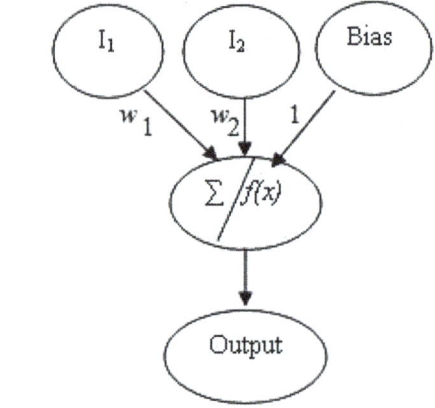

Fig. 15.20 Two-class
classification using
single-layer perceptron

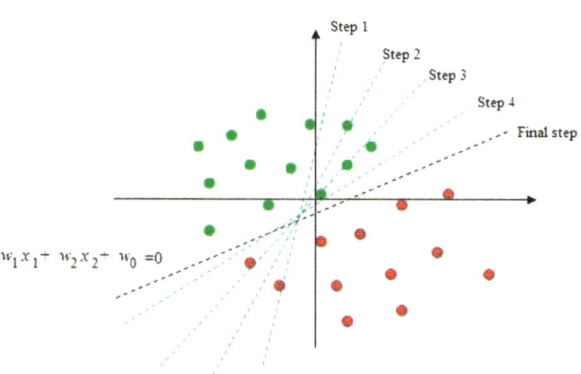

The perceptron algorithm makes a classification function that looks like:

$$f(x_1, x_2) = \begin{cases} +1 \ (\text{Green Class}), \ if \ w_1x_1 + w_2x_2 + w_0 \geq 0 \\ -1 \ (\text{Red Class}), \quad if \ w_1x_1 + w_2x_2 + w_0 < 0 \end{cases}. \qquad (15.9)$$

The perceptron learning algorithm decreases the weights of neurons in the output
layer that output 1 rather than 0, while in the opposite case, it raises these weights
following the delta rule. The algorithm is as follows.

- Input a pattern into the network and calculate the output.
- For all output neurons if output $o = t$, then no correction of weights;
  here o is the output value generated by a neuron, and t is the true class
  information.

(continued)

- If not, then if output = 0, then

$$w_i \Longleftarrow w_i + o_i$$

i.e., increase weight toward one class; and if output = 1, then

$$w_i \Longleftarrow w_i - o_i$$

i.e., decrease weight toward another class.
- While there are patterns available from the training set and the error remains large, the whole operation is iterated.

It has been shown that the aforementioned algorithm completes the task in a finite number of steps. Hence, within a finite number of steps, the perceptron can learn anything it can represent. However, unfortunately, the percentage of the linearly separable problems rapidly decrease as dimensionality of the input pattern grows. Also, for nonlinearly separable cases like the XOR problem, we need something more powerful than a single-layer perceptron.

15.7.1 XOR Problem

It can be mathematically proved that the perceptron algorithm can find the decision hyperplane for any problem that is originally linearly separable within finite time. The perceptron algorithm fails if the problem is not linearly separable. For example, let us consider the classical exclusive-OR (XOR) problem as shown in Fig. 15.21. The points belonging to two classes cannot be separated using one hyperplane. But it could be done with a convex region instead. One single perceptron gives a decision hyperplane, so two layers create a combination of multiple hyperplanes, thus, a convex polygon. The XOR problem is the one that needs such polygon.

The truth table for XOR can be written as:

x_1	x_2	Output
0	0	0
0	1	1
1	0	1
1	1	0

The output is 1 if x_1 is on and x_2 is off or x_2 is on and x_1 is off. Otherwise, the output is zero. This problem is not linearly inseparable.

Fig. 15.21 XOR pattern
space

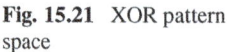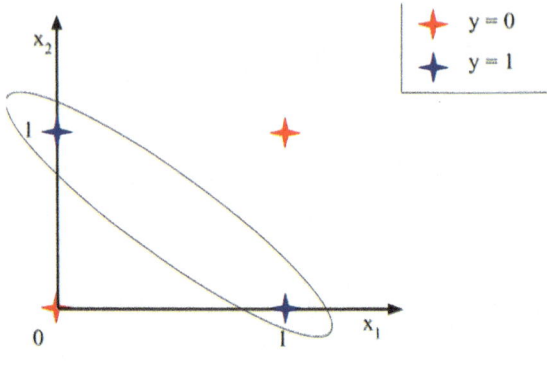

Fig. 15.22 Single-layer
perceptron model with bias
and weights

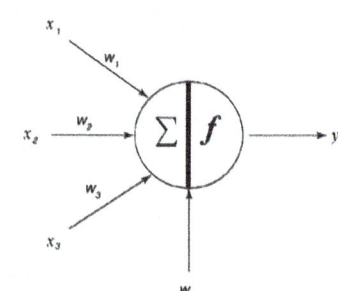

15.7.2   Bias and Weights

The perceptron can have another input known as **bias** (denoted by w_0). It is a
normal practice to treat the bias as just another input. The bias allows us to shift the
transfer function horizontally along the input axis while leaving the shape unaltered.
The weights determine the slope of the curve. Changing the bias will result in a
horizontal shift of the decision boundary while change in weights alters the slope.
The stepwise alteration of both bias and weights together will finally yield the
correct decision boundary. The bias and weights will be adjusted after each training
example, and finally, a generalized boundary is obtained as shown in Fig. 15.20. The
single-layer perceptron model with bias and weights is shown in Fig. 15.22. The bias
is a mathematical simplifying strategy to represent threshold values as connection
weights with input value $x = 1$.

15.7.3   Learning Rate

The weights and bias are required to be updated in order to minimize the error. The
learning rate helps us control the pace at which we control the weight and bias in
order to get the optimal boundary. Random change of bias and weights can result in

Fig. 15.23 Learning rate curve

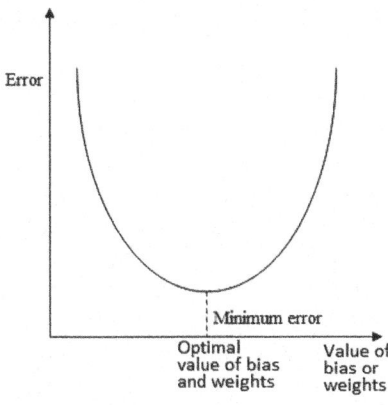

Fig. 15.24 Single-loop feedback system

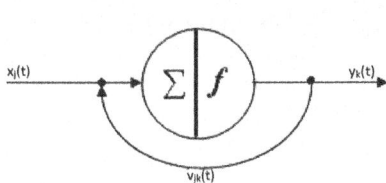

high error. As we can see from the curve in Fig. 15.23, the error is least at the optimal value of weight and bias. The change of bias or weight is zero at the optimal point. So, basically if we have n variables, then we need to find $n + 1$ weight values (n weights and one bias). These will be the coefficients in the equation of the separation line (or hyperplane for multidimensional space).

$$w_{t+1} = w_t + \Delta w; \qquad (15.10)$$

15.7.4   Feedback

Feedback is defined to be present within a dynamic system in which an element's output in the system partially influences the next value for input provided to that same element, leading to cyclic closed path(s) when the signal is transmitted through the whole system. It becomes a salient trait in studying the mechanisms of associative as well as feedback neural networks. Figure 15.24 shows a single-loop feedback system for which the signal being input, $x_j(t)$, feedback $v_{jk}(t)$, and the output $y_k(t)$ are functions of discrete-time variable t. The forward channel gives an output, which influences, in part, its own next output via the feedback channel.

$$x'_j(t) = x_j(t) + v_{jk}(t) \qquad (15.11)$$

where, x'_j is the modified input after feedback.

15.8   Learning Paradigms in Artificial Neural Networks

Whether or not it requires supervision in learning is an important aspect of an ANN design. Based on their way of learning, all artificial neural networks can be classified into two groups—supervised and unsupervised. Most of these neural network architectures are focused on learning, and therefore, selecting a learning technique is a key issue in the design of the artificial neural network. Learning means the ability of a processing unit to adjust input/output actions based on the modifications in its environmental conditions. Since the activation rules typically are defined when the network is initially formed and the input/output vectors cannot be adjusted in any way, the weights determining the input to output function must be modified to adjust the input/output behavior. Thus, a system is required that can adjust weights in response to input/output conditions at least during a training phase. For neural network models, a variety of these learning rules are available. In the neural network, learning can be guided by providing the correct response for the output during the training process (supervised) or learning without the intervention of any external instructor (unsupervised).

A suitable output outcome is expected for each input vector if the network is equipped with supervised learning. The target output is used by an ANN, trained using supervised learning mechanisms, to control neural parameters development. The neural network can thus learn the behavior of the experimental process.

The network training is entirely driven by data and no target outputs are given for the input vectors in unsupervised learning. An unsupervised ANN can be used to group input data and to recognize inherent features in the dataset.

15.9   Learning Rules in Neural Network

The word "learning" here is an umbrella term denoting activities related to the design of neural networks, including single-layer binary classifiers to multilayered models, which are relatively advanced systems. The training process starts with a random matrix of weights or more often with a zero weight matrix and adjusts the synaptic weights regularly and ideally improves the state until a stable weight matrix has been identified. The process of learning continues from one neuron to another (each one of which learns individually) either sequentially or in a random fashion. There are five basic learning rules: Error Correction Learning, Memory-Based Learning, Hebbian Learning, Competitive Learning, and Boltzmann Learning. The method of learning chosen controls the way in which parameters get changed.

15.9.1 Error-Correction Learning

Error-correction learning process is a supervised method. The method is used to compare the machine's output value with the expected or target output value, and the training is implemented according to the error between the two. In the most straightforward way, a method such as a backpropagation algorithm may be used to change the connecting weights directly. If the system output is $y_k(t)$ at a particular neuron k for a particular pattern vector $\mathbf{x}(t)$, and the desired system output is known to be $d_k(t)$ for that particular neuron, the error signal can be defined as:

$$e_k(t) = d_k(t) - y_k(t). \tag{15.12}$$

$e_k(t)$ operates a control mechanism so as to lessen the difference between the output signal $y_k(t)$ and the desired response $d_k(t)$ in gradual steps. A goal of learning is to keep tweaking the weights so that the following is achieved for the new output:

$$y_k(t + 1) = y_k(t) + e_k(t). \tag{15.13}$$

Put in words, this means that the new output and the desired output are equal. The algorithm must continue tweaking the weights until this is achieved. This is done by Widrow–Hoff [7, 8] delta rule

$$w_0(t + 1) = w_0(t) + \eta e_0(t); \tag{15.14}$$

and

$$w_i(t + 1) = w_i(t) + \eta e_i(t)x_i(t) \tag{15.15}$$

where, $i = 1, 2, 3, \ldots, d$, $\mathbf{w}(t) = \{w_0, w_1, w_2, \ldots, w_d\}$, $\mathbf{x}(t) = \{x_1, x_2, x_2, \ldots, x_d\}$ and η is the learning rate (equal to 1 for threshold neuron). The learning mechanism minimizes a cost function or index of performance. Here it is the mean square error (M.S.E) calculated over all the patterns.

$$\varepsilon(t) = \frac{1}{N} \sum_{n=1}^{N} \{e_k(t)\}^2; \tag{15.16}$$

where $e_k(t)$ denotes the error for the tth pattern, and N denotes the total number of samples. This cost function gives the instantaneous value of the error. Minimization of this yields a steady state, i.e., synaptic weights are stabilized. The learning rate α determines the speed of convergence.

15.9.2   Memory-Based Learning

In memory-based learning, the whole set of data points in the training data is essentially saved in memory. Whenever test data come, they are classified according to the similar training data points using some local model (the idea is inspired by nearest neighbor algorithm) and making predictions based on this model. There are four main components in a memory-based learning system:

(i) **Distance metric** is needed to find the similarity between two points.
(ii) **Number of nearest neighbors** is crucial for the algorithm. The value of k in k-NN, for example, is the key hyperparameter for classification.
(iii) **Weight function** or a weighted average of the neighbors is used to smooth out the discontinuities at the boundaries of the regions.
(iv) **Local model** is a function the algorithm will use to fit the local data.

An example of memory-based learning system is k-nearest neighbor algorithm, shown in Fig. 15.25.

15.9.3   Hebbian Learning

The change in the value of weights connecting two neurons is a function of the pre- and post-synaptic neural activities—Hebbian learning rule. According to Hebb, "When an axon of cell/neuron A is near enough to excite a cell/neuron B and repeatedly or persistently takes part in firing it, some growth process or metabolic change takes place on one or both cells such that A's efficiency as one of the cells firing B, is increased."

Fig. 15.25 Memory-based learning: k-NN with k = 3

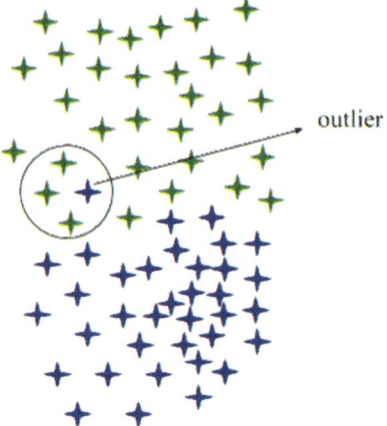

outlier

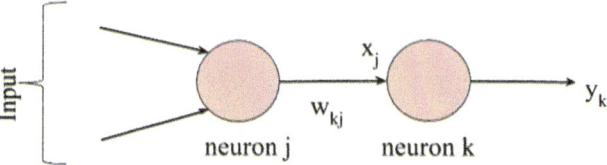

Fig. 15.26 Hebbian learning rule

The simplest form of Hebbian learning rule where the weight change is given as:

$$\Delta w_{kj} = \eta y_k x_j, \quad (\eta > 0);$$

η is the learning rate, y_k denotes the output of the kth neuron, and x_j is the output of the jth neuron (Fig. 15.26).

15.9.4 Competitive Learning

In this type of learning, more than one output neuron compete with each other to become active (or to get fired) at a time. So, only one neuron fires at a time, and it is called the "winner-takes-all" neuron. In order to determine the winner-takes-all neuron, the following rule is used (Fig. 15.27).

$$y_k = \begin{cases} 1, & \text{if } v_k > v_j, \ \forall j, j \neq k \\ 0, & \text{otherwise;} \end{cases}$$

with $\sum_j w_{jk} = 1, \forall k$.

In the competitive learning, the synaptic weights of the winning neuron are updated using the following learning rule, and no learning takes place for the neurons other than the winning neuron for that particular input pattern. In other words, only the neuron that gets fired by an input pattern learns from the corresponding input pattern, the other neurons do not learn anything from that input pattern.

$$\Delta w_{jk} = \begin{cases} \eta(x_j - w_{jk}), & \text{if neuron k is the winner} \\ 0, & \text{otherwise.} \end{cases}$$

A simple competitive learning neural network (Fig. 15.28) has an input layer along with an output layer. The input neurons are fully connected to the output neurons, and the output neurons might have lateral connections among themselves.

Neural networks using competitive learning techniques may be used as salient feature detectors (Fig. 15.29). Each neuron is fired by a different subset of data; hence, the different neurons act like different feature detectors.

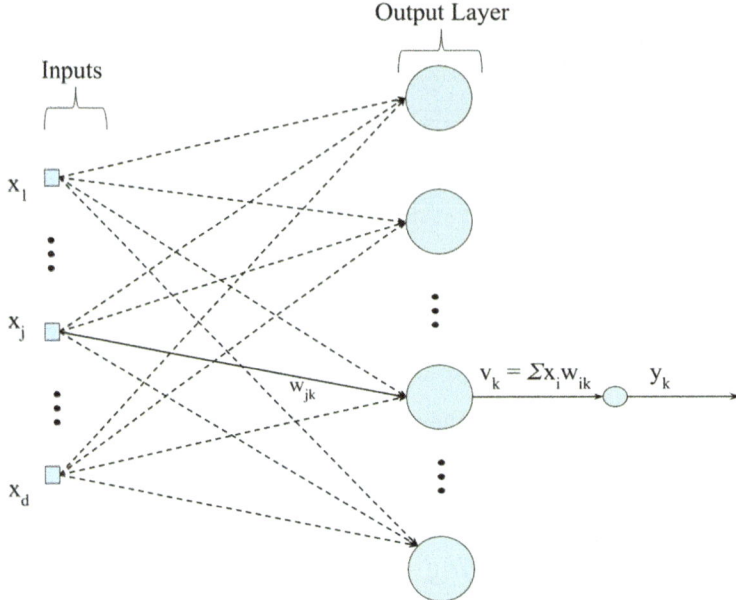

Fig. 15.27 Winner-takes-all strategy

Fig. 15.28 A simple
competitive learning network

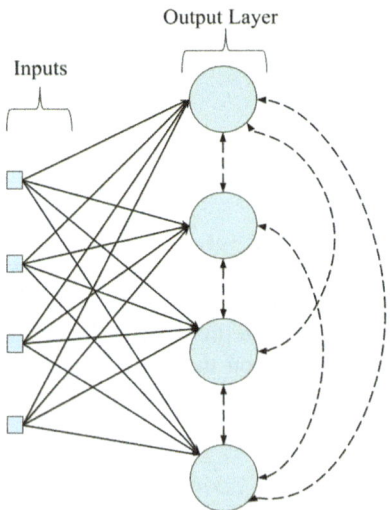

15.9.5   Boltzmann's Learning

The idea of Boltzmann's learning comes from statistical mechanics, and it is named
after Ludwig Boltzmann. When a neural network uses Boltzmann's learning as the
learning rule, then it is called Boltzmann's machine. The states of the neurons in the

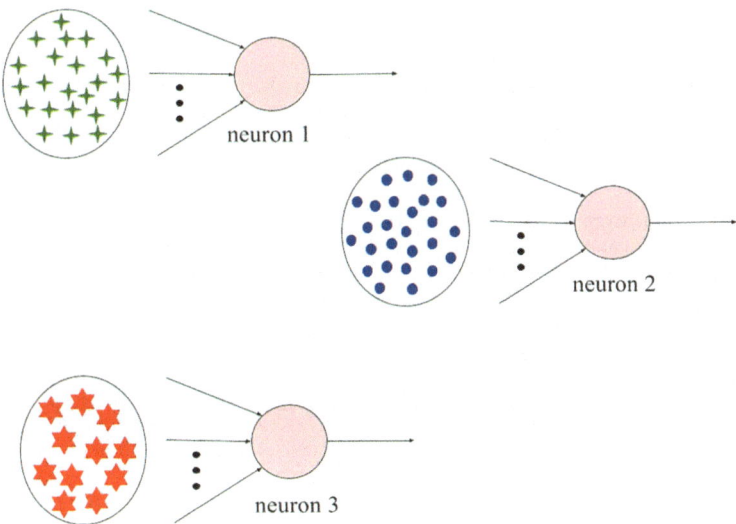

Fig. 15.29 Competitive learning used to detect features

Boltzmann's machine are always binary. What sets the Boltzmann's machine apart is the use of an energy function

$$E = -\frac{1}{2} \sum_{j} \sum_{k, k \neq j} w_{kj} x_k x_j;$$

where x_j is the state of neuron j, x_k is the state of neuron k, and w_{kj} is the weight of the connection between neuron x_j and x_k. The neurons in a Boltzmann's machine are divided into two groups: visible and hidden. Visible neurons become a path for the hidden neurons and environment to interact with each other; so operation of the visible neurons depends on the environment. Hidden neurons are independent of the environment. The two modes of operations for the neurons of the Boltzmann's machine are as follows.

- **Clamped condition:** Visible neurons are fixed at a specific state determined by the environment. ρ_{kj}^{+} is the correlation between neuron j and neuron k at clamped condition.
- **Free-running condition:** All neurons (visible and hidden) are permitted to function without restrictions. ρ_{kj}^{-} is the correlation between neuron j and neuron k at free-running condition.

Boltzmann's learning rule induces the change Δw_{kj} to the weight w_{kj} (the synaptic connection between neuron k and neuron j), where,

$$\Delta w_{kj} = \eta(\rho_{kj}^{+} - \rho_{kj}^{-}), k \neq j;$$

with η as the learning rate.

References

1. Anderson, J. A. (1995). *An introduction to neural networks*. A Bradford Book. MIT Press.
2. Coates, T. D. (2008). *Neural interfacing: Forging the human-machine connection*. Synthesis lectures on biomedical engineering. Morgan and Claypool.
3. Cochocki, A., & Unbehauen, R. (1993). *Neural networks for optimization and signal processing*. Wiley.
4. Hornik, K., Stinchcombe, M., & White, H. (1989). Multilayer feedforward networks are universal approximators. *Neural Networks, 2*(5), 359–366.
5. Rojas, R., & Feldman, J. (2013). *Neural networks: A systematic introduction*. Springer.
6. Rosenblatt, F. (1958). The perceptron: A probabilistic model for information storage and organization in the brain. *Psychological Review, 65*(6), 386.
7. Widrow, B., & Hoff, M. E. (1960). Adaptive switching circuits (pp. 96–104).
8. Widrow, B., Lehr, M., et al. (1990). 30 years of adaptive neural networks: Perceptron, madaline, and backpropagation. *Proceedings of the IEEE, 78*(9), 1415–1442.

Chapter 16
Multilayer Perceptron

As a computational model, perceptrons are too poor, because they can only learn linear functions. The generalized neural networks have several layers of hidden units to improve their computational abilities. The goal is to find a training framework for such multilayered networks. A multilayered feed-forward network [6] tries to model a function of the given input and the corresponding network weights. A network that is feed-forward in nature and has one or multiple layers in the middle of the input and output layers is often known as a multilayer perceptron [3] (Fig. 16.1). Feed-forward implies the data moves in a unidirectional fashion from the input layer toward the output layer (i.e., forward). In order to train such a network, an algorithm called "backpropagation" is used. MLPs are widely used to classify, recognize patterns, forecast, and approximate nonlinear functions. MLP can address problems that cannot be separated linearly. The basic components of a multilayer perceptron are the following:

- **Input units**
- **Output units**
- **Hidden units**

Single-layer perceptrons have input units followed by one layer of output units, i.e., no hidden units. As already shown, it is not possible to learn weights which will enable it to deal with the XOR problem. However, multilayer perceptrons can handle the nonlinear separability issues [9]. Thus, the main motive that arises now is to find the correct combination of the weights. As there are several connections and several layers, finding the correct combination of weights is a question of computational effort. A popular learning method therefore emerged that enabled handling of such situations—the **Backpropagation Algorithm(BP)**.

© The Author(s), under exclusive license to Springer Nature Singapore Pte Ltd. 2025   265
A. Ghosh, *Data Science and Cases in Sustainability*, Mathematics for Sustainable
Developments, https://doi.org/10.1007/978-981-96-8362-8_16

Fig. 16.1 Multilayer
perceptron

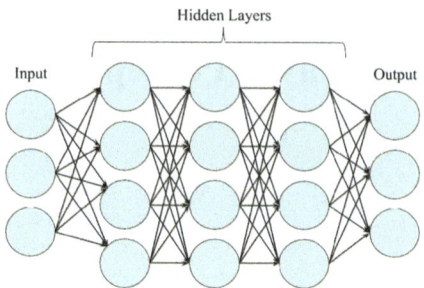

16.1 Backpropagation Algorithm

The backpropagation algorithm deploys a gradient descent technique to find out the
lowest error value in the global weight space. The choice of weights which gives the
lowest value of the error function is regarded as a possible solution to the learning
process. As such an approach has to measure the error value gradient at each step,
the continuity and differentiability of the error function must be ensured [11]. Thus,
we cannot use any activation function which is discontinuous or non-differentiable.
Hence, using a step function would not suffice. The "Sigmoid" function, however,
grew in favor as it included the beauty of a step function, is continuous, and
differentiable. Several alternate types of activation functions which have been
discovered, for which the backpropagation algorithm works. The sigmoid's output
is always more than 0 and less than 1. Thus, 0 and 1 can be achieved asymptotically.
For fixed weights $w_1, w_2, \ldots, w_d$ along with a bias w_0, a sigmoidal unit gives the
following value for the input $\mathbf{x} = \{x_1, x_2, \ldots, x_d\}$ as

$$f(\mathbf{x}) = \frac{1}{1 + e^{-\left(\sum\limits_{i=1}^{d} w_i x_i - w_0\right)}}. \tag{16.1}$$

For training, the operation of the network is initially started by choosing small
random values for its weights (e.g., in the numbers ranging from $-\epsilon$ to $+\epsilon$;
where ϵ is a small positive quantity). Then, a sequence of inputs is presented to
the network, and the output is determined (this is called the forward pass). The
estimation generates an output that is entirely different from what we are aiming
for (the target), because all the weights in the network are randomly initialized. We
then determine the error produced for each output neuron, which effectively means
the difference between target output and predicted output (i.e., the output we expect
and the output we see). This error is useful in changing the weights of the network,
so that the error is minimized and the predicted output becomes close to the target
output. In other terms, each neuron's output value will be nearer to the target value
(this is referred to as the reverse pass). This process continues till convergence; i.e.,
the minimum error function is achieved.

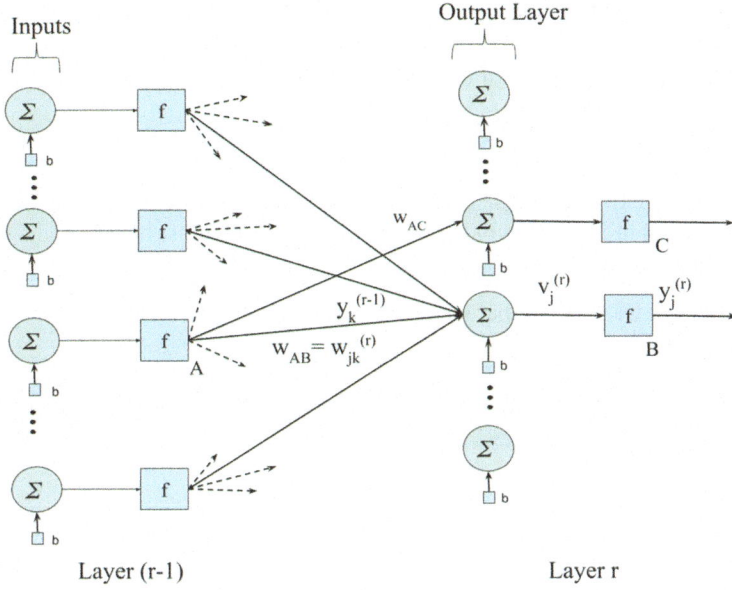

Fig. 16.2 A single connection learning in a back propagation network

Let us consider a simple example with a single connection learning in a network (Fig. 16.2) with a BP algorithm. We are interested in an arbitrary connection from neuron A in the hidden layer to neuron B in the output layer. The weight connecting the two is denoted by w_{AB}. The algorithm works as follows:

(I) First provide an input vector to the neural network and determine the output. This value of output in the beginning will be random, as the setting of the weights at starting point were random numbers.

(II) Error for neuron B:

$$\text{Error}_B = \text{Output}_B(1 - \text{Output}_B)(\text{Target}_B - \text{Output}_B).$$

The sigmoid function necessitates the introduction of the "Output(1 – Output)" term in the equation

(III) The weight is then updated:

$$w'_{AB} = w_{AB} + (\text{Error}_B \times \text{Output}_A).$$

The weights are adjusted for each training pattern, so that the mean squared error between the network's predicted output and the target output value is minimized. These adjustments are incorporated during the "backward" phase. It means that the flow of error originates at the output layer, goes in the backward direction through the hidden layers, and finally reaches the first hidden layer (which is how it came to be called backpropagation).

(IV) Now, we calculate error in neuron A by taking the errors from B and C

$$\text{Error}_A = \text{Output}_A(1 - \text{Output}_A)(\text{Error}_B \times w_{AB} + \text{Error}_C \times w_{AC}).$$

(V) After calculating the error for each of the hidden layer neurons, the algorithm proceeds as in Step III to tweak the hidden layer weights. Iterating this method can thus train a network of arbitrarily many layers.

So, in summary, we can say that backpropagation algorithm works in the following manner in general:

- Initialize the weights with some random values, and choose and fix the learning rate η.
- While the network is being trained for all the samples and the error has not yet been minimized,

  - For each training sample, a forward pass is made to generate an output. Fixing J as the number of hidden layer nodes and d as the number of inputs for a two-layer MLP, the final output $y_k^{(2)}$ at the kth node of the output layer is

$$y_k^{(2)} = f(\sum_{j=0}^{J} w_{jk}^{(2)} y_j^{(1)}); \qquad (16.2)$$

  where w's represent the corresponding weights and $y_j^{(1)}$ is the output given by hidden node j:

$$y_j^{(1)} = f(\sum_{i=0}^{d} w_{ij}^{(1)} x_i). \qquad (16.3)$$

  - For each output unit k, compute errors

$$\delta_k^{(2)} = (y_{\text{target}} - y_k^{(2)}) y_k^{(2)} (1 - y_k^{(2)}). \qquad (16.4)$$

(continued)

- For each hidden unit j (considering the hidden layers in backward order, when multiple hidden layers are present) generate the error

$$\delta_j^{(1)} = y_j^{(1)}(1 - y_j^{(1)}) \sum_k w_{jk}^{(1)} \delta_k^{(2)} . \tag{16.5}$$

- For all weights on connections between input layer unit i and hidden layer unit j, change weight using gradient descent

$$\Delta w_{ij}^{(1)} = \eta \delta_j^{(1)} x_i; \tag{16.6}$$

and for all weights on connections between last hidden layer's unit j and output layer unit k, change weight using the equation:

$$\Delta w_{jk}^{(2)} = \eta \delta_k^{(2)} y_j^{(1)} . \tag{16.7}$$

A flowchart representation of the backpropagation algorithm is depicted in Fig. 16.3.

Thus, in summary, the network aims to reduce the total error to below some predetermined low target value by repeatedly training all the patterns, following which it ends the process. After the error becomes low for the validation or the test set, the network stops. Figure 16.4 shows the idea. Overtraining the network (becoming too accurate) only for the training patterns can start to raise the validation set error. This hampers its performance with noisy data.

16.2   Number of Hidden Layers

The conceptual assumption is that every layer in a multilayer input perceptron contributes its own degree of nonlinearity, which cannot be expressed by only one layer. The input of each layer is only combined linearly, and therefore the nonlinearity which is recognizable across several layers cannot be generated. The concept of getting a decision hyperplane at each neuron at the first layer and a decision polygon (found by combining the decision hyperplanes at the 1st layer of the network) at the next layer of the network is valid only if the activation function is a step function. It is evident that the same argument is essentially not valid if the activation function is not a step function (like sigmoid, tanh, etc.).

Considering the classical XOR problem, we can see that linear separation is not possible. So, each "decision boundary" represents one layer in MLP (Fig. 16.5).

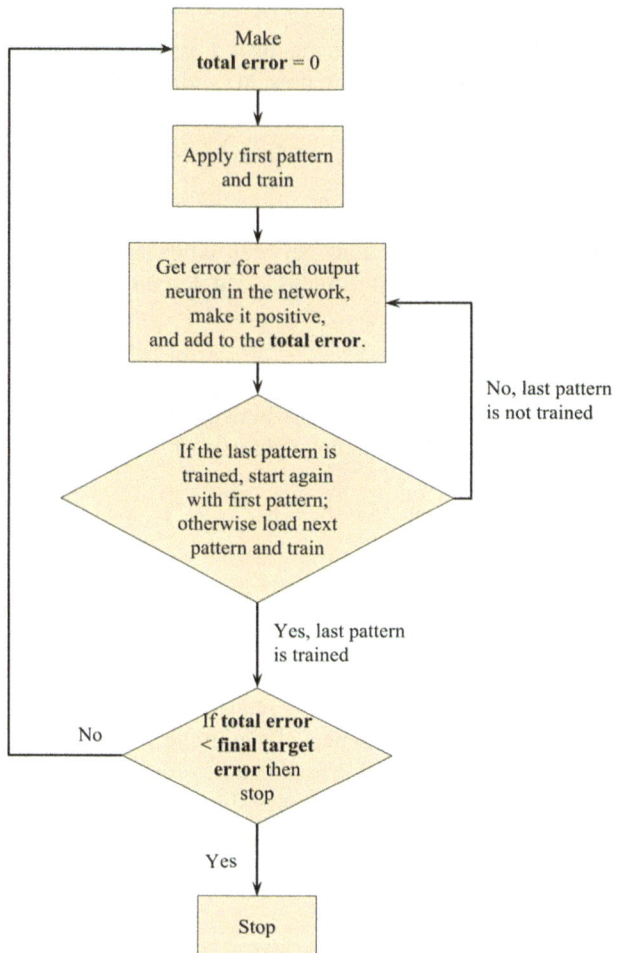

Fig. 16.3 A characterization of the backpropagation algorithm using a flowchart

The size of a neural network depends on the number of hidden layer neurons, the complexity of the functions that are aimed to be approximated by the network as a result of multiple layers, and their interconnections. If the network is tiny, a particular task or part of it cannot be entirely solved by it. This is referred to as underfitting case. On the other hand, a network that is too large tends to memorize the data and lose generalization capabilities. Such a network focuses too much on the presented data and attempts only to model them. Such situations are referred to as overfitting.

A linearly separable data needs no hidden layers to be solved. Multiple layers provide complexity levels that cannot be expressed through a single layer, even if they can contain the same number of parameters [8]. However, the complexity of the

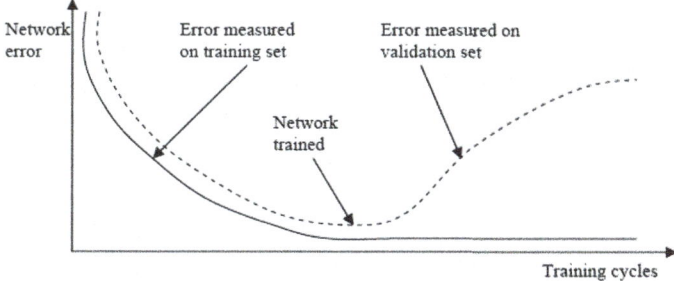

Fig. 16.4 A single connection learning in a back propagation network

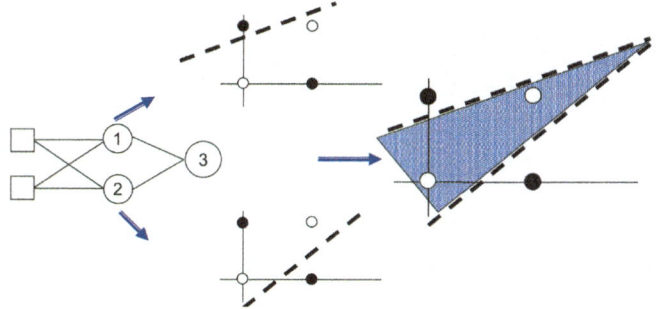

Fig. 16.5 Solving the XOR problem with one hidden layer MLP

network can increase unnecessarily by increasing the number of hidden layers. To render MLPs as a universal approximator, one hidden layer is enough [2]. However, the concept of deep learning indicates that the addition of hidden layers contributes significantly.

For determining the correct number of neurons also, to determine the correct number of neurons that are to be used in the hidden layers, there are many rule-of-thumb methods, such as the below-mentioned ones:

- The number of hidden neurons should lie in the range defined by the numbers of neurons in the input layer and of those in the output layer as margins.
- The number of hidden neurons should be 2/3rd the size of the input layer, plus the number of neurons in the output layer.
- The number of hidden neurons should not exceed twice the size of the input layer.

The selection of an optimal network size is still a big issue, since there is no analytical solution to that problem available.

Fig. 16.6 Basic overview of
an RBFN

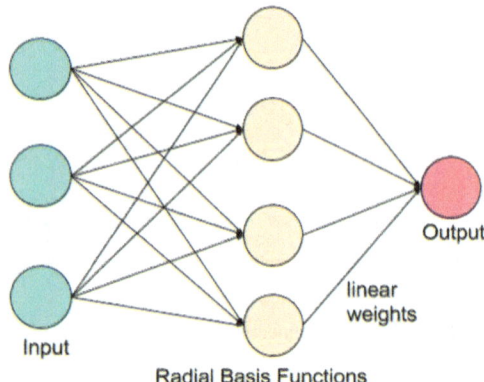

16.3   Radial Basis Function Networks

Radial basis function (RBF) networks are three-layer networks. They have input layer, a hidden layer, and an output layer (Fig. 16.6). They are mainly used for function approximation. This means their output is real valued. Each neuron in the input layer represents a feature of the pattern. Each neuron in the hidden layer consists of a radial basis function like Gaussian, multi-quadric, etc. centered at a point with the same dimensions as the number of features. Overall functioning of the network is similar to multilayer networks discussed previously.

16.4   Radial Basis Functions

A radial basis function (RBF) can be defined formally as a real-valued function which is a function solely of the radial distance between the point of origin and the input point x being considered, so that $\phi(\mathbf{x}) = \phi(||\mathbf{x}||)$. This may otherwise also signify the distance between a center (a point c) and the point x, such that $\phi(\mathbf{x}, \mathbf{c}) = \phi(||\mathbf{x}-\mathbf{c}||)$. Every function ϕ which satisfies the property $\phi(\mathbf{x}) = \phi(||\mathbf{x}||)$ may be regarded as a radial function. The usual norm is Euclidean distance, but other distance functions are equally acceptable (Fig. 16.7).

Usually, the summation of radial basis functions is used to approximate the specified function. In 1988, David Broomhead and David Lowe's works [1], which was inspired by groundbreaking studies by Michael J. D. Powell from 1977 [10], state that this approximating mechanism can be viewed as a basic form of a neural network. RBFs are also used as a kernel in support vector machines (discussed in Chap. 18). Commonly used radial basis functions include (here $r = ||\mathbf{x} - \mathbf{x}_i||$) the following:

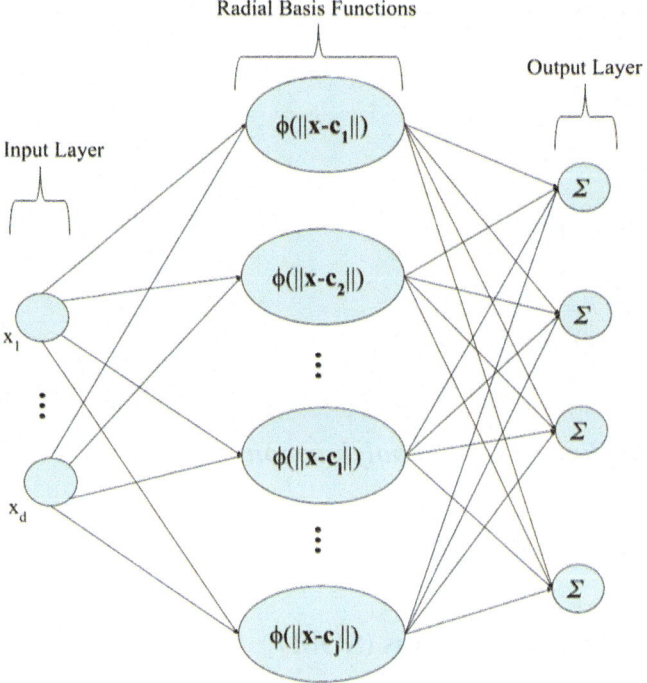

Fig. 16.7 Architecture of an RBFN

Gaussian

The first term that is used for normalization of the Gaussian is missing, because every Gaussian has a weight in our sum; thus, the normalization is not necessary.

$$\phi(r) = e^{-(\varepsilon r)^2}.$$

Multiquadric

$$\phi(r) = \sqrt{1 + (\varepsilon r)^2}.$$

Inverse quadratic

$$\phi(r) = \frac{1}{1 + (\varepsilon r)^2}.$$

Inverse multiquadric

$$\phi(r) = \frac{1}{\sqrt{1 + (\varepsilon r)^2}}.$$

Polyharmonic spline

$$\phi(r) = r^k, \, for \, k = 1, 3, 5, \ldots$$
$$\phi(r) = r^k \ln(r), \, for \, k = 2, 4, 6, \ldots$$

Thin plate spline (a special polyharmonic spline)

$$\phi(r) = r^2 \ln(r).$$

16.5   Self-Organizing Maps (SOM)

Kohonen's self-organizing maps (SOM) or self-organizing feature maps (SOFM) [4] fall under the category of unsupervised learning network. The objective is detecting some intrinsic structure in the data. The desired form of structure, however, is highly distinct from the dimensionality reduction techniques, for example, PCA or vector quantization.

Kohonen's SOM is referred to as a topology-preserving map, since the nodes of the network are subject to imposition of topological structure. A topological map is just a map that retains relationships with the neighborhood.

We largely neglected the structural configurations of the output nodes in the neural networks we have examined so far. Every node belonging to the same layer has been identical to each other, because they are connected to every other node in the upper and/or lower layers. The spatial arrangement of the nodes will now be taken into consideration. Communication between "close" and "far" away nodes will be of different nature.

What do the terms "close" and "far" here signify? The output nodes can be arranged in a linear or a planar structure. The target is training of the network to suit neighboring inputs; i.e., similar inputs will be mapped to those neurons which are in close vicinity to them.

For example, if two vectors $\mathbf{x}_1$ and $\mathbf{x}_2$ are presented as input with winning output nodes positioned at $\mathbf{t}_1$ and $\mathbf{t}_2$, respectively, now if $\mathbf{x}_1$ and $\mathbf{x}_2$ are similar, then $\mathbf{t}_1$ and $\mathbf{t}_2$ should be close. This kind of mapping is performed by a network which is known as a feature map.

Neurons in the brain have a propensity to aggregate together into groups. Such neurons have a preference to form greater in-group connections than out-group ones. Kohonen's network attempts to mimic a simplified version of this.

The algorithm for Kohonen's self-organizing map is given below:

Assumption: Connection exists between every possible pair of an input layer node and an output layer node, and there is also connection between the output nodes in an array (usually one or two dimensional). The competitive learning algorithm is used as described below:

- Pick a random vector $\mathbf{x}$ as input for the network.
- Ascertain which output node i is the "winning" one. We denote by $\mathbf{w}_i$ the weight vector contained in the connection between the input nodes and output node i. Thus

$$|\mathbf{w}_i - \mathbf{x}| \leq |\mathbf{w}_j - \mathbf{x}| \text{ for all } j. \tag{16.8}$$

- Having ascertained the winning node i, the weight's value is revised according to the rule:

$$\mathbf{w}_j(\text{new}) = \mathbf{w}_j(\text{old}) + \mu \mathcal{H}(i, j)(\mathbf{x} - \mathbf{w}_j); \tag{16.9}$$

where $\mathcal{H}(i, j)$ denotes the neighborhood function maps to 1 when $i = j$ and keeps reducing the further the distance $dist_{ij}$ is between units i and j in the output array. Hence, relatively high magnitude of changes in weights is observed in the group of units close to the winning unit including the winner itself. This is in contrast to units far away from the winner, whose weights do not change by any high magnitudes relative to the closer ones. Thus the topological information exerts a significant influence on the changes brought about in the network's weights. Units that are close together have their weights changed by similar amounts which are why they tend to respond to input patterns corresponding to locations nearby.
- The above rule tends to bring each of the weight vector $\mathbf{w}_i$ of the winner along with those of nearby units and the input $\mathbf{x}$ closer.

The neighborhood function $\mathcal{H}(i, j)$ in Eq. 16.9 can be of the form

$$\mathcal{H}(i, j) = e^{(-|d_{ij}|^2)/(2\sigma^2)}; \tag{16.10}$$

where σ^2 denotes the width parameter that is reduced with every iterations.

Kohonen's self-organizing feature map (SOFM) network [4, 5] includes two layers, one for input, the other for output. Connection exists between every possible

Fig. 16.8 Rectangular
topological neighborhood of
neuron *i* over *ep*

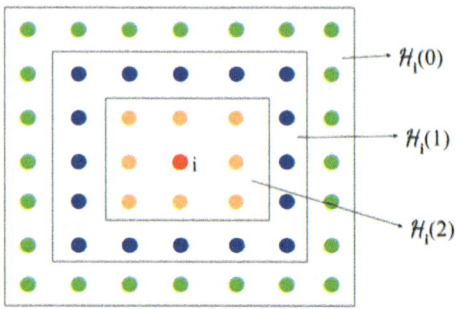

pair of a neuron in the output layer and one in the input layer; i.e., the *d*-dimensional input vector $\mathbf{x} = [x_1, x_2, \ldots, x_d]$ can get transferred to any given output neuron. Assume that the synaptic weight vector of an output neuron j is $\mathbf{w}_j = \{w_{1j}, w_{2j}, \ldots, w_{dj}\}^T$, $j = 1, 2, \ldots \Omega$, Ω being the number of output layer neurons and w_{pj} being the weight of the connection between the jth output unit with the pth component of the input. Assume that the input vector $\mathbf{x}$ corresponds to the synaptic weight vector $\mathbf{w}_i$ belonging to output neuron i as the best match; then $\sum_{p=1}^{d} w_{pj}.x_p$, $\forall j$ will be maximized for $j = i$ with this maximum value being $\sum_{p=1}^{d} w_{pi}.x_p$. This neuron i is known as the *winning neuron* corresponding to the input vector $\mathbf{x}$. A topological neighborhood of cooperating units is centered around the *winning neuron*. Denote this topological neighborhood of *winning neuron i* for epoch number *ep* as $\mathcal{H}_i(ep)$. Topological neighborhood can be defined using several ways [7] such as rectangular, Gaussian *etc*. With increase in *ep*, the size of the topological neighborhood gets shrinked. Figure 16.8 demonstrates the reduction of the width/area of the topological neighborhood of the *winning neuron i* as successive epochs are executed, that is, as *ep* keeps increasing (this is presuming a rectangular topological neighborhood).

The term *self-organization* refers to the ability to learn the structure of the input data even when there is no available information about it. Self-organization is a basic principle of sensory paths of the human visual system. In the feature mapping algorithm of Kohonen, the same input is given to all the neurons to perform self-supervised mapping.

16.5.1   SOM Initialization

Choice of a decent beginning guess is a surely understood issue for every single iterative technique for learning neural systems. Kohonen utilized random start of SOM weights, yet as of late the principal component initialization, a technique where the space of the first principal component is used as a domain for selecting the initial map weights, turned out to be somewhat prominent in the light of precise reproducibility of the outcome.

Cautiously comparing such random initiation approach with the principal component initialization for one-dimensional SOM (models of principal curves) illustrated the widely accepted presumption about pros of the principal component SOM initialization as not being ubiquitous. The optimal initialization is dependent on the dataset geometry. In the case that the principal curve approximating the dataset is being univalently and linearly projected on the first principal component (quasilinear sets), the principal component initialization should be used if the dimension is one. The random initiation performs better for essentially nonlinear datasets.

16.5.2 Pros and Cons

Likely the best thing about SOMs is that they are straightforward, since they are governed by simple rule-closer units connected by gray value which is set as similar. In contrast, the existence of a dark gorge between them forces them to be set as distinctive. Unlike Multidimensional Scaling or N-land, effective use of SOMs is learned relatively fast. They are also known to have vastly good performance. They classify data well and after that the quality of their own output is also effortlessly evaluated. Thus, the accuracy of a map along with the similarities between objects are calculated spontaneously.

One noteworthy disadvantage of using SOMs is the need for acquiring the appropriate data. That is, the value of each dimension of each member of samples is required as input to output a map. At times, this basically is impractical, and frequently it is exceptionally hard to procure every one of this information, which is why it is a constraining component to the utilization of SOMs regularly alluded to as missing data.

The second issue with SOMs is that they differ from each other and find different similarities among the sample vectors. For two samples to get grouped together, it is usually necessary for them to be similar; however, it is not a sufficient condition, and it does not guarantee their grouping together. For example, a lot of shades of purple will not always get grouped together with all the purples in that cluster; it is possible for such a group to get broken into, say, two groups of purple in the output. While for such a case of samples of colors, it might be easy to conclude that those two groups are actually similar and manually rescinding the split, but with most data, those two clusters will look completely different. So, in order to get one final good map, building of several maps becomes necessary.

References

1. Broomhead, D. S., & Lowe, D. (1988).   Radial basis functions, multi-variable functional interpolation and adaptive networks. Technical report, Royal Signals and Radar Establishment Malvern, United Kingdom.
2. Cybenko, G. (1989). Approximation by superpositions of a sigmoidal function. *Mathematics of Control, Signals and Systems, 2*(4), 303–314.
3. Gardner, M. W., & Dorling, S. R. (1998).   Artificial neural networks (the multilayer perceptron)- a review of applications in the atmospheric sciences. *Atmospheric Environment, 32*(14), 2627–2636.
4. Kohonen, T. (2001). *Self-organizing maps* (2nd ed.), Springer Nature.
5. Kohonen, T. (1982). Self-organized formation of topologically correct feature maps. *Biological Cybernetics, 43*, 59–69.
6. Kollias, S. (2006). Artificial neural networks. In *Proceedings of 16th international conference*, Athens, Greece ICANN. Springer.
7. Lo, Z. P., Yu, Y., & Bavarian, B. (1993). Analysis of the convergence properties of topology preserving neural networks. *IEEE Transaction on Neural Networks, 4*, 207–220.
8. Melin, P., Castillo, O., Ramírez, E., & Pedrycz, W. (2007). *Analysis and design of intelligent systems using soft computing techniques*.   Advances in Intelligent and Soft Computing. Springer Berlin Heidelberg.
9. Pal, S. K., & Mitra, S. (1992). Multilayer perceptron, fuzzy sets, and classification. *IEEE Transactions on Neural Networks, 3*(5), 683–697.
10. Powell, M. J. D., & Sabin, M. A. (1977). Piecewise quadratic approximations on triangles. *ACM Transactions on Mathematical Software (TOMS), 3*(4), 316–325.
11. Van Ooyen, A., & Nienhuis, B. (1992). Improving the convergence of the back-propagation algorithm. *Neural Networks, 5*(3), 465–471.

Chapter 17
Evolutionary Computing for Machine Learning

The collective name for a variety of problem-solving methods based on biological evolutionary principles, such as natural selection and genetic traits, is evolutionary computing [2]. Such methods are widely used on a range of tasks varying from practical applications in manufacturing and commerce to state-of-the-art scientific research. An initial set of candidate solutions is generated and iteratively modified in evolutionary simulations. In each new generation, seemingly inferior alternatives are stochastically excluded, and small random adjustments are introduced. In biological language, natural selection (or artificial selection) and mutation are inevitably incorporated to a population of solutions. In this scenario, the fitness function of the algorithm will slowly increase and improve the population. Evolutionary concepts have been used in the 1950s to simplify problem-solving. Three separate interpretations of the theory began to be established at three different locations only in the 1960s. Lawrence J. Fogel proposed evolutionary programming [3] in the USA, while John Henry Holland named his technique as a genetic algorithm [6]. Evolution strategies have been pioneered in Germany by Ingo Rechenberg and Hans Paul Schwefel [9]. These fields continued establishing independently over nearly 15 years. From the early 1990s, all these ideas were combined and started being collectively called evolutionary computation. In the early 1990s, a fourth stream of called the genetic programming [1] also developed, which followed the general ideas. Nature-based algorithms have become a substantial part of evolutionary computation since the 1990s. These terminologies describe evolutionary computation that includes subareas like evolutionary programming, evolutionary strategies genetic algorithms, and genetic programming. In the following sections, we will only concern ourselves with genetic algorithms. Let us first look at the clearer description of how genetic algorithms can be used in machine learning.

A. Ghosh, *Data Science and Cases in Sustainability*, Mathematics for Sustainable Developments, https://doi.org/10.1007/978-981-96-8362-8_17

17.1   The Chromosome

The chromosomes provide representations for candidates of the solution. Chromosomes have the ability to mate, produce offsprings, and undergo mutations. Sometimes they perish because of the survival of the fittest rule or they are permitted to mate and have offspring with more desirable characteristics and stick to a natural selection process. For example, we can consider a simple phrase "Hello, World!" and understand how GAs can be evolved or taught how to respond correctly. In the presented example, each chromosome is represented by a string of characters. We can assume for fairness that the lengths of all chromosomes are fixed at 13 characters each (the same as "Hello, World!"). Solutions to our problem could be presented by some candidates among the entire population of the chromosomes:

Geimo+%xosmd!

Geiln, worle"

Gelln, woslf!

Gello, worlx!

Hello, world!

Clearly, the "correct" (or globally-optimum) chromosome is the only true solution. However, then, the question on the measurement of the degree of how optimal a choice of a chromosome as "correct" is arises.

17.2   Cost Function

The optimality of a chromosome can be measured by a cost function (or error function, or fitness function as the inverse). If we are concerned about the "fitness function," then we should look out for higher scores, and if we are concerned about the "cost function," then we should try to maintain low scores.

In this case, we can always define a cost function as described below:

For every character in the chromosome string that does not match the target character (character of the target chromosome at the same position), "cost" is incremented by one. For instance, the presence of a character "H" (ASCII 72) in a position of the string that is supposed to actually be a capital "D" (ASCII 68) causes an increase in the cost for that position by 1.

Again, the reason we are using such a cost function is that it cannot become negative. We can also use the square or the absolute value of the difference between the two ASCII values. This gives us the values (in parentheses) of the cost function for the five example chromosomes:

Geimo+%xosmd! (8)

Geiln, worle" (6)

Gelln, woslf! (5)

Gello, worlx! (3)

Hello, world! (0)

This is a very simple, easy, and contrived case as we already know that we are aiming to reduce our cost function to zero, which is our stopping criterion. However, in practical situations, it is not possible to ascertain the precise optimum value beforehand. So, most of the time we just search for the minimum possible cost achievable, and figuring out the different criteria to terminate the calculations becomes our concern. Other times we search for the maximum "fitness score" and similarly face a requirement for a criterion to decide when to terminate the process.

The evaluation of a cost function is highly crucial to GAs. In this example, we are just concentrating on the characters. However, in real-world scenarios, like developing an app which provides driving directions, we need to consider every aspect like toll booths, distances, speed of the vehicle, traffic lights, bad neighborhoods, presence of bridges, etc. These are completely independent parameters which we can definitely club together into one compact cost function in order to plan a route by giving different weights to the parameters.

17.3 Mating and Death

Mating is an inevitable reality in the life cycle of many species of living creatures and is very commonly used in GAs. Mating is referred to as the process by which two chromosomes share their traits among themselves. The jargon that is used for mating is "crossover."

If one performs a GA, one chromosome is not examined at a moment, but an entire chromosomal population is analyzed. We can have 20 or 100 or 5,000 chromosomes and all constituting a single population, and we need to analyze them all at once. Just as humans have evolved, we may be motivated to have the best and most successful traits of our populations, in the expectation of a healthier child than any parent.

Mating chromosomes, as given in "Hello, World!" instance is quite straightforward and simple to understand. Two candidate chromosomes (in our case strings) can be chosen along with a pivot point somewhere in the length of the string. This point could, say, be the exact midpoint or can also be any random point.

Consider the following two strings, for instance:

Hello, akrlt+ (4)

Peolv, world! (3)

If we halve them and make two new strings (offsprings) by exchanging the halves among these strings, we get the following "children":

Peolv, akrlt+ (7)

Hello, world! (0)

It can thus be observed that the population of offsprings include one extremely unhealthy child which has inherited the most undesirable traits from its parents but also a healthy child that has inherited the most desirable traits from its parents. Mating process controls the evolution of the population of one generation of genes to that of the next.

17.4   Mutation

Mating exclusively by itself is a grave problem as it causes in-breeding and propagation of genetic defects. If all one does is just mate the candidates present to propagate genes through generations, one will get trapped inside the neighborhood of a "local optimum": a solution that looks fine but is not the best or the most desirable in most cases.

We may think of the world as a physical environment in which these genes exist. There are many peaks and valleys in the optimization function. The lowest valley is the best solution out of all. But one can get trapped in any valley that appears deep as compared to the surrounding hills (not the optimum position). Such a strategy is like initiating a set of balls in random places on the slopes of a mountain range having multiple hills and valleys, causing the balls to go downhill. The balls should ideally stop at the deepest valleys. However, in the event, most of the balls will top at some local valleys which are still pretty high up the hill (local optimum). The goal is then to ensure that a minimum of one ball stops at the lowest point on the map: the minimum globally. Because the balls begin at random locations, it is difficult to know where which ball is caught up. However, what we can do is to arbitrarily kick a number of balls and push them out of the local minima. It will be like giving a momentum to the stuck balls.

This is what is known as mutation. This is a completely random operation, by which a random chromosome is selected and treated with only enough stimulation for a random modification to one of its characters.

Here is an illustration—suppose the following two chromosomes show up:

Hfllr, worlb!

Hfllr, worlb!

This is an abstract instance again, but it occurs. The two chromosomes have become exactly identical to each other. It means that their offspring remain just the same as their parents, and there is no evolution. But in the case that one out of every hundred chromosomes has a character that undergoes random mutation, then given sufficient time, the second chromosome will inevitably and eventually turn into "Ifllr, worlb!" after which the evolution continues, since the offspring will finally be different from the parents. Mutation enables progress in the evolution.

17.5 The Population

Multiple groups of chromosomes are present in a population. It generally retains the same size and gradually undergoes evolution over time to better average cost rates. We have the right population size to choose from. The evolution has "generations." A standard generation comprises the following:

(i) Calculation of the cost/fitness function for every chromosome in the population
(ii) Sorting the chromosomes by their cost/fitness score
(iii) Forcefully killing a specified number of the weakest chromosomes
(iv) Mating a certain number of the strongest chromosomes to produce offsprings
(v) Mutating random chromosomes using a mutation function
(vi) Performing a specific chosen completeness test—i.e., identifying whether the task at hand has yet been "solved"!

A population is easier to start, i.e., only filled with arbitrary chromosomes. Once we agree on the requirement to interrupt population growth, problems get complicated. In the above state particular case, it is somewhat easy: If we reach 0 cost, we terminate the process. However, this thumb rule is only applicable sometimes. Most of the times, we have no foresight on final minimal cost that would be appropriate to be expected. And sometimes even if we use fitness function rather than cost function, we cannot achieve the optimal level fitness. In these cases, a prerequisite of completeness should be specified. For example, we terminate the process if the highest score has not improved in thousands of generations.

Now that we have a simpler intuition of how evolutionary algorithms work, let us proceed to genetic algorithms.

17.6   Genetic Algorithms

Genetic algorithms (GAs) were first designed by John Holland in 1975 [7] to model the natural evolution—development of highly complex, highly fitted organisms from lower complex ones. Genetic algorithms are versatile and efficient systems that rely on natural genetic manipulation. These act as evolutionary models and try to emulate those natural evolutionary processes.

(i) Natural evolution operates on encoding of biological entities in the form of genes. Similarly, GAs operate on string representation of possible solutions (individual/chromosomes).
(ii) Selection obeys Darwinian survival of the fittest strategy.
(iii) Nature acts as environment. Objective function information plays the role of environment.
(iv) Variation is introduced mainly through recombination (crossover) and mutation.

However the original GAs developed by Holland had distinct features:

(i) a bit string representation,
(ii) proportional selection and
(iii) crossover as the primary method to produce new chromosomes.

Goldberg et al. [6] have done several changes to the original Holland's GA, which use different representation schemes, selection, crossover, mutation, and elitism operators. A simple genetic algorithm that yields good results in many practical problems is composed of three operators:

(i) Reproduction,
(ii) Crossover,
(iii) Mutation.

17.6.1   Reproduction

Reproduction is a mechanism in which entities (i.e., chromosomes) are replicated based on their specific functional values f (also known as the fitness function). Intuitively, the function f is something we want to optimize for gain or perfection. Chromosomes having higher scores are more likely to contribute one or more relatives in the next generation, as per their health values. In reality, this operation is a heuristic form of natural selection, which reflects Darwinian theory of the survival of the fittest. Based on the above idea, the simple genetic algorithm for function minimization can be summarized in the following algorithm.

(a) randomly initialize population(t) at any instant t
(b) determine fitness of population(t)

(c) repeat

> (i) select parent pairs from population(t)
> (ii) perform crossover on selected parent pairs creating population(t+1)
> (iii) perform mutation on individual of population(t+1)
> (iv) determine fitness of population(t+1)

(d) until best individual is good enough.

17.6.2 Chromosome Encoding

This is among the most significant aspects of a genetic algorithm. There are several methods to depict individual genes such as binary encoding, permutation encoding real-valued encoding, etc. Here we will only discuss the aforesaid three encoding schemes.

17.6.2.1 Binary Encoding

This is the most commonly used encoding scheme in GA. Holland (1975) worked mainly with binary encoding for solving optimization problem (Table 17.1).

For instance, if we want to code the variables of a function having two parameters assuming four bits are used for each variable, we represent the two variables x_1 and x_2 as $< 1010\ 0010 >$. In optimization problem, each variable x_i will have a fixed range given by $x_i^L \leq x_i \leq x_i^U$.

As we know, a four bits string bijectively maps to the integers 0–15 (16 search points) in an obvious linear fashion, and hence (0000 0000) and (1111 1111) correspond to the points for x_1 and x_2 as $(x_1^L, x_2^L); (x_1^U, x_2^U)$, respectively, the reason being that sub-strings (0000) and (1111) are the lowest and the highest permitted decoded values. Hence each n-bit string corresponds to a unique integer from 0 to $2^n - 1$, i.e., 2^n integers, assuming that x_i is encoded as a sub-string b_i of length n_i. The decoded form of the binary sub-string b_i is computed to be:

$$\sum_{j=0}^{n_i-1} 2^j b_{ij};$$

b_{ij} being either 1 or 0 and the sub-string is represented as

$$b_{ij-1}, \ldots, b_{i2}, b_{i1}, b_{i0}.$$

Table 17.1 Binary-encoded chromosomes	
Chromosome 1	100101010101010
Chromosome 2	010000100011110

Knowing that x_i^L and x_i^U are encoded as (0000) and (1111), the equivalent value for any 4-bit string is thus:

$$x_i = x_i^L + \frac{x_i^u - x_i^L}{2^{n_i} - 1} \times \text{(Decoded value of String)}.$$

Assume for a variable x_i $x_i^L = 1$, and $x_i^u = 31$, to find what value of four-bit string of $x_i = (1010)$ would encode. First we get the decoded value for bias 10.

$$x_i = 1 + \frac{31 - 1}{2^4 - 1} \times 10 = 21.$$

Therefore, the performance achieved by a four-bit code is just $1/16^{th}$ of the entire search space. The resultant performance is exponentially increased to $1/32^{th}$ of the search area, as a consequence of an expansion in string size by one. All variables may not be mandatorily coded with the same length. The size of a sub-string that represents a specific variable depends on how significant the variable is. By generalizing this concept, we can say that the accuracy of a variable approximation is $(x_i^u, x_i^l)/2^{n_i}$ for n_i bit length coding.

17.6.2.2 Permutation Encoding

Permutation encoding is used in problems named ordering problems such as task ordering problem or traveling sales person problem. In a permutation encoding, every individual/chromosome is basically a string of numbers, which represents the number in the sequence as shown in Table 17.2.

Even for ordering problems after applying for sometimes the genetic operators, rectifications become necessary for the sake of consistency of the chromosome.

17.6.2.3 Value Encoding

In this scheme, each chromosome is a string of some characters, and these characters can be anything associated with the problem. The different types of value-encoded chromosomes are given in Table 17.3.

Table 17.2
Permutation-encoded
chromosomes

Chromosome 1	1 5 3 7 8 2 4
Chromosome 2	2 3 1 6 4 5 7

Table 17.3 Value-encoded
chromosomes

Chromosome 1	5.432.7891.22356.78
Chromosome 2	5.671.2346.782.134

Value encoding is very good for some special problems like data clustering—one of the primary tasks of pattern recognition. On the other hand, this encoding is often necessary to develop new genetic operators specific to the problem.

17.6.3 Selection

The first operator to be introduced to population is typically reproduction. Chromosomes are chosen to be parents and reproduce within the population to generate offsprings. The strongest should sustain and produce new descendants, as stated in Darwin's theory of natural selection. This is why it is often regarded as the operator for selection. There are many reproductive operators in GA, but the basic idea is that the above-average strings are selected from the existing population and their many copies are probabilistically placed in the pairing pool. There are various techniques to select chromosomes as parents to crossover are:

   (i) Roulette-wheel selection
  (ii) Tournament selection
 (iii) Rank selection
  (iv) Steady-state selection
   (v) Boltzmann selection
  (vi) Elitism

17.6.3.1 Roulette-Wheel Selection/Proportionate Selection

In roulette wheel selection, the fitness value determines the selection of parents. Better are the chromosomes, the more is the chance of them getting selected. A strategy to accomplish this selection process is to picture a disk with its boundary that is directly proportional to the fitness for each chromosome. The fitness of the entire population is determined by spinning the disk (also called roulette wheel) 'n' times. Every time the roulette wheel is rotated, the string instance is chosen to which the pointer points. Figure 17.1 illustrates the mechanism with five fitness values.

As the third individual has the highest fitness value, it is also the most likely to get selected by the roulette wheel selection.

Such a process can be simulated by the following algorithm:

   (i) *Calculate $S = \sum_{i=1}^{n} F_i$. F_i is the functional value of chromosome i.*
  (ii) *Generate a random number $r = (0, S)$*
 (iii) *For $i = 1$ to n*
  (iv) *sum = sum+ F(i);*
   (v) *if (sum > r)*
  (vi) *return i;*
 (vii) *The selected chromosome is i.*

Fig. 17.1 Roulette wheel
marked for five individuals

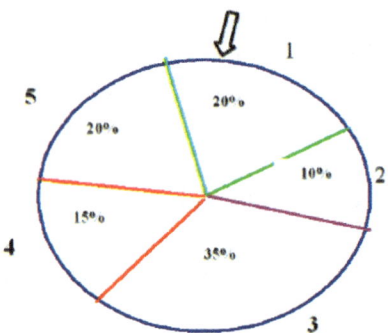

17.6.3.2   Tournament Selection

The evolution process of genetic search faces two important issues, *population diversity* and *selective pressure* [4]. When the genes from the already recognized good individuals are utilized, it is defined as population diversity, while the promising unexplored regions of the search space are scanned through. The extent of favoritism toward better individuals for selection is quantified by what is termed as selective pressure.

Greater selective pressure favors the better individuals more, thus directing GA toward more desirable population fitness in the futurex generations. The selective pressure and population diversity largely determine the convergence rate of GA. By and large, increasing selective pressure increases convergence rates. Unfortunately, this also increases the chance of prematurely converging to a local minimum, especially if the selective pressure is set too high, as much of population diversity in a given generation will not be able to survive or pass on their genes into the next generation to explore the search space.

On the other hand, the GA decreasing the selective pressure increases the time to converge to an optimal solution as it slows down the convergence rate as more genes are explored in the search space. An optimal selection strategy thus needs to incorporate fine-tuning GA search performance so as to enable adjusting selective pressure and population diversity.

The fitness proportional selection (e.g., roulette wheel selection) pointed out by Darrel Whitley [10] likely leads two problems, namely,

(i) Stagnation of search because it lacks selection pressure
(ii) Premature convergence of the search because it causes the search to narrow down too quickly.

Goldberg and Deb [5] contributed one-selection strategies to the GA community known as tournament selection to overcome this difficulty. The selection scheme works as: Choose t number of individuals chromosomes from the population, and copy the best individual to the intermediate population, and repeat this procedure until the population pool is full. Here t is called tournament size. If tournaments are held only between two individuals, then it is treated as binary tournament. A simulation of this scheme is as follows:

Tournament_Selection (t, pool)

(i) *Select t individuals from the pool*
(ii) *Copy best individuals to the intermediate pool*
(iii) *Until the pool is full repeat from i*

17.6.4   Rank Selection

Rank-based fitness assignment is performed in the following way:

(i) According to rank, the population is sorted.
(ii) According to some function, usually linear but not necessarily, the fitness to individuals is assigned by interpolating from the best (rank 1) to the worst (rank n $\leq$ N) in the usual way
(iii) The fitness of individuals with the same rank is to be averaged, so that all of them will be sampled at the same rate. This procedure keeps the global population fitness constant while maintaining appropriate selection pressure, as defined by the function used.

17.6.4.1   Steady-State Selection

This strategy attempts to make larger part of chromosome survive to the next generation. Each generation has a few good individuals with higher fitness that are chosen for passing on their genes to the next generation via their offspring. Subsequently, some undesirable low fitness chromosomes are eliminated, and new offsprings are inserted in the population.

17.6.4.2   Boltzmann Selection

Simulated annealing is a strategy that is a type of functional optimization. This strategy simulates the slow cooling of molten metal to score the lowest possible functional value in a minimization problem. The physics of cooling is simulated by actuating a parameter that represents temperature, paired with the idea of Boltzmann probability distribution, so that a system in thermal equilibrium with temperature T has a random energy with probability distribution as given below:

$$P(E) = e^{-E/kT};$$

where k is Boltzmann constant. This expression suggests that a system lowering temperature reduces the probability of a high energy state, but at higher temperatures, the probability acquires a uniform distribution across all energy states. Therefore, it is possible to control the convergence of the algorithm by manipulating the temperature T and enforcing a Boltzmann probability distribution on the search process.

17.6.4.3 Elitism

In this method, at first, the best chromosomes or few best chromosomes (obtained until now from the beginning) are copied to new population. The rest of the process is done in the classical way. The performance of GA can be very rapidly increased by Elitism, because it prevents losing the best-found individuals.

17.6.4.4 Generation Gap and Steady-State Replacement

The fraction of the population which is replaced in every generation is defined by the generation gap. So far, we have been doing reproduction with a generation gap equal to 1, i.e., the entire population is replaced every generation. However, more recently, steady-state replacement has been gaining more popularity, which is given by Whitley in 1987 and 1989 [8]. This is in stark contrast to the reproduction strategies we have been following so far, as it only replaces a small fraction of the population each generation. This may be an improved model, because it considers the aspect of nature. Short-lived species including insects exemplify the former reproduction strategy as parents lay eggs and then pass away before their eggs hatch. So the parents are unable to nurture their offspring. But in longer-lived species including mammal's offspring, the offsprings and parents can live simultaneously. This enables parents to take care of and teach their kids but also themselves contribute to competition among them. Generation gap can be classified as

$$G_p = \frac{P}{N_P};$$

where N_p is the population size and P is the number of individuals that will be replaced. Several schemes are possible. Some of which are the following:

(i) Selection of parents according to fitness and selection of replacement at random.
(ii) Selection of parents at random and selection of replacement by inverse fitness.
(iii) Selection of both parents as well as replacements according to fitness/inverse fitness.

Generation gap can be gradually increased as the evolution takes place to widen exploration space and may lead to better results.

17.6.5 Crossover

The population is filled with better candidates after the reproductive process is over. Reproduction produces good string cloning, but no new strings are generated. The crossover operator is used with the hope of producing better strings inside the mating pool. The crossover operator is supposed to scan into the parameter space. Furthermore, the scan should be carried out to maximize the preservation of the information stored in this string, because both the parent strings are examples of desirable strings chosen during the past generation.

Three steps are taken by the crossover operator (also referred to as the recombination operator). First, the operator must pick a pair of two different strings for arbitrary reproduction. Then, a cross-point will be randomly selected along the string length, and the positional values will be modified in both the strings following the cross-point.

There exist many types of crossover operations in genetic algorithm, which are discussed in the following sections.

17.6.5.1 Single-Point Crossover

A cross-point is chosen randomly over a single point overlap lengthwise of the parent strings and bits are swapped across cross-point, as shown in Fig. 17.2. When a suitable point is selected, a combination of good parent factors will produce better offsprings. Given that it is not known and the point of crossover was randomly selected, it can lead to better kids only if the position of the selected point is appropriate. If not, the quality of the string may be severely impaired. Nevertheless, stronger children are created because of the crossover of parents, and this will proceed even in the next generations. But if crossover does not create good strings, the reproduction will reject those strings from entering the subsequent mating pool, and thus, they will not survive beyond next generation.

17.6.5.2 Two-Point Crossover

A two-point crossover operator chooses two arbitrary cross-points and swaps the parts of these cross-points between two parents. If the first cross-point is at two and

Fig. 17.2 Single-point crossover

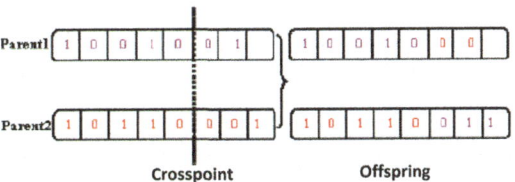

Fig. 17.3 Two-point crossover

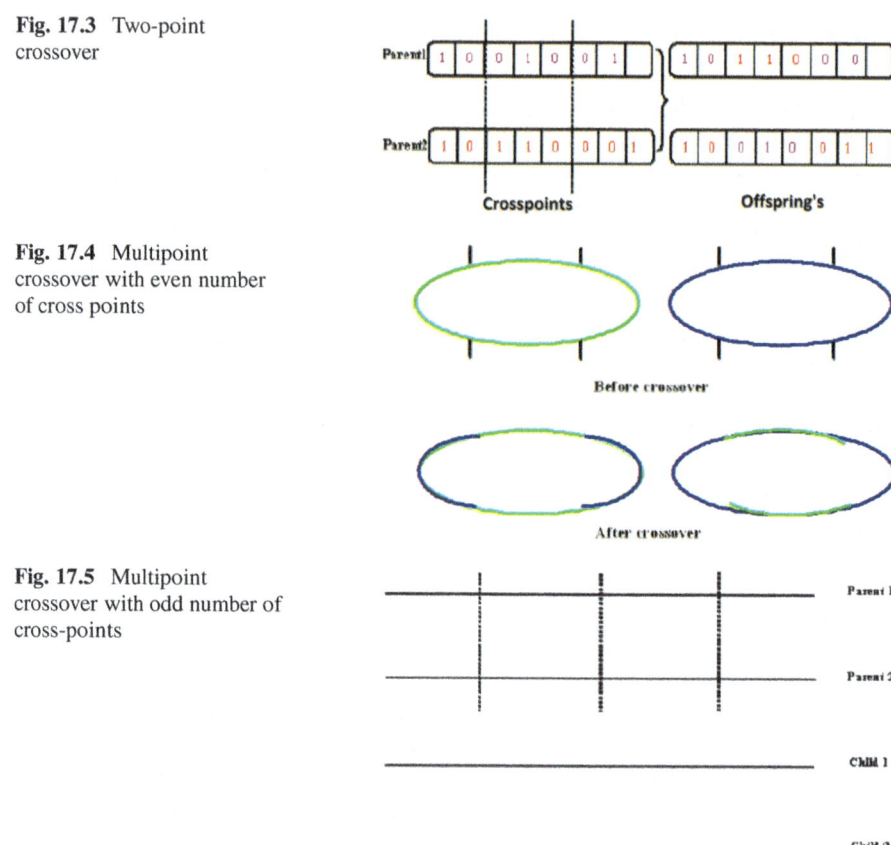

Fig. 17.4 Multipoint crossover with even number of cross points

Fig. 17.5 Multipoint crossover with odd number of cross-points

the second cross-point is at five, the strings between two and five are swapped as shown in Fig. 17.3.

17.6.5.3 Multipoint Crossover

In a multipoint crossover, once more, there may be two scenarios, that is, the number of cross-points can either be even or odd. In the former case, the string has no start or end; it is treated as a ring. The cross-points are selected at random uniformly around the circle. Now as shown in Fig. 17.4, the information between alternative pairs of points are interchanged. Now, when the number of cross-points is odd, then at the beginning of the string, a different cross-point is always assumed. The information (genes) is shared by alternate pairs with each other as shown in Fig. 17.5.

Fig. 17.6 Uniform crossover with odd number of cross-points

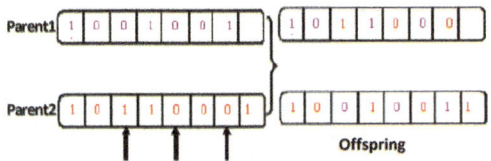

17.6.5.4   Uniform Crossover

The uniform crossover operator is an extreme case of multipoint crossover. In a uniform crossover operator, as shown in Fig. 17.6, each bit is chosen from either parent with a probability of 0.5 and then interchanged. It has been seen that uniform crossover is unlike anything one-point crossover. Sometimes the offspring's gene is a duplication of the corresponding gene from one of the parents selected according to a randomly generated crossover mask. The presence of one or zero in the mask makes the algorithm duplicate the gene from the second or the first parents, respectively. The genes for the second offspring are decided by reversing the order of the parents in this process and repeating it. For each pair of parents, a new crossover mask is randomly generated. Therefore, a mixture of genes from each of the parent is contained in the offspring.

17.6.5.5   Crossover Rate

In GA literature, the term crossover rate is normally represented by P_c, the probability of crossover. The probability lies within a range of 0–1, computed in GA using the ratio of the number of pairs to be crossed to some fixed population. Usually, crossover lies between 0.5 and 1 for a population having size of 30–200.

It has been seen that with random cross-points, the cross-point getting assigned to the appropriate position influences the desirability of the sub-strings from parent strings the combination of which gives the children's string. But there is nothing to be worried about, because if crossover has created good strings, the reproduction operator will generate more duplicates. But in case good strings are not produced by the crossover, they will be swiftly wiped out from the population, since reproduction will not favor selection of such strings in latter generations. Thus this discussion makes it clear that the crossover has either detrimental effect or it has beneficial effect, which is why some strings are not used in the crossover so that some of the good strings already there in the mating pool are conserved.

For the crossover probability being equal to P_c, the crossover operation is applied on $100P_c$ percent of the strings in the population and the rest $100(1-P_c)$ percentage of the population is conserved as it is. Despite the fact that it is possible to copy deterministically the best $100(1 - P_c)\%$ of the current population to the new population, this is normally done randomly. A crossover operation is primarily in charge of looking for new strings.

17.6.6   *Mutation*

Post the crossover operation, mutation operation is applied to the strings. Bit mutation is the process of flipping the bit, changing the bit value from 0 to 1 and vice versa with a small mutation probability P_m. By flipping a coin with a probability of P_m, the bit-wise mutation operation is performed bit by bit. Flipping a coin with a probability of P_m is simulated as follows. A number is chosen randomly between 0 and 1. If the chosen random number is smaller than P_m (the mutation probability), then the outcome is true; otherwise, the outcome is false. If at any bit the outcome comes as true, then the bit is altered; otherwise, the bit is kept as it is (unchanged). The bits of the strings undergo independent mutation, that is, when one bit goes through the mutation process, it has no effect on the probability of mutation of other bits. A simple genetic algorithm uses the mutation only as a secondary operator with the purpose of bringing back genetic materials no longer available. Consider the instance where every string in a population has conveyed to a zero at a given position and the optimal solution has a one at that position, then crossover cannot regenerate a one at that position, while a mutation could. Thus against the irreversible loss of genetic material, mutation is an insurance policy. New genetic structures are introduced in the population by the mutation operator introduces by randomly modifying some of its building blocks. Since the modification is not related to any of the previous genetic structure of the population, it helps the search algorithm to escape from the traps of local minima. It creates different structures representing other section of the search space. Mutation is also useful in maintaining the diversity in the population.

17.6.6.1   Mutation Rate

The probability of mutation is termed as the mutation rate, which calculates the number of bits that are to be mutated. Preserving the diversity among population is very important for the search, and the mutation operator helps in doing that. In natural populations, the mutation probabilities are generally smaller from which it can be inferred that mutation is appropriately considered as a secondary mechanism of genetic algorithm adoption. Usually, a simple genetic algorithm restricts the population size to a range of 30–200 with the mutation rates lying within the range 0.001–0.5.

References

1. Banzhaf, W., Nordin, P., Keller, R. E., & Francone, F. D. (1998). *Genetic programming: An introduction* (vol. 1). Morgan Kaufmann.
2. Eiben, A. E., Smith, J. E. (2003). *Introduction to evolutionary computing* (vol. 53). Springer.

3. Fogel, D. B., Fogel, L. J., & Atmar, J. W. (1991). Meta-evolutionary programming. In *IEEE conference Record of the twenty-fifth Asilomar conference on signals, systems and computers* (pp. 540–545).
4. Gen, M., & Cheng, R. (2000). *Genetic algorithms and engineering optimization* (vol. 7). Wiley.
5. Goldberg, D. E., & Deb, K. (1991). A comparative analysis of selection schemes used in genetic algorithms. *Foundations of Genetic Algorithms, 1*, 69–93.
6. Goldberg, D. E., & Holland, J. H. (1988). Genetic algorithms and machine learning. *Machine Learning, 3*(2), 95–99.
7. Holland, J. H. (1975). *Adaptation in natural and artificial systems. An introductory analysis with application to biology, control, and artificial intelligence.* University of Michigan Press.
8. Schaffer, J. D., Whitley, D., & Eshelman, L. J. (1992). Combinations of genetic algorithms and neural networks: A survey of the state of the art. In *International workshop on combinations of genetic algorithms and neural networks, COGANN-92* (pp. 1–37).
9. Schwefel, H.-P. P. (1993). *Evolution and optimum seeking: The sixth generation.* Wiley.
10. Whitley, D. (1994). A genetic algorithm tutorial. *Statistics and Computing, 4*(2), 65–85.

Chapter 18
Support Vector Machine

Support vector machines (SVMs) were introduced in COLT-92 by Boser, Guyon, and Vapnik [3]. SVM is related to statistical learning theory [11]. SVM is now seen as an illustration of "kernel techniques," among the most significant topics of machine learning. A support vector machine (SVM) is a classifier based on the discriminative concepts which generate a class separating hyperplane. This means that the algorithm generates an efficient hyperplane which categorizes new unlabeled instances after being trained with labeled data (supervised learning).

In a simple minimum distance classifier (MDC discussed in Chap. 8), the decision boundary is the perpendicular bisector of the straight line joining the mean of the two classes. Minimum distance classifier (MDC) performs well if the two classes are nonoverlapping and the spread or variance of the classes are nearly equal or comparable as shown in Fig. 18.1a. If one of the classes has large variance compared to the other, the minimum distance classifier misclassifies some data points (Fig. 18.1b). This is a drawback of the minimum distance classifier; SVM can handle this issue given that the classes are nonoverlapping.

SVM finds out the maximum separation between two classes and draws the decision boundary or separating hyperplane so that it falls along the middle of the region of separation. So, position of the decision boundary (separating hyperplane) is not dependent on the variances of the individual classes as shown in Fig. 18.1c.

18.1 Linear SVM

The data points which are positioned nearest to the decision boundary are known as support vectors. These are the most complicated to identify as a member of any one of the classes. Their effects directly relate to the ideal positioning of the decision surface. For example, we can consider a two-class linearly separable classification problem.

© The Author(s), under exclusive license to Springer Nature Singapore Pte Ltd. 2025   297
A. Ghosh, *Data Science and Cases in Sustainability*, Mathematics for Sustainable
Developments, https://doi.org/10.1007/978-981-96-8362-8_18

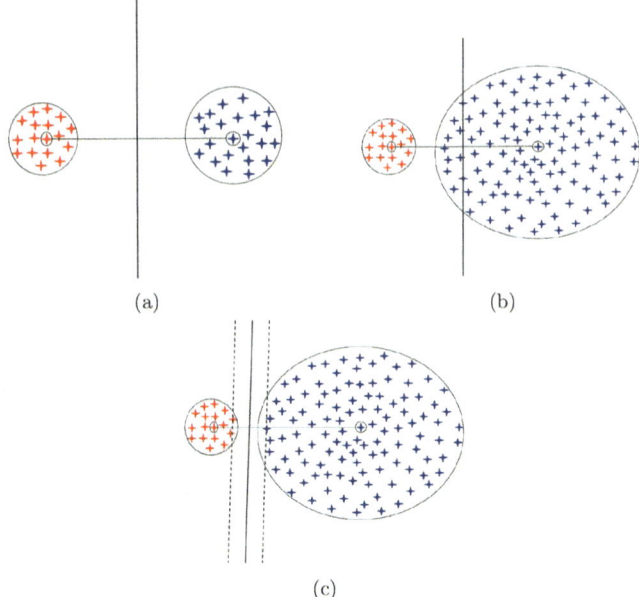

Fig. 18.1 Advantage of SVM over MDC (**a**) MDC applied on separable classes with similar/comparable variances, (**b**) MDC applied on separable classes with different variances, (**c**) SVM applied on separable classes with different variances

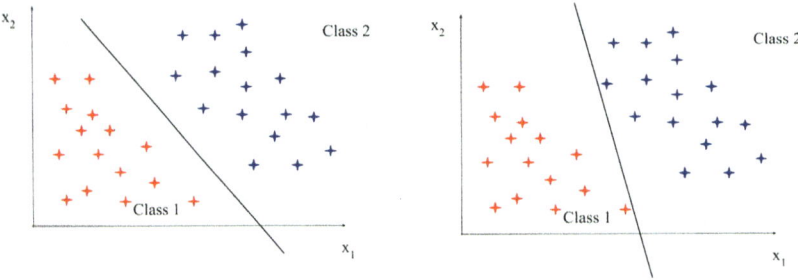

Fig. 18.2 A two-class, linearly separable classification problem

There may be many decision boundaries based on different algorithms (Fig. 18.2). But the question that arises now is whether all decision boundaries are equally good or not! A line is bad if it passes too close to the boundary points, because it will be noise sensitive and it will not generalize properly.

Therefore, our goal should be to find the line passing as far as possible from all boundary points. The best decision boundary is the one that is far away from the boundary data points of both classes (Fig. 18.3). Support vector machine (SVM) finds such an optimal boundary.

Fig. 18.3 Optimal
hyperplane

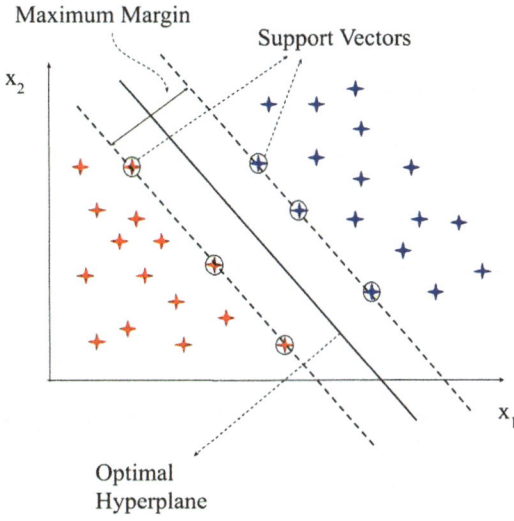

In Fig. 18.4, the black hyperplane gives smaller margin, and the green hyperplane is the optimal one as it maximizes the margin. The SVM algorithm is focused on finding the hyperplane which provides the highest minimum distance to all boundary points among the training samples. As per the theory of SVM, the double of this distance is known as the **margin**. The optimum separating hyperplane thus maximizes the margin for the training set. Support vectors are these boundary input vectors, and they satisfy

$$\mathbf{w}^T \mathbf{x} + w_0 = 1 \tag{18.1}$$

or,

$$\mathbf{w}^T \mathbf{x} + w_0 = -1. \tag{18.2}$$

where $\mathbf{w}$ denotes the weight vector for the decision hyperplane in Fig. 18.5. The decision function is completely specified by the crucial subset of training samples, which are known as the support vectors. Support vectors are therefore the critical samples in the training set.

Optimized Hyperplane

For an optimal separating plane, the margin is maximum. The optimization problem is suitably chosen as the following:

$$\text{minimize } J(\mathbf{w}, w_0) = \frac{1}{2}||\mathbf{w}||^2; \tag{18.3}$$

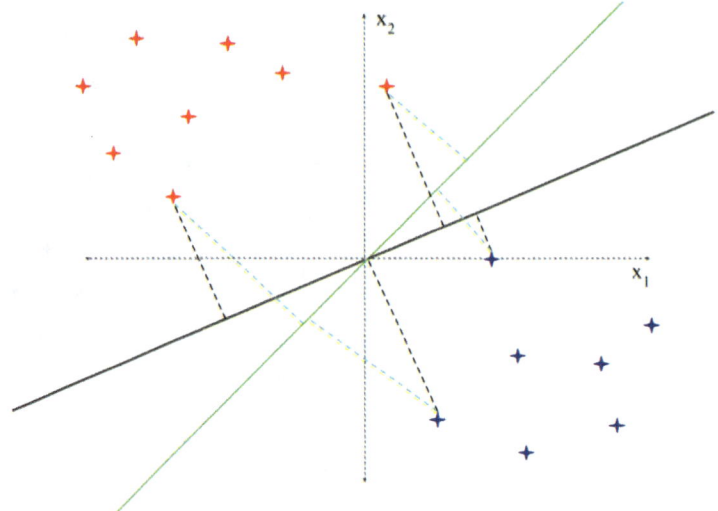

Fig. 18.4 Choosing the hyperplane that maximizes the margin

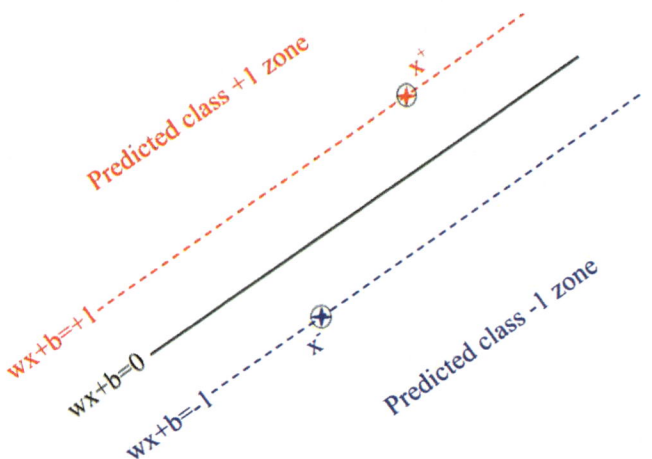

Fig. 18.5 Solving the optimization problem

subject to, $y_i(\mathbf{w}^T + w_0) \leq 1$, $i = 1, 2, \cdots, N$, where $J(.)$ is the cost function. From the equation, we see that this is accomplished by minimizing $|\mathbf{w}|$ (subject to the distance constraints). This minimization problem is what is a quadratic programming problem [7]. Fortunately, many techniques have been developed to solve them [6].

Figure 18.5 nicely explains the goal of the SVM that is to maximally separate the data points belonging to two different classes. To achieve this, SVM finds out two parallel lines at maximum margin $||\mathbf{w}||$ from each other that separates the data points belonging to two different classes and then finds out the line in the middle of these two lines. The solution comprises of engineering a dual problem [8] where a Lagrange multiplier α_i [2] is put together with every constraint in the primary problem [9] thus combining it into one expression. The solution has the form:

$$\mathbf{w} = \sum \alpha_i y_i \mathbf{x}_i; \tag{18.4}$$

where y_i is the predicted class for the ith training data and

$$w_0 = y_k - \mathbf{w}^T \mathbf{x}_k; \tag{18.5}$$

for any $\mathbf{x}_k$ such that $\alpha_k \neq 0$. Each nonzero α_i indicates that the corresponding $\mathbf{x}_i$ is a support vector. Then the classifying function will have the form

$$f(\mathbf{x}) = \sum \alpha_i y_i \mathbf{x}_i^T \mathbf{x} + w_0. \tag{18.6}$$

It heavily relies on an inner product of the test point $\mathbf{x}$ and the support vectors $\mathbf{x}_i$. In order to solve the optimization problem, one would need to compute the inner products $\mathbf{x}_i^T \mathbf{x}$ between all possible pairs of training points.

18.2 Nonlinear SVM

SVM is a linear classifier that gives an optimal hyperplane to separate data points from two classes. Data sets that are linearly separable with some noise can be efficiently classified using SVMs. If some dataset is not linearly separable by an SVM in d-dimensions, then there exists a mapping like

$$\mathbf{x} \in \mathbb{R}^d \rightarrow \mathbf{y} \in \mathbb{R}^k, k > d$$

which maps the input feature space into a k-dimensional space, where the classes can be separated by a hyperplane.

The general idea is that the original input space can always be mapped to some higher-dimensional feature space where the training set is separable (Fig. 18.6).

Kernel Functions

Kernel is essentially a mapping function—one that transforms data from a given space into some other (usually very high dimensional) space satisfying certain conditions. An architecture of SVM using kernel functions is shown in Fig. 18.7.

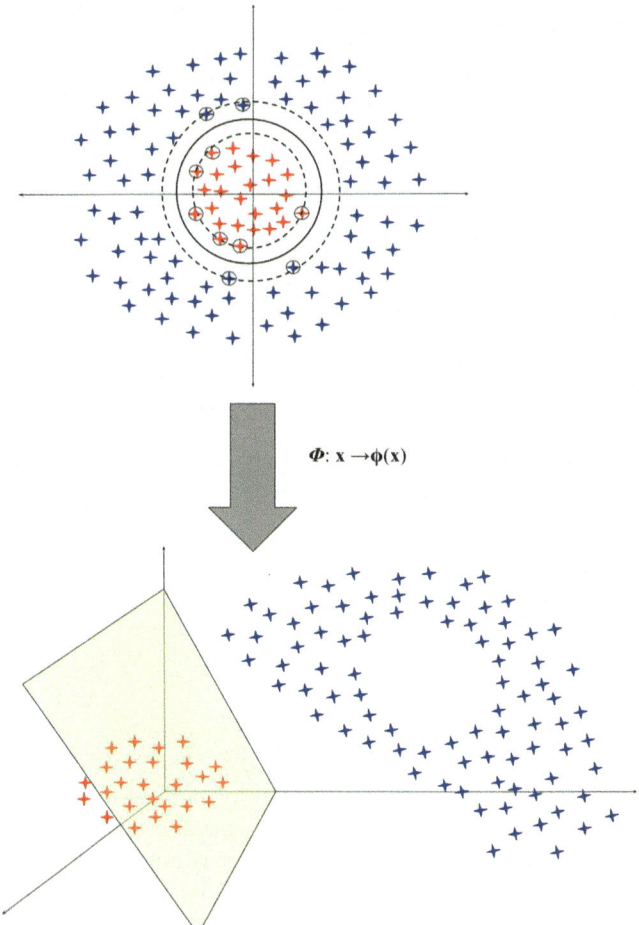

Fig. 18.6 Nonlinear SVMs: Feature spaces

The assignment rule is as follows:

Assign $\mathbf{x}$ to ω_1 if $\sum_{i=1}^{N_s} \lambda_i y_i \phi(\mathbf{x_i}, \mathbf{x}) + w_0 > 0$;
assign $\mathbf{x}$ to ω_2 otherwise.
Here, N_s is the number of support vectors, and ω_1 and ω_2 are the two classes.

Such kernel functions can be of different types:

(i) **Polynomial**: $\phi(\mathbf{x}, \mathbf{z}) = (\mathbf{x}^T \mathbf{z} + 1)^q$, $q > 0$.

(ii) **Radial Basis Function**: $\phi(\mathbf{x}, \mathbf{z}) = \exp\left(-\dfrac{||\mathbf{x} - \mathbf{z}||^2}{\sigma^2}\right)$.

(iii) **Hyperbolic Tangent**: $\phi(\mathbf{x}, \mathbf{z}) = \tanh\left(\beta \mathbf{x}^T \mathbf{z} + \gamma\right)$.

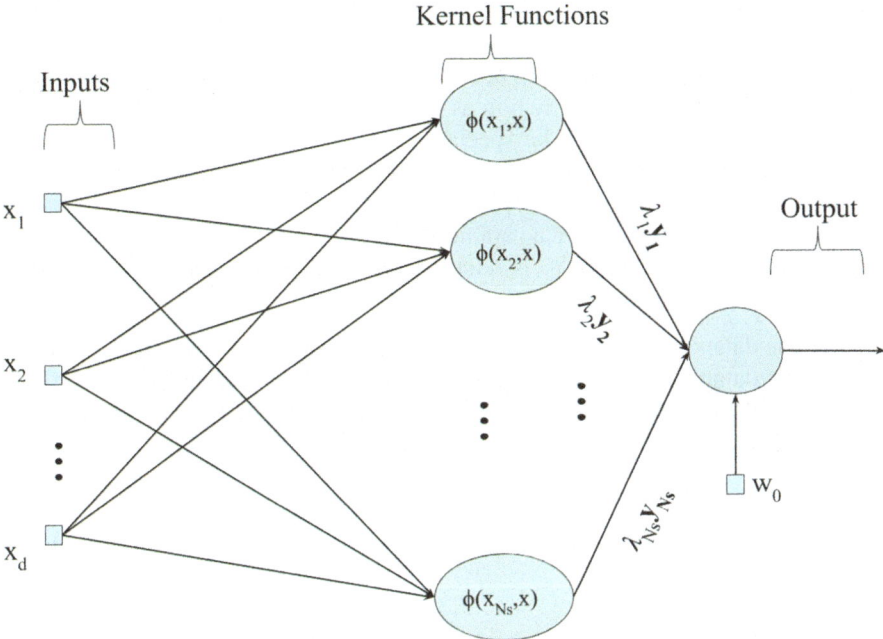

Fig. 18.7 SVM architecture using kernel functions

18.3   Properties of SVM

The clear connection to statistical learning theory, efficient performance, and simple usage of SVMs has attracted many researchers to investigate its properties. Many literatures have shown that SVMs are the state-of-the-art method in solving classification and regression problems. It has various applications in text (and hypertext) categorization [10], image classification [12], bioinformatics (protein classification, cancer classification) [4, 5], hand-written character recognition [1], etc. They have gained popularity due to many attractive features some of which are listed below.

(i) Flexibility in choosing a similarity function.
(ii) When dealing with large data sets, there is sparseness of solutions, i.e., only support vectors are used to specify the separating hyperplane. While dealing with large datasets (N is very high), the solution that SVM provides is of very sparse nature. This is because only the support vectors are used to represent the hyperplane and the number of support vectors (N_s) is very less in comparison with the number of data points (N) in the dataset.
(iii) Ability to handle large feature spaces as complexity does not depend on the dimensionality of the feature space. The time complexity of SVM is $O(N^3)$, where N is the number of data points in the training set. So it has the ability to

handle large feature space as the complexity does not depend on the number of dimensions d.

(iv) Overfitting can be controlled. The inherent nature of the data is sometimes non-separable at the boundary region as shown in the figure below. For handling such cases, the SVM can be specially modified to avoid overfitting. In this new setting, the mixture of the data from both the classes is allowed inside the region of margin but not outside the margin. The size of the margin becomes variable here depending on a parameter C'.

A new slack variable ξ_i is adopted for the ith datapoint. A datapoint is divided into three categories:

(a) the datapoints lying outside the band and are correctly classified, $\xi_i = 0$.
(b) the datapoints lying inside the band and correctly classified, $0 < \xi_i \le 1$.
(c) the datapoints lying inside the band and wrongly classified, $\xi_i > 1$.

The modified cost function here is as follows:

$$\text{minimize } J(\mathbf{w}, w_0, \boldsymbol{\xi}) = \frac{1}{2}||\mathbf{w}||^2 + C' \sum_{i=1}^{N} \xi_i; \tag{18.7}$$

subject to $y_i \left[\mathbf{w}^T \mathbf{x_i} + w_0 \right] \ge 1 - \xi_i, i = 1, 2, \cdots, N$.

The classifier can be tuned by changing the value of C' as shown in Fig. 18.8.

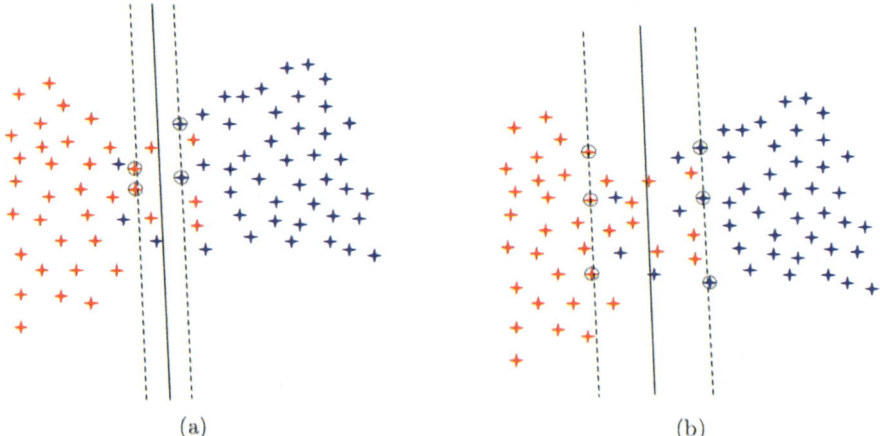

(a)   (b)

Fig. 18.8 Tuning the SVM classifier (a) C'=0.5, the band size decreased and less points are allowed inside it, (b) C'=1000, the band size increased and more points are allowed inside it

References

1. Bahlmann, C., Haasdonk, B., & Burkhardt, H. (2002). Online handwriting recognition with support vector machines-a kernel approach. *Eighth international workshop on frontiers in handwriting recognition* (pp. 49–54).
2. Bellman, R. (1956). Dynamic programming and lagrange multipliers. *Proceedings of the National Academy of Sciences, 42*(10), 767–769.
3. Boser, B. E., Guyon, I. M., & Vapnik, V. N. (1992). A training algorithm for optimal margin classifiers. *Proceedings of the fifth annual workshop on computational learning theory* (pp. 144–152).
4. Byvatov, E., & Schneider, G. (2002). Support vector machine applications in bioinformatics. *Applied Bioinformatics, 2*(2), 67–77.
5. Dixit, P., & Prajapati, G. I. (2015). Machine learning in bioinformatics: A novel approach for DNA sequencing. *Fifth international conference on advanced computing and communication technologies (ACCT), 2015* (pp. 41–47).
6. Frank, M., & Wolfe, P. (1956). An algorithm for quadratic programming. *Naval Research Logistics Quarterly, 3*(1–2), 95–110.
7. Hildreth, C. (1957). A quadratic programming procedure. *Naval Research Logistics Quarterly, 4*(1), 79–85.
8. Lemke, C. E. (1954). The dual method of solving the linear programming problem. *Naval Research Logistics (NRL), 1*(1), 36–47.
9. Mowrer, O. (1947). On the dual nature of learning a reinterpretation of conditioning and problem solving. *Harvard Educational Review, 17*, 102–148.
10. Taira, H., & Haruno, M. (1999). Feature selection in SVM text categorization. *Innovative applications of artificial intelligence conference (IAAI) sponsored by the association for the advancement of artificial intelligence (AAAI)* (pp. 480–486).
11. Vapnik, V. (2013). *The nature of statistical learning theory.* Springer.
12. Zhou, S. Liang, P., & Qin, J. (2015). Attention region latent SVM for image classification. *2015 International symposium on computers and informatics.*

Chapter 19
Kernel Machines

The raw data interpretation should explicitly be translated into feature vector representations through a user-defined feature mapping for many algorithms which tackle machine learning challenges. On the other side, kernel approaches only allow a user-specified kernel (the function of similarity between pairs of raw data points) and aim to bring two similar data objects much closer and two different data points even further by projecting the data points to a higher dimension.

A kernel can be regarded a sort of a similarity function which is provided by the domain experts to a machine learning algorithm. It inputs two data points or vectors and outputs the extent of similarity they have. For example, if we wanted to classify images, we would have many (image, label) pairs as training data. We would extract features by various means from the images and then feed them into a learning algorithm. Kernels make our task easier. Instead of defining a huge lot of unpredictable features, we can define a single kernel to compute the similarity between two images. Kernels are used to represent the data points in a higher-dimensional space in order to achieve better separability between the data points from different classes. In other words, the main aim behind using a kernel is to move the less similar data points even far away and the more similar data points even more close in higher-dimensional space, so that better separability can be reached at the cost of moving the data points to a higher-dimensional space. The main intuition behind this is the Cover's theorem which states, "*a complex pattern classification problem cast in high dimensional space nonlinearly is more likely to be linearly separable than in low dimensional space*"[1].

19.1   Kernel

Intuitively, a kernel is just a mapping function that transforms the input data from one space to another such that the data can be processed. As shown below in (Fig. 19.1), imagine that we have the toy problem of separating the red circles from the blue circles on a plane

Given below is the example of a kernel function as an inner product similarity measure:

$K : X \times X \rightarrow \mathbb{R}; (x, x') \rightarrow K(x, x')$ where $x, x' \in X$,
$K(x, x') =< \phi(x), \phi(x') >$

where ϕ maps into some inner product space (sometimes called as feature space). The similarity measure K is usually the kernel, and ϕ is called its feature map.

The curve drawn on the left figure would be our separating surface in the original input space (Fig. 19.1). However, it would be far simpler to transform the data into a three-dimensional space by creating the mapping, as the points can now be divided by a simple plane (right side figure of Fig. 19.1). This embedding of input data on a greater dimension is known as the kernel trick.

19.2   Feature Space Mapping

Let us consider a data set which is nonlinearly separable in the original space. As a remedy, the kernel function becomes highly useful at this point, as it is an implicit function mapping the input space to a linearly separable feature space, where the linear classifiers are useful. Let us take an example. Let us suppose a nonlinear mapping function

$$\phi : (I =)\mathbb{R}^2 \rightarrow (F =)\mathbb{R}^3$$

from the two-dimensional input space I into the three-dimensional feature space F, which is defined in the following way:

$$\phi(\mathbf{x}) = (x_1^2, \sqrt{2x_1x_2}, x_2^2)^T. \tag{19.1}$$

Fig. 19.1   Kernel trick

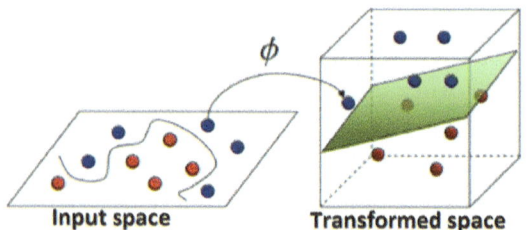

Input space   Transformed space

Considering the above equation, by taking into account a separating hyperplane, we get a linear function in $\mathbb{R}^3$:

$$\mathbf{w}^T \phi(\mathbf{x}) = w_1 x_1^2 + w_2 \sqrt{2 x_1 x_2} + w_3 x_2^2 = 0. \tag{19.2}$$

This is an elliptic function and evaluated in $\mathbb{R}^2$. Hence, by using an appropriate mapping function and not much effort, we can use the linear classifier in F on a transformed version of the data to get a nonlinear classifier in I. As seen in (Fig. 19.1), a linear separating hyperplane can be found after mapping our nonlinearly separable data into a higher-dimensional space.

Kernels are the functions which are used to project the data from lower dimensions to higher dimensions with an expectation to separate the data linearly. The input space (lower dimensional space) is usually denoted by X space, and feature space is usually denoted by Z space. The idea behind the kernel is that we want to go to higher-dimensional Z space without paying actual price for it. In nonlinear transformation, we still have to transform the points and perform the inner product between two z vectors with high dimensions. The only thing we need from Z space is the result of the inner product.

$$D(\boldsymbol{\alpha}) = \sum_{i=1}^n \alpha - \frac{1}{2} \sum_{i=1}^n \sum_{j=1}^n \alpha_i \alpha_j y_i y_j (\mathbf{z_i.z_j}).$$

Example using polynomial kernel is shown below:

Let $\mathbf{x}$ be two-dimensional feature vector $\mathbf{x} = (x_1, x_2)$. Let K be a polynomial kernel with degree two given by

$$K(\mathbf{x}, \mathbf{x'}) = (1 + \mathbf{x.x'})^2 = (1 + x_1 x_1' + x_2 x_2')^2;$$

which will be reduced to

$$K(\mathbf{x}, \mathbf{x'}) = 1 + x_1^2 x_1'^2 + x_2^2 x_2'^2 + 2 x_1 x_1' + 2 x_2 x_2' + 2 x_1 x_1' x_2 x_2'.$$

Now we got everything in terms of x and x', but the coefficients are extra. So our z and z' will become as follows

$$\mathbf{z} = (1, x_1^2, x_2^2, \sqrt{2} x_1, \sqrt{2} x_2, \sqrt{2} x_1 x_2).$$

$$\mathbf{z} = (1, x_1'^2, x_2'^2, \sqrt{2} x_1', \sqrt{2} x_2', \sqrt{2} x_1' x_2').$$

This example is for degree 2 polynomial kernel with two features. In this way, we can map from input space X to feature space Z.

19.3   Why Kernels?

Instead of feature vectors, here more stress is put on kernels. A big reason is that in many situations, it is easy to compute your kernel, but it is very challenging to compute the feature vector for the kernel. Even the basic kernels can have the feature vectors which are of very big dimensions. For kernels, like the RBF kernel

$$K(\mathbf{x}, \mathbf{y}) = \exp\left(-(\mathbf{x} - \mathbf{y})^2\right),$$

the corresponding feature vector is infinite dimensional. However, the kernel computation process is almost trivial.

Several machine learning algorithms can be written using dot products, and then, dot products can be replaced by kernels. By that, we may not at all be using the feature vector. This allows us not to manage the vector with a vast, seemingly infinite dimension, with extremely complex, efficient, and yet powerful kernels. Therefore we are left with relatively lower-dimensional low-performance features if we are not allowed to use kernel functions straight away. Often not only it is computationally expensive but is also impossible to go into higher dimensions. The kernel can be thought of as a mapping from some low dimension to some infinite dimension which we may have very little idea about how to deal with. Then kernel gives us a pretty decent shortcut.

The use of kernels has recently received significant attention in learning systems. The main reason for this is that the kernels enhance the computational power of linear machines by mapping the data into a high-dimensional feature space. This is a way to extend linear theories to nonlinear models, which can be indirectly applied.

One of the attractiveness of the support vector machines is that its computational complexity makes it possible to use a kernel function, as stated in the last section. This is sometimes referred to as the kernel trick. This efficient instrument regardless of the dimensionality of the original feature space is the construction of a linear classifier in the high-dimensional space. The practical success of SVMs [10] led to a research effort to extend a range of linear classifiers to nonlinear ones through the integration of the kernel trick into their design. This is feasible to use whenever all the calculations can be done using inner products.

19.4   Constructing Kernels

We need to construct a valid kernel function to successfully substitute feature vector computation. A kernel function $K : X \times X \to \mathbb{R}$ is a symmetric function in its arguments for which the following property holds.

$$K(\mathbf{x}, \mathbf{z}) = f(\mathbf{x}).f(\mathbf{z}); \tag{19.3}$$

for some feature map f. It is a type of metric measuring the similarity of input objects, which is why it should only be natural to put together these somehow different similarity measures to generate new kernels. Following closure properties are defined over kernels, assuming that: K_1 and K_2 are kernels over $X \times X$, $X \subseteq \mathbb{R}^n$,

$a \in \mathbb{R}^+$,
$f()$: a real-valued function,
$\phi : X \rightarrow \mathbb{R}^m$ with K_3 a kernel over $\mathbb{R}^m \times \mathbb{R}^m$ and
$\mathbf{B}$ as a symmetric semi-definite $n \times n$ matrix.

The following properties can be used to construct new kernels:

(i)

$$K(\mathbf{x}, \mathbf{z}) = c.K_1(\mathbf{x}, \mathbf{z}). \tag{19.4}$$

(ii)

$$K(\mathbf{x}, \mathbf{z}) = c + K_1(\mathbf{x}, \mathbf{z}). \tag{19.5}$$

(iii)

$$K(\mathbf{x}, \mathbf{z}) = K_1(\mathbf{x}, \mathbf{z}) + K_1(\mathbf{x}, \mathbf{z}). \tag{19.6}$$

(iv)

$$K(\mathbf{x}, \mathbf{z}) = K_1(\mathbf{x}, \mathbf{z}).K_1(\mathbf{x}, \mathbf{z}). \tag{19.7}$$

(v)

$$K(\mathbf{x}, \mathbf{z}) = f(\mathbf{x})f(\mathbf{z}). \tag{19.8}$$

(vi)

$$K(\mathbf{x}, \mathbf{z}) = K_1(\phi(\mathbf{x}), \phi(\mathbf{z})). \tag{19.9}$$

(vii)

$$K(\mathbf{x}, \mathbf{z}) = \mathbf{x}^T \mathbf{B} \mathbf{z}. \tag{19.10}$$

(viii)

$$K(\mathbf{x}, \mathbf{z}) = K(\mathbf{z}, \mathbf{x}). \tag{19.11}$$

19.5 Types of Kernels

The linear SVM classifier depends on the dot product between vectors,

$$K(\mathbf{x}_i, \mathbf{x}_j) = \mathbf{x}_i^T \mathbf{x}_j. \tag{19.12}$$

Assuming that every data point is mapped into high-dimensional space using the following transformation:

$$\Phi : \mathbf{x} \to \phi(\mathbf{x}), \tag{19.13}$$

the dot product can be expressed as

$$K(\mathbf{x}_i, \mathbf{x}_j) = \phi(\mathbf{x}_i)^T \phi(\mathbf{x}_j). \tag{19.14}$$

A *kernel function* is some function that corresponds to an inner product in some expanded feature space. For example, let us consider two-dimensional vectors, $\mathbf{x} = [x_1, x_2]$, and let

$$
\begin{aligned}
K(\mathbf{x}_i, \mathbf{x}_j) &= (1 + \mathbf{x}_i^T \mathbf{x}_j)^2 \\
&= 1 + x_{i_1}^2 x_{j_1}^2 + 2x_{i_1} x_{j_1} x_{i_2} x_{j_2} + x_{i_2}^2 x_{j_2}^2 + 2x_{i_1} x_{j_1} + 2x_{i_2} x_{j_2} \\
&= \begin{bmatrix} 1 \\ x_{i_1}^2 \\ \sqrt{2} x_{i_1} x_{i_2} \\ x_{i_2}^2 \\ \sqrt{2} x_{i_1} \\ \sqrt{2} x_{i_2} \end{bmatrix} \begin{bmatrix} 1 & x_{j_1}^2 & \sqrt{2} x_{j_1} x_{j_2} & x_{j_2}^2 & \sqrt{2} x_{j_1} & \sqrt{2} x_{j_2} \end{bmatrix} \\
&= \phi(\mathbf{x}_i)^T \phi(\mathbf{x}_j).
\end{aligned}
$$

There is no need to explicitly represent the space, simply by defining a kernel function. The role of the dot product in the feature space is played by the kernel function. In practice, for most of the common settings, a couple of kernels turned out to be appropriate. Some of the popular kernels are listed in Table 19.1.

Given below are some detailed discussion on various types of kernels and application of some of them.

19.5.1 Linear Kernel

The simplest kernel function is represented by the linear kernel. It is defined as equal to the inner product $< \mathbf{x}, \mathbf{y} >$ added to an optional constant c.

Table 19.1 Summary of Inner-Product Kernels

Type of Kernel	Inner Product Kernel	Comments
Polynomial Kernel	$K(\mathbf{x}, \mathbf{y}) = (\mathbf{x}^T \mathbf{y} + \theta)^p$	Power p and threshold θ is specified a priori by the user
Gaussian Kernel	$K(\mathbf{x}, \mathbf{y}) = e^{-\frac{1}{2\sigma^2} \|\mathbf{x}-\mathbf{y}\|^2}$	Width σ^2 is specified a priori by the user
Spectrum Kernel for strings	Count the number of substrings in common	It is a kernel, since it is a dot product between vectors of indicators of all the substrings

$$K(\mathbf{x}, \mathbf{y}) = \mathbf{x}^T \mathbf{y} + c.$$

Kernel-based algorithms using a linear kernel are often equivalent to their non-kernel counterparts.

19.5.2 Polynomial Kernels

In machine learning, the polynomial kernel is basically a kernel function which is frequently used with support vector machines (SVMs) as well as other kernelized models that divide feature vectors of training data into groups of similar data in a feature space over polynomials of the variables initially provided, thus making it possible to learn nonlinear models. Roughly speaking, the polynomial kernel uses the given individual features of input samples as well as their combinations when dividing data into groups based on similarity.

For degree-d polynomials, the polynomial kernel is shown below:

$$K(\mathbf{x}, \mathbf{y}) = (\mathbf{x}^T \mathbf{y} + c)^d;$$

where $\mathbf{x}$ and $\mathbf{y}$ are vectors in the input space, i.e., vectors of features computed from training or test samples and $c \geq 0$ is a free parameter controls the influence of higher-order versus that of lower-order terms in the polynomial. The kernel is said to be homogeneous, when $c = 0$.

As each kernel, K is associated with its own inner product in a feature space expressed as the mapping $\boldsymbol{\phi}$

$$K(\mathbf{x}, \mathbf{y}) = < \boldsymbol{\phi}(\mathbf{x}), \boldsymbol{\phi}(\mathbf{y}) > .$$

The correspondence between ϕ and its kernel can be demonstrated as below—the special case of the quadratic kernel is found when $d = 2$.

$$K(\mathbf{x}, \mathbf{y}) = \left(\sum_{i=1}^{n} x_i y_i + c \right)^2 = \sum_{i=1}^{n} (x_i^2)(y_i^2) + \sum_{i=2}^{n} \sum_{j=1}^{i-1} (\sqrt{2} x_i x_j)(\sqrt{2} y_i y_j)$$

$$+ \sum_{i=1}^{n} (\sqrt{2} c x_i)(\sqrt{2} c y_i) + c^2.$$

From this, it follows that the feature map is given by:

$$\phi(\mathbf{x}) = < x_n^2, \cdots, x_1^2, \sqrt{2} x_n x_{n-1}, \cdots, \sqrt{2} x_n x_1, \sqrt{2} x_{n-1} x_{n-2}, \cdots,$$
$$\sqrt{2} x_{n-1} x_1, \cdots,$$
$$\sqrt{2} x_2 x_1, \sqrt{2} c x_n, \cdots, \sqrt{2} c x_1, c > .$$

This kernel is most widely used in natural language processing (NLP). Since higher degrees tend to overfit the NLP problems, the degree that is most commonly used is $d = 2$ (quadratic).

19.5.3   Radial Basis Function Kernel

The (Gaussian) radial basis function kernel, or RBF kernel, is a popular kernel function that is used in various kernelized learning algorithms in the field of machine learning. To be more specific, it is commonly used in support vector machines.

The RBF kernel on two samples $\mathbf{x}$ and $\mathbf{x}'$, represented as feature vectors in some input space, is defined as

$$K(\mathbf{x}, \mathbf{x}') = \exp \left(-\frac{||\mathbf{x} - \mathbf{x}'||^2}{2\sigma^2} \right)$$

where

$$||\mathbf{x} - \mathbf{x}'||^2.$$

It is recognized as the squared Euclidean distance between the two feature vectors, and σ is a free parameter. The feature space of the kernel has an infinite number of dimensions. For $\sigma = 1$, it gets expanded as

$$
\exp\left(-\frac{1}{2}||\mathbf{x} - \mathbf{x}'||^2\right) = \sum_{j=0}^{\infty} \frac{\mathbf{x}^T \mathbf{x}'^j}{j!} \exp\left(-\frac{1}{2}||\mathbf{x}||^2\right) \exp\left(-\frac{1}{2}||\mathbf{x}'||^2\right)
$$

$$
= \sum_{j=0}^{\infty} \sum_{\sum n_i = j} \exp\left(-\frac{1}{2}||\mathbf{x}||^2\right) \frac{\mathbf{x_1}_1^n \cdots \mathbf{x_k}_k^n}{\sqrt{n_1! \cdots n_k!}}
$$

$$
\exp\left(-\frac{1}{2}||\mathbf{x}'||^2\right) \frac{\mathbf{x}_{11}'^n \cdots \mathbf{x}_{kk}'^n}{\sqrt{n_1! \cdots n_k!}} \tag{19.15}
$$

19.5.4 Fisher Kernel

The Fisher kernel has been named after Ronald Fisher. Based on the sets of measurements for each object and a statistical model, this function measures the similarity of two objects. In a classification problem scenario, this is used to estimate the class for a new object (whose real class is unknown).This is achieved by minimizing, across classes, an average of the Fisher kernel distance from the new object to each known member of the given class.

The Fisher kernel makes use of the Fisher score, defined as

$$
U_{\mathbf{x}} = \Delta_\theta \log(P(\mathbf{x}|\theta));
$$

where θ is a set (vector) of parameters. The function $\log[P(X|\theta)]$ is the log likelihood of the probabilistic model. The Fisher kernel is defined as

$$
K(x, x') = U_x I^{-1} U_{x'};
$$

where I represents the Fisher information matrix and U_x is Fisher score. Fisher's kernel is widely used in information retrieval.

19.5.5 Exponential Kernel

The exponential kernel is similar to the Gaussian kernel, differing only in the aspect that the square of the norm is missing in its formulation.

$$
K(\mathbf{x}, \mathbf{y}) = \exp\left(-\frac{||\mathbf{x} - \mathbf{y}||}{2\sigma^2}\right).
$$

19.5.6   Laplacian Kernel

The Laplace kernel is completely equivalent to the exponential kernel, except that it is less sensitive for changes in the sigma parameter.

$$K(\mathbf{x}, \mathbf{y}) = \exp\left(-\frac{||\mathbf{x} - \mathbf{y}||}{\sigma}\right).$$

It should be noted that the observations that had been made about the sigma parameter related to the Gaussian kernel are equally applicable to the Exponential and the Laplacian kernels.

19.5.7   Hyperbolic Tangent (Sigmoid) Method

The hyperbolic tangent kernel is also called the sigmoid kernel or the multilayer perceptron (MLP) kernel. From the domain of neural networks comes the sigmoid kernel, where the bipolar sigmoid function is often used as an activation function for artificial neurons.

$$K(\mathbf{x}, \mathbf{y}) = \tanh(\alpha \mathbf{x}^T \mathbf{y} + c).$$

A SVM model using a sigmoid kernel function is the same as to a two-layer perceptron neural network in terms of functionality. The slope alpha and the intercept constant c are the two adjustable parameters in the sigmoid kernel. A common value for α is $1/d$, where d is the data dimension.

Although some kernels are domain specific; there is in general no best choice for it. Each of the kernel suffers from some degree of variability, so in practice, the best solution is to experiment with different kernels and adjust their parameters. This is done via model search such that the error on a test set gets minimized. Generally, a low polynomial kernel or a Gaussian kernel has shown to be good initial try, and they generally beat conventional classifiers in terms of performance [4, 8].

19.6   Utility of Kernel

The use of kernels in learning system has achieved appreciable attention. The main reason behind that is, they help in increasing the computational power of linear machines, because the kernels allow to map the data into a high-dimensional feature space. Thus, in a way, it is an extension of linear hypotheses to nonlinear ones, and this step can be performed implicitly.

Support vector machines [2], kernel principal component analysis [9], kernel Gram-Schmidt [5], Bayes point machines [3], and Gaussian processes [7] are just few of the algorithms that make significant use of kernels for problems of classification, regression, density estimation, and clustering.

Kernels are the similarity functions which are usually applied to nonlinearly separable data. This function converts the data in X space (i.e., the given space) to the data of Z space which provides better separability for the dataset that was not initially easily separable in the X space.

Applications of some of the popular kernels are discussed below. Kernels like Fisher Kernel [6] are used in information retrieval. It forms a bridge between generative and probabilistic model of documents. The Fisher kernel can also be applied to image representation for classification or retrieval problems. The Fisher kernel can create a compact and dense representation, which is more desirable for image classification and retrieval problems.

Polynomial kernel is quite popular in natural language processing (NLP). Kernels are widely used in SVM, because this kernel trick makes the linear classifier design in the high-dimensional space independent of the dimensionality of this space. The success of SVM has also led the research to extend the linear classifiers to nonlinear ones using the powerful kernel trick.

References

1. Cover, T. M., & Ordentlich, E. (1996). Universal portfolios with side information. *IEEE Transactions on Information Theory, 42*(2), 348–363.
2. Hearst, M. A., Dumais, S. T., Osuna, E., Platt, J., & Scholkopf, B. (1998). Support vector machines. *IEEE Intelligent Systems and Applications, 13*(4), 18–28.
3. Herbrich, R., Graepel, T., & Campbell, C. (2001). Bayes point machines. *Journal of Machine Learning Research, 1*, 245–279.
4. Joachims, T. (1998). *Text categorization with support vector machines: Learning with many relevant features*. Springer.
5. Leon, S. J., Björck, Å., & Gander, W. (2013). Gram-schmidt orthogonalization: 100 years and more. *Numerical Linear Algebra with Applications, 20*(3), 492–532.
6. Perronnin, F., Sánchez, J., & Mensink, T. (2010). Improving the fisher kernel for large-scale image classification. *Computer Vision–ECCV 2010* (pp. 143–156).
7. Rasmussen, C. E., & Williams, C. K. I. (2006). *Gaussian processes for machine learning*, vol. 1. MIT Press Cambridge.
8. Schólkopf, B., Burgest, C., & Vapnik, V. (1995). Extracting support data for a given task. *Proceedings of the First International Conference on Knowledge Discovery and Data Mining*, 252–257.
9. Schólkopf, B., Smola, A., & Múller, K.-R. (1997). Kernel principal component analysis. In *International conference on artificial neural networks* (pp. 583–588).
10. Xue, H., Yang, Q., & Chen, S. (2009). Svm: Support vector machines. In *The top ten algorithms in data mining* (pp. 51–74). Chapman and Hall/CRC.

Chapter 20
Extreme Learning Machines

Numerous forms of neural networks exist. One of the most widely used form of neural networks is feedforward neural networks. The feedforward neural network is composed of one input layer that receives the inputs from external environments, one or more hidden layers, and one output layer that sends the results to external environments. In general, the training of feedforward networks utilizes three primary strategies:

(i) Gradient descent based (e.g., backpropagation for multilayer feedforward neural network) methods.
(ii) Standard optimization methods (e.g., support vector machines).
(iii) Least square-based methods (e.g., radial basis function).

The learning speed of feedback networks is generally much slower than necessary. In recent decades, this has been a limiting aspect in their use. This may be attributed to two main reasons:

(i) Slow gradient learning algorithms for the training of neural networks are widely used, and
(ii) All network parameters are designed iteratively using these learning algorithms..

A new system known as the extreme learning machines (ELM) was developed to tackle these problems [7] for increasing training rate and overfitting issues and achieving better generalization performance. Challenges faced by classic gradient descent, such as local minima, incorrect learning rate, etc. are also addressed. ELM can also be used to train single hidden layer feedforward networks (SLFN) on a wide range of non-differentiable functions, while the conventional gradient-based algorithms operate on differentiable functions. For a generalized single hidden layer feedforward neural network, the extreme learning machine was built for where the hidden layer doesn't have to be like neuron.

© The Author(s), under exclusive license to Springer Nature Singapore Pte Ltd. 2025   319
A. Ghosh, *Data Science and Cases in Sustainability*, Mathematics for Sustainable
Developments, https://doi.org/10.1007/978-981-96-8362-8_20

20.1   Brief Overview

Extreme learning machines (Fig. 20.1) are feedforward neural network for regression or classification with a single layer of hidden nodes. The weights that map inputs to hidden nodes are assigned haphazardly and never revised. The hidden nodes in ELM are generated randomly, while the neuron activation functions are nonlinear piecewise continuous, unlike other neural networks with back propagation. The weights between the hidden layer and the output layer are calculated analytically.

ELM offers effective coherent solutions to generalized feedforward networks which include but are not limited to (both single and multiple hidden layer) neural networks, the RBF networks, and kernel machines as emerging learning techniques. Hypotheses of ELM indicate that hidden neurons are necessary but can be generated randomly and are application-free and that ELM have both large guessing and classification capability. They also link different speculations (especially edge relapse, optimization, neural network performance generalization, linear system stability, and matrix theory) immediately. Therefore, ELM may be biologically inspired and can provide tremendous advantages such as quick learning, ease of operation, and limited human intervention. As a viable alternative for large-scale computing and machine learning, they thus hold a great potential. ELM can be applied easily where it tends to produce the lowest training error, achieves the smallest weight norm, good generalization performance, and works extremely efficiently. It is named the "extreme learning machine " to separate it from the other common SLFN learning algorithms.

One does not need to tune the input weights and first hidden layer biases in applications, contrary to popular belief and most practical applications where all the

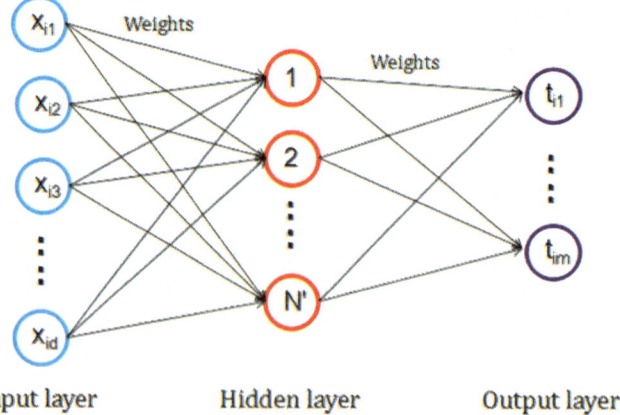

Fig. 20.1 Basic structure of an ELM

Fig. 20.2 Representation of
an ELM

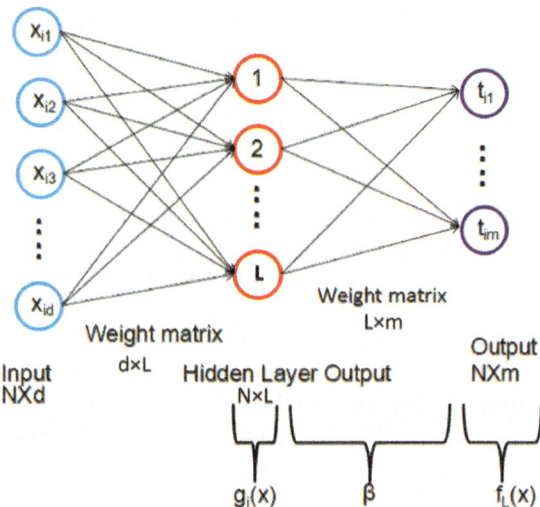

parameters of the feedforward networks are adjusted. This way, learning is not only extremely quick but also facilitates strong generalization efficiency.

The crucial difference of the ELM is that the neurons of a single, hidden layer feedforward network (SLFN) need not be modified, and this makes ELM vary from the common understanding of learning. The use of random computational nodes in a hidden layer, independent of the training data, is one of the standard implementations of ELMs. In comparison to conventional neural network learning algorithms, ELM appears to attain not only the least training error but also, over a very short timeframe, the smallest level of output weight (Fig. 20.2).

20.2   Interpolation Theorem

We present the constructs of a standard ELM and its variations for supervised classification and regression in this section. The output function of SLFNs with L hidden nodes can be represented by

$$f_L(\mathbf{x}) = \sum_{i=1}^{L} \boldsymbol{\beta}_i g_i(\mathbf{x}) \tag{20.1}$$

$$= \sum_{i=1}^{L} \boldsymbol{\beta}_i G_i(\mathbf{a}_i, b_i, \mathbf{x}) \tag{20.2}$$

where $\mathbf{x} \in \mathbb{R}^d$, $\boldsymbol{\beta}_i \in \mathbb{R}^m$ and $G_i(\mathbf{a}_i, b_i, \mathbf{x})$ of the ith hidden node. For additive nodes with activation function g, g_i is defined as

$$g_i = G_i(\mathbf{a}_i, b_i, \mathbf{x}) = g(\mathbf{a}_i.\mathbf{x} + b_i); \qquad (20.3)$$

where $\mathbf{a}_i \in \mathbb{R}^d$ and $b_i \in \mathbb{R}$.

For RBF nodes with activation function g, g_i is defined as

$$g_i = G_i(\mathbf{a}_i, b_i, \mathbf{x}) = g(b_i \|\mathbf{x} - \mathbf{a}_i\|); \qquad (20.4)$$

where $\mathbf{a}_i \in \mathbb{R}^d$ and $b_i \in \mathbb{R}^+$.

SLFNs with a maximum of L hidden nodes as well as any specific nonlinear activation function that has a limit at infinity can learn any N arbitrary distinct zero-error samples. The threshold, sigmoid, ramp and radial functions, as well as radial basis, cosine squasher and many nonregular functions are included in these activation functions. Theoretically, it was proved that given an activation function $g(x)$ obeying certain rather general conditions, there exists a sequence of network functions $\{f_L\}$ approximating to any given continuous target function f with an expected learning error $\varepsilon > 0$. According to conventional neural network theories, hidden layer parameters $(\mathbf{a}_i, b_i)$ must be properly tuned and appropriate values of network parameters (e.g., $(\mathbf{a}_i, b_i)$ and b_i) must be sought for any given target function f. In such traditional learning models, hidden layer parameters must be modified on the basis of training samples at least once. On the other hand, all hidden layer parameters within the ELMs should not be changed and can be isolated from the training samples. For N arbitrary distinct samples $(\mathbf{x}_i, \mathbf{t}_i) \in \mathbb{R}^d \times \mathbb{R}^C$, (with C being the number of classes) SLFNs having L hidden nodes can be mathematically modeled as

$$\sum_{i=1}^{L} \boldsymbol{\beta}_i g_i(\mathbf{x}_j) \qquad (20.5)$$

$$= \sum_{i=1}^{L} \boldsymbol{\beta}_i G_i(\mathbf{a}_i, b_i, \mathbf{x}_j) = \mathbf{o}_j; \qquad (20.6)$$

where $j = 1, 2, \ldots N$.

The SLFNs can approximate these N samples with zero error means; that $\sum_{j=1}^{N} \|\mathbf{o}_j - \mathbf{t}_j\| = 0$, i.e., there exists $(\mathbf{a}_i, b_i)$ and $\boldsymbol{\beta}_i$ such that

$$\sum_{i=1}^{L} \boldsymbol{\beta}_i G_i(\mathbf{a}_i, b_i, \mathbf{x}_j) = \mathbf{t}_j. \qquad (20.7)$$

The above equations can be compactly written as

$$\mathbf{H}\boldsymbol{\beta} = \mathbf{T}; \tag{20.8}$$

where

$$\mathbf{H} = \begin{bmatrix} h(\mathbf{x}_1) \\ h(\mathbf{x}_2) \\ \vdots \\ h(\mathbf{x}_N) \end{bmatrix}. \tag{20.9}$$

$$= \begin{bmatrix} G(\mathbf{a}_1, b_1, \mathbf{x}_1) & \dots & G(\mathbf{a}_L, b_L, \mathbf{x}_1) \\ G(\mathbf{a}_1, b_1, \mathbf{x}_2) & \dots & G(\mathbf{a}_L, b_L, \mathbf{x}_2) \\ \vdots & \dots & \vdots \\ G(\mathbf{a}_1, b_1, \mathbf{x}_N) & \dots & G_i(\mathbf{a}_L, b_L, \mathbf{x}_N) \end{bmatrix} \tag{20.10}$$

$$\boldsymbol{\beta} = \begin{bmatrix} \boldsymbol{\beta}_1^T \\ \boldsymbol{\beta}_2^T \\ \vdots \\ \boldsymbol{\beta}_N^T \end{bmatrix} \tag{20.11}$$

and

$$\mathbf{T} = \begin{bmatrix} \mathbf{t}_1^T \\ \mathbf{t}_2^T \\ \vdots \\ \mathbf{t}_N^T \end{bmatrix}. \tag{20.12}$$

$\mathbf{H}$ is called the hidden layer output matrix of the SLFN. The ith column of $\mathbf{H}$ is the ith hidden node output with respect to inputs $\mathbf{x}_1, \mathbf{x}_2, \dots, \mathbf{x}_N$. $h(\mathbf{x}_i) = G(\mathbf{a}_1, b_1, \mathbf{x}_i) \dots G(\mathbf{a}_L, b_L, \mathbf{x}_i)$ is known as the hidden layer feature mapping. The ith row of $\mathbf{H}$ is the hidden layer feature mapping with respect to the ith input $\mathbf{x}_i : h(\mathbf{x}_i)$. It has been shown [7] that from the interpolation capability point of view, if the activation function g is infinitely differentiable in any interval, it is possible to generate the hidden layer parameters randomly.

20.3   Basic ELM

The hidden node parameters $(\mathbf{a}_i, b_i)$ are kept fixed after random generation. SLFN training is equivalent to finding a least-square solution $\hat{\boldsymbol{\beta}}$ of the linear system $\mathbf{H}\boldsymbol{\beta} = \mathbf{T}$, and

$$||\mathbf{H}\hat{\boldsymbol{\beta}} - \mathbf{T}|| = min_{\boldsymbol{\beta}}||\mathbf{H}\boldsymbol{\beta} - \mathbf{T}||. \qquad (20.13)$$

If $L = N$, that is, the number L of hidden nodes becomes equal to the number N of distinct training samples, then the matrix H becomes square and invertible [7] when hidden node parameters $(\mathbf{a}_i, b_i)$ are randomly selected, and SLFNs can therefore estimate the samples with zero error. In most of the cases, the number of hidden nodes is, however, remarkably lower than that of distinct training samples, that is, $L << N$. In this case, $\mathbf{H}$ is a nonsquare matrix, and there may not exist $\mathbf{a}_i, b_i, \boldsymbol{\beta}_i$ $(i = 1, 2, \ldots L)$ such that $\mathbf{H}\boldsymbol{\beta} = \mathbf{T}$. The smallest norm least-squares solution of the above linear system is

$$\hat{\boldsymbol{\beta}} = \mathbf{H}^{\dagger}\mathbf{T};$$

where $\mathbf{H}^{\dagger}$ is the Moore-Penrose generalized inverse of matrix $\mathbf{H}$ [18, 21]. ELM algorithm can therefore be summarized as follows:

Given: A training set of N samples $S = \{(\mathbf{x}_i, \mathbf{t}_i) | \mathbf{x}_i \in \mathbb{R}^d, \mathbf{t}_i \in \mathbb{R}^m, i = 1, 2, \ldots, N\}$, hidden node output function $G(\mathbf{a}_1, b_1, \mathbf{x})$ and L being the number of hidden nodes

  (i) Randomly generate hidden node parameters $(\mathbf{a}_i, b_i)$, $i = 1, 2, \ldots, L$
  (ii) Calculate the hidden layer output matrix $\mathbf{H}$
  (iii) Calculate the output weight vector $\boldsymbol{\beta} = \mathbf{H}^{\dagger}\mathbf{T}$.

A wide range of activation functions can function with ELM algorithm. Many prominent algorithms do not interact directly with threshold networks. Rather, other analog networks are being used for approximating threshold networks in order to finally use the gradient-descent method. ELM can however be used for the direct training of threshold networks. Various methods can be used for the computation of the Moore Penrose generalized inverse of a matrix: orthogonal projection method, iterative method, orthogonalization method, and singular value decomposition (SVD) [18].

20.4 Adaptations of ELM

In addition to the ELM basic model, a variety of other ELM implementations were studied by researchers. In this section, we discuss few of them.

20.4.1 Random Hidden Layer Feature Mapping-Based ELM

The ridge regression theory [6] suggests [10, 24] that a positive value $\frac{1}{\lambda}$ be added to the diagonal of $H^T H$ or $H H^T$ while calculating the output weights β. The solution thus obtained is stabler and is likely to have better generalization performance. In other words, so as to increase the stability of ELM, we can have

$$\beta = H^T \left(\frac{I}{\lambda} + H H^T \right)^{-1} T \qquad (20.14)$$

and the associated output function of ELM can be written as

$$f(x) = h(x)\beta = h(x)H^T \left(\frac{I}{\lambda} + H H^T \right)^{-1} T. \qquad (20.15)$$

Or, following equation

$$\beta = \left(\frac{I}{\lambda} + H H^T \right)^{-1} H^T T; \qquad (20.16)$$

and the corresponding output function of ELM as

$$f(x) = h(x)\beta = h(x) \left(\frac{I}{\lambda} + H H^T \right)^{-1} H^T T. \qquad (20.17)$$

Huang et al. [10] show that the Eqs. (20.14) and (20.16) are actually consistent in minimizing $||H\beta - T||^2 + \lambda ||\beta||^2$, which is the essential target of ELM. Thus, ELM algorithm can be rewritten as

Given: a training set N, the hidden node output function $G(a_i, b_i, x)$, and L as the hidden node number:

(i) Randomly generate hidden node parameters (a_i, b_i), $i = 1, \ldots, L$.
(ii) Calculate the hidden layer output matrix H.

(continued)

(iii) Calculate the output weight vector β as:

$$\beta = H^T \left(\frac{I}{\lambda} + H H^T \right)^{-1} T \qquad (20.18)$$

or

$$\beta = \left(\frac{I}{\lambda} + H^T H \right)^{-1} H^T T. \qquad (20.19)$$

The number of hidden nodes that are required in these implementations can be moderate; it has no direct dependency on the number of training samples N. It works for both the cases $L < N$ and $L \geq N$. It is distinct from the theorem of interpolation which needs $L \leq N$, yet consistent with the universal approximation theorem. Toh [24] and Deng et al. [2] studied such regularization enhancement under sigmoid additive type of SLFNs. Deng et al. [2] and Man et al. [13] concentrated on acquiring the analytical solution (21) based on optimization methods. Toh [24] came up with a corresponding total error rate-based multi-class solution of ELM (TER-ELM). Miche et al. [14] studied ELM with a cascade of two regularization penalties. Huang et al. [10] further contributed to this study on generalized SLFNs with varieties of hidden nodes (feature mappings) as well as kernels and showed that for regression, a basic integrated ELM algorithm can be achieved, binary and multi-label classification cases which, however, have to be handled separately by SVMs and its variations.

20.4.2   Kernel-Based ELM

Kernel-based ELM is another implementation of the ELM [10]. If it is not known by the users what the hidden layer mapping function h(x) is, an ELM kernel matrix can be specified as follows:

$$\Omega_{ELM} = H H^T : \Omega_{ELMij} = h(x_i).h(x_j) = K(x_i.x_j). \qquad (20.20)$$

Then the output function of ELM can be written as

$$f(x) = h(x) H^T \left(\frac{I}{\lambda} + H H^T \right)^{-1} T \qquad (20.21)$$

$$= \begin{bmatrix} h(\mathbf{x}_1) \\ h(\mathbf{x}_2) \\ \vdots \\ h(\mathbf{x}_N) \end{bmatrix}^T \left(\frac{I}{\lambda} + \Omega_{ELM}^{-1} T \right). \tag{20.22}$$

In this particular kernel implementation of ELM, users do not need to learn about hidden layer mapping $h(x)$ but rather about the corresponding kernel $K(u, v)$ (e.g., $K(u, v) = exp(-\gamma ||u - v||^2)$, which is provided. There is no need for finding out the amount of hidden nodes L (the dimensionality of the hidden layer feature space). The ELM algorithm can therefore be changed as follows for the kernel case:

Given a training set and kernel $K(u, v)$:
   The output function can be calculated as

$$f(x) = \begin{bmatrix} h(\mathbf{x}_1) \\ h(\mathbf{x}_2) \\ \vdots \\ h(\mathbf{x}_N) \end{bmatrix}^T \left(\frac{I}{\lambda} + \Omega_{ELM}^{-1} T \right). \tag{20.23}$$

It can be seen that an ELM algorithm based on kernel can be implemented in just one single learning step. Frenay and Verleysen [3, 4] examined ELM's kernel implementation when users became aware of $h(x)$. If the users have the idea of the hidden layer feature mapping $h(x)$, then the ELM kernel can be defined as [3]

$$K(u, v) = \lim \frac{1}{L} h(u).h(v). \tag{20.24}$$

For the regression method for SVM, a parameter-insensitive kernel having analytical structure can be acquired which significantly lessens the computational complexity. We have speculated that Frenay and Verleysen's ELM kernel [3] can work for SVM and its variants. All of these methods can be used straight away for regression, binary, and multi-label classification systems. For complex environments, ELMs can be implemented.

20.4.3   Fully Complex ELM

Equalizers are very often used in high-speed digital communication networks on recipients to restore the initial signals [9]. For the resolution of equalization problems, two conventional approaches are usually employed.

 (i) Real-valued neural network models like feedforward neural networks, RBF networks, and recurrent neural networks.
(ii) Complex-valued neural networks: Split-complex activation (basis) functions consisting of two real-valued activation functions, one processing the real part and the other processing the imaginary part, have been traditionally employed in these complex-valued neural networks.

Instead of using a split-complex activation function, fully complex activation functions can be used explicitly in the extreme learning machines.

20.4.4   Online Sequential ELM (OS-ELM)

Most ELM algorithms only learn samples after all samples are completed. In many commercial applications, the training data can be provided one-by-one or in chunk-by-chunk manner. In those situations, sequential online learning algorithms get favored over batch learning algorithms, because no retraining is needed when new data are obtained. Most of the usual online sequential learning algorithms have to define multiple parameters, and it takes a great deal of time to adjust them. OS-ELM [12] is a simple and efficient online sequential learning algorithm that is able to study training data not only in one-by-one fashion but also in chunk-by-chunk fashion(with fixed or variable length) and can delete the processed training data. The results are sent to the learning algorithm sequentially (one-by-one or chunk-by-chunk with fixed or variable chunk length). Once the learning process is full for the individual observation(s), a single observation or a chunk of training observations will be scrapped and no longer useable. One of the solutions of the output weight vector β is

$$\beta = (H^T H)^{-1} H^T T. \tag{20.25}$$

The OS-ELM, using the recursive least square algorithm of the above equation, succeeds in the sequential application of the least square solution [11].

When $rank(H_0) = L$, then OS-ELM and ELM can attain the same learning performance (in terms of training error and generalization accuracy). Additionally, if $N_0 = N$, OS-ELM also becomes the batch ELM. In OS-ELM, the chunk size of incoming training data has no requirement of being constant.

20.4.5 Incremental ELM (I-ELM)

Using an incremental learning approach by inserting hidden nodes one by one, ELMs were realized for their universal approximation potential. The proof is indeed an innovative, practical method that essentially demonstrates a technique to build an incremental feedforward network (known as I-ELMs). Unlike other incremental learning algorithms which only operate with certain types of hidden nodes (like the resource allocation network and its variants work only for RBF networks), I-ELM works well with a broad range of activation functions, where it becomes immaterial whether they are sigmoidal or non-sigmoidal, continuous or noncontinuous, and differentiable or non-differentiable. The classical gradient-descent-based learning algorithms are not available on networks with non-differential activation functions such as the threshold networks, due to unavailability of needed derivatives. These classical methods may also suffer from problems like local minima. Although various other incremental learning algorithms had been suggested in the literature, the general potential to approximate earlier learning algorithms has not been proven in contrast to I-ELM.

In contrast to other traditional incremental learning algorithms, I-ELM does not have any parameters to be defined, other than maximum network architecture and targeted accuracy. The findings of studies demonstrate that I-ELM beats other learning algorithms (including support vector regression (SVR), stochastic gradient-descent BP, and incremental RBF networks on the basis of parameters like performance outside of training set and learning speed. I-ELMs may be applied in different ways:

(i) Basic I-ELM: Every time, only one hidden node is created and connected to the existing network randomly.
(ii) Enhanced I-ELM: Every time, k hidden nodes are generated randomly. But among the randomly generated k hidden nodes, only the most suitable hidden node will be added to the existing network.

The improved version, in comparison to the original I-ELM, has a more lightweight network architecture, and learning can be achieved more easily. When $k = 1$, I-ELM becomes a specific case of EI-ELM.

20.4.6 ELM Ensembles

The performance of a single network can be anticipated to increase with an ensemble of neural networks having a plurality consensus system. Bagging and boosting ensembles were the most significant techniques for training neural networks [20]. An integration of several ELMs was proposed by Sun et al. [22] for the prediction of the future sales amount. Multiple ELM networks were linked concurrently, and the final estimated sales volume is the average of the outcomes of ELM. The subsequent ensemble had shown better result in generalization scenario. Heeswijk et al. [25] investigated the adaptive ensemble models of ELM on the application of one-step ahead prediction in (non-)stationary time series. It was reported that the method worked on fixed time series and the method was checked on nonstationary time series. The empirical studies found that a reasonable test error was made with strong adaptability with the adaptive ensemble model. Heeswijk et al. [26] also studied ELM ensemble for large-scale regression applications. In fact, network ensembles are potentially valuable tools for sequential learning [15]. Network ensemble comprises of some single networks which can adapt to new data differently. Few networks in the system will adapt to new data more easily and faster than others to solve the issue of networks that cannot fit the new data. Lan et al. [11] proposed an integrated network structure, which is called ensemble of online sequential ELM (EOS-ELM). EOSELM comprises of several OS-ELM networks. As the final calculation of network performance, the average output value of each OS-ELM within the ensemble was used. The findings of the simulation revealed that EOS-ELM is more robust in comparison to the original OS-ELM for most problems in every simulation trial.

20.5 Controversy

There are several problems with the ELMs, which have caused much debate. Some say that an ELM is exactly what Minsky and Papert [16] called a Gamba perceptron (a perceptron whose first layer is a bunch of linear threshold units) [5]. The original 1958 Rosenblatt perceptron [19] was an ELM in that the first layer was randomly connected. Some people also say that the worst thing to do is to just connect the first layer randomly. Nearly 60 years have passed since the discovery of perceptron in developing better schemes to extend the dimension of the input vector nonlinearly in order to separate the data. Setting the weight of a layer one arbitrarily (if properly done) can be successful if the function we are learning is very easy and there is a limited amount of labeled data. The advantages are close to the benefits of an SVM (although in fewer ways): The number of parameters to be trained is limited (as the first layer has been fixed) and easy to regularize (as it constitutes a linear classifier). But the issue then is, "Why don't you first use a SVM or an RBF network?". The field of basic classification problems can be very limited with small data sets, which can do well with this kind of two-layer network with a random first layer. Yet you can never see them overcome records on complex operations like ImageNet [1] or speech recognition [17].

A number of types of ELMs have emerged to combat such controversies and are able to handle several machine learning tasks. ELMs were used in numerous forms effectively like kernel ELM [27], incremental ELM [8], evolutionary ELM [29], etc. Advances in data science also motivated researchers to build ELM to suit other big data handling systems and deep learning techniques. Multilayer ELM [23], ELM-based autoencoder [28], hierarchical ELM [30], etc. were created, in this respect.

References

1. Deng, J., Dong, W., Socher, R., Li, L. J., Li, K., & Fei-Fei, L. (2009). Imagenet: A large-scale hierarchical image database. In *IEEE conference on computer vision and pattern recognition, CVPR* (pp. 248–255).
2. Deng, W., Zheng, Q., & Chen, L. (2009). Regularized extreme learning machine. In *IEEE symposium on computational intelligence and data mining, CIDM'09* (pp. 389–395).
3. Frénay, B., & Verleysen, M. (2011). Parameter-insensitive kernel in extreme learning for non-linear support vector regression. *Neurocomputing, 74*(16), 2526–2531.
4. Frénay, B., Verleysen, M. (2010). Using SVMs with randomised feature spaces: An extreme learning approach. *European symposium on artificial neural networks (ESANN)*
5. Gamba, A. (1961). Optimum performance of learning machines. *Proceedings of the Institute of Radio Engineers, 49*(1), 349.
6. Hoerl, A. E., & Kennard, R. W. (1970). Ridge regression: Biased estimation for nonorthogonal problems. *Technometrics, 12*(1), 55–67.
7. Huang, G., Zhu, Q., & Siew, C. (2006). Extreme learning machine: Theory and applications. *Neurocomputing, 70*(1), 489–501.
8. Huang, G.-B., & Chen, L. (2007). Convex incremental extreme learning machine. *Neurocomputing, 70*(16), 3056–3062.
9. Huang, G.-B., Li, M.-B., Chen, L., & Siew, C.-K. (2008). Incremental extreme learning machine with fully complex hidden nodes. *Neurocomputing, 71*(4), 576–583.
10. Huang, G.-B., Zhou, H., Ding, X., & Zhang, R. (2012). Extreme learning machine for regression and multiclass classification. *IEEE Transactions on Systems, Man, and Cybernetics, Part B (Cybernetics), 42*(2), 513–529.
11. Lan, Y., Soh, Y. C., & Huang, G.-B. (2009). Ensemble of online sequential extreme learning machine. *Neurocomputing, 72*(13), 3391–3395.
12. Liang, N.-Y., Huang, G.-B., Saratchandran, P., & Sundararajan, N. (2006). A fast and accurate online sequential learning algorithm for feedforward networks. *IEEE Transactions on Neural Networks, 17*(6), 1411–1423.
13. Man, Z., Lee, K., Wang, D., Cao, Z., & Miao, C. (2011). A new robust training algorithm for a class of single-hidden layer feedforward neural networks. *Neurocomputing, 74*(16), 2491–2501.
14. Miche, Y., Van Heeswijk, M., Bas, P., Simula, O., & Lendasse, A. (2011). TROP-ELM: A double-regularized ELM using LARS and tikhonov regularization. *Neurocomputing, 74*(16), 2413–2421.
15. Minku, F. L., Inoue, H., & Yao, X. (2009). Negative correlation in incremental learning. *Natural Computing, 8*(2), 289–320.
16. Minsky, M. (1963). Neural nets and theories of memory. *Artificial Intelligence, 48*, 301–310.
17. Rabiner, L. R., Juang, B.-H., & Rutledge, J. C. (1993). *Fundamentals of speech recognition* (vol. 14). PTR Prentice Hall Englewood Cliffs.
18. Rao, C. R., & Mitra, S. K. (1972). Generalized inverse of matrices and its applications. In *Proceedings of the sixth berkeley symposium on mathematical statistics and probability, theory of statistics 1* (pp. 601–620).
19. Sandler, J., & Rosenblatt, B. (1962). The concept of the representational world. *The Psychoanalytic Study of the Child, 17*(1), 128–145.
20. Schapire, R. E. (1990). The strength of weak learnability. *Machine Learning, 5*(2), 197–227.
21. Serre, D. (2002). Matrices: Theory and applications. Springer-Verlag New York.
22. Sun, Z.-L., Choi, T.-M., Au, K.-F., & Yu, Y. (2008). Sales forecasting using extreme learning machine with applications in fashion retailing. *Decision Support Systems, 46*(1), 411–419.
23. Tang, J., Deng, C., & Huang, G.-B. (2015). Extreme learning machine for multilayer perceptron. *IEEE Transactions on Neural Networks and Learning Systems, 27*(4), 809–821.
24. Toh, K.-A. (2008). Deterministic neural classification. *Neural Computation, 20*(6), 1565–1595.

25. Van Heeswijk, M., Miche, Y., Lindh-Knuutila, T., Hilbers, P. A. J., Honkela, T., Oja, E., & Lendasse, A. (2009). Adaptive ensemble models of extreme learning machines for time series prediction. *International conference on artificial neural networks* (pp. 305–314).
26. Van Heeswijk, M., Miche, Y., Oja, E., & Lendasse, A. (2011). GPU-accelerated and parallelized ELM ensembles for large-scale regression. *Neurocomputing, 74*(16), 2430–2437.
27. Wang, J., Cai, L., & Zhao, X. (2017). Multiple-instance learning via an RBF kernel-based extreme learning machine. *Journal of Intelligent Systems, 26*(1), 185–195.
28. Wang, Y., Xie, Z., Xu, K., Dou, Y., & Lei, Y. (2016). An efficient and effective convolutional auto-encoder extreme learning machine network for 3D feature learning. *Neurocomputing, 174*, 988–998.
29. Zhu, Q.-Y., Qin, A. K., Suganthan, P. N., & Huang, G.-B. (2005). Evolutionary extreme learning machine. *Pattern Recognition, 38*(10), 1759–1763.
30. Zhu, W., Miao, J., Qing, L., & Huang, G.-B. (2015). Hierarchical extreme learning machine for unsupervised representation learning. In *2015 International joint conference on neural networks (IJCNN)* (pp. 1–8).

Chapter 21
Deep Learning

A well-trained multilayered neural network could learn many complex decision regions. It was proved that any three-layered neural network could successfully represent any polynomial function [2, 7]. However, the number of neurons needed by the network to approximate any given function was uncertain. A high-dimensional polynomial function could need a thousand neurons too. Due to this uncertainty, these networks failed to develop an accurate model, often ending up with either much more or much lesser neurons than actually needed. Deep learning promises to develop the accurate model by adding more layers and thereby reducing the effective number of neurons. Back-propagation was however slower in deep networks and with many layers these networks performed poorly.

21.1 Neural Coding

Neuroscience attempts to answer the big question: How is information processed and represented in the human brain? Almost 50 years ago, Herbert Simon had written an essay entitled "The architecture of complexity" [13]. He concluded in his glorious research that most dynamic functional processes, such as financial, biological or physical, are hierarchically ordered. He introduced the concept of "nearly-decomposable systems," that is, structures with elements communicating (in some way) mostly with a group of elements similar to them and less with elements out of the group. The near-decomposability proposed by Simon in conventional contemporary terminology is very much similar to the principle of topological modularity: Nodes in the same module have extensive intra-modular connectivity and minimal inter-modular connections with nodes in other modules [9].

The neurons use the axon, a tubular structure, to transfer their influence to the various parts of the brain over extended ranges. If a neuron fires, an electrical activity known as the action potential travels down the axon, which is further divided

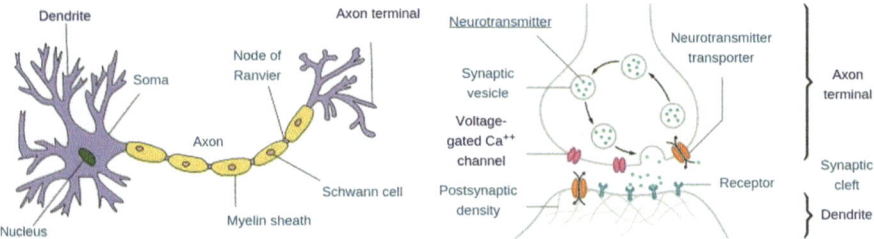

Fig. 21.1 Neuron in human brain

into branched tree-like structure called the axon terminals. At the bottom of each axon terminal, there are some proteins that translate this electrical signal into a chemical. Small particles (synaptic vesicles) that contain a set of neurotransmitters are distributed into an outer neighborhood of the neuron known as a synaptic cleft. This area isolates the axon terminal from the beginning of the next neuron (a synapse) and allows the neurotransmitter to move freely to go after different tasks. The synapses are mostly found at the roots of the branched tree-like structure which resembles a root system of any plant. This is a branching of the dendritic tree which splits into multiple branches and even form longer arms. It finally reaches the core of the cell named soma, reflecting the neural connections in neural networks. These dendrites contain nearly all of the synapses connecting one neuron to another and forming the main connections. Thousands of receptors may be retained in a synapse, to which neurotransmitters can bind (Fig. 21.1).

This compound of axon terminal and synapses at a dendrite can be conceptualized as the (dense) input layer of a deep neural network. Each neuron may have less than five dendrites or as many as a few hundred thousands. Thus, the idea of hierarchically organized artificial neural network sprang up. Similar to human brain, the deep networks had many hidden layers that could learn a given function.

21.2   Deep Neural Networks

Neuroscience attempts to answer the big question how is information processed and represented in the human brain. The theory that complex systems have a centralized hierarchical structure emerged in the early 1960s has recently gained new perspective from large-scale real-life network computational research. This had led to some major advancements in the field of artificial neural networks that were thought of mimicking the human brain in a simpler fashion. McCulloch-Pitts model and the perceptron model were used for various pattern recognition tasks. Multilayered neural networks have emerged as a need to model nonlinear decision regions. It was proved [2, 7] that there always exists a three-layer network (with some desired numbers of neurons) that can approximate (with a desired level of accuracy) any arbitrary nonlinear, continuous, multidimensional function f. The theorem does not, however, reveal how many neurons are required to approximate a specific function. The insufficient quantity of hidden neurons will mean insufficient learning and can thus be linked to the failure to develop an exact model. A high-dimensional polynomial function could need a thousand neurons too. Due to this uncertainty, these networks failed to develop an accurate model, often ending up with either much more (causing overfit) or much less (causing underfit) neurons than actually needed. In practice, one or two hidden layers (that is three-layered or four-layered neural networks, respectively) are commonly used in many pattern recognition tasks.

It was proved that if a circuit with depth d needed n nodes to model a function, one with depth $d - 1$ will need 2^n nodes [4, 5]. Intuitively, a four-layer neural network would perform better in modeling nonlinear problems with effectively lesser neurons than a three-layer network. Structural optimization algorithms that determine the optimal number of layers have also been investigated [3].

Machines are becoming increasingly more intelligent—able to see, speak and even think like us. Some people even claim it to be "deep learning" which has revolutionized machine learning. The computers need only a series of algorithms to perceive and recognize objects. What differentiates deep learning from other forms of machine learning is the inspiration that it took from the fundamental framework of our perception of the brain. Models can learn patterns from the data if they are subjected to a large amount of data. Such types of data include pictures, videos, audio, human language, etc.

21.2.1   What Is Deep?

Depth is a dimension in neural networks, and it measures the number of hidden layers. Shallow networks have one hidden layer. Deep neural networks have more than one. Each hidden layer can have a different number of nodes, where the features of the previous layer are recombined. This series of recombination is the basis of a feature hierarchy. In a feature hierarchy, a neural network starts by working with low-level, relatively concrete features and moves on to more complex and abstract features. One can gain an intuition of this by imagining a neural network that processes facial images. In its visible or input layer, it takes a pixel in each node. Each pixel is represented as a feature in the input layer. In the subsequent, hidden layer, those pixels are recombined, and the nodes detect edges, or lines of pixels (this might be the angle of your upper eyelid). In the layer after that, the edges are recombined to form intersections, or combinations of edges (this might be the outer corner of your left eye). And so on and so forth as nodes recognize whole eyes, regions of a face, and whole faces. That is a feature hierarchy, and deep neural networks can learn them automatically.

21.2.2 Why Going Deep?

Neural networks are not really a new concept. It originated during the 1950s, and many of the significant algorithmic contributions were introduced in the 1980s and 1990s. The transition that has occurred is that computer scientists are today finally leveraging both the immense computing power and the huge data storage facilities which include photos, videos, audio, and text files distributed over the Internet. In order to make a neural network function efficiently, it seems that such massive amounts of data are necessary. Deep learning can be interpreted as a subset of the neural network learning. "Artificial intelligence" covers a broad range of technologies, such as conventional logic and rule-based systems, allowing computers and robots to make them solve problems in respects which resemble (at least superficially) human thinking and reasoning process. A subgroup in this vast range is called machine learning. These are powerful mathematical approaches that enable computers to enhance their skills in carrying out tasks through experiences. Furthermore, the smaller subcategory under the realm of machine learning is called deep learning.

Now, MLPs have been around for 25 years already. But the algorithms were not good at learning weights for networks with more hidden layers. The new thing in deep learning is the algorithms for training many-layer networks. An algorithm is deep if the input is passed through several nonlinearities before being output.

However, many researchers found that having more than two hidden layers provided only diminishing effects (slowed the training time significantly but did not improve results). This is because these first hidden layers are too far from the output (which is the source of learning by back-propagation) and are basically more influenced by initial random setting than by real training data.

21.2.3 Vanishing and Exploding Gradients

Vanishing gradient [6] problem is a difficulty faced while training some artificial neural networks with gradient-based methods (e.g., back-propagation). To be more specific, this problem makes it really hard while learning and tuning the parameters of the earlier layers in the network. This problem worsens as the number of layers in the architecture increases. This is not a fundamental problem with neural networks. It is a problem with gradient-based learning methods caused by certain activation functions. Gradient-based approaches study how a small change in the value of a parameter influences the performance of the network. When a variation in the value of the parameter affects the network performance only marginally, the network actually cannot fully learn the parameter, which seems to be a concern.

This is exactly what happens with the issue of the vanishing gradient. The network output gradients are extremely small compared to the parameters in the early layers. This is an excellent way to say that even a significant change in the value of early layer parameters is not of great importance for output.

The problem with the gradient shrinking depends on what kind of activation function is selected. Some specific activation functions (e.g., sigmoid or tanh) squeeze the input into a very small range of outputs in a nonlinear manner. Sigmoid, for example, maps the real numbers into a restricted range of [0, 1]. As a consequence, large sections of the input space are restricted to a very small range. Even a major change in the input would produce a small difference in the output in these regions of the input space—i.e., the gradient is minimal.

If we add more layers of such nonlinearities on top of each other, that becomes even worse. For example, the first layer would map a large area of input to a smaller region of output, and the second layer will map it to an even smaller area. This function is mapped by the third layer to an even smaller region, etc. As a consequence, the output does not vary much even with large changes to the parameters of the first layer.

The explosion of gradients [11] is more of a concern in recurrent networks where the reverse happens because of a Jacobian with a determinant above 1. More commonly, the gradient of deep neural networks works out to be unstable and tends to explode or vanish in initial layers. Such inconsistency is a basic problem for deep neural networks' gradient-based learning.

21.3 Building Blocks of Deep Architecture

Most of the deep neural networks have very simple level-wise building blocks. In this section, we will explore some of these building blocks of deep neural networks which can be stacked to form a deep architecture.

21.3.1 Restricted Boltzmann Machines

A restricted Boltzmann machine (RBM) [8] is an algorithm that is used for dimensionality reduction, classification, regression, collaborative filtering, feature learning, and topic modeling. Thanks to its relative simplicity and historic significance, the first neural network we will explain in this chapter is the restricted Boltzmann machine.

Initially, RBMs had been coined as Harmonium [14]. RBMs are shallow neural networks of two layers, which serve as a basis for deep-belief networks. The first layer of the RBM is termed as the visible layer (or input layer), and the second layer is the hidden layer. Each circle in the above diagram (Fig. 21.2) represents a neuron or node. Nodes perform calculations. The nodes are connected together across layers, but there are no connections between two nodes of the same layer. In a restricted Boltzmann machine, this is the restriction. Each node is a computation

Fig. 21.2 The layers in an RBM

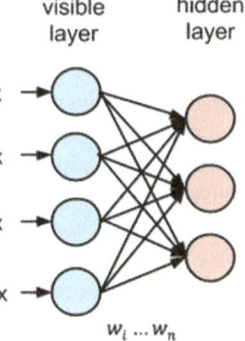

Fig. 21.3 A two-layer RBM

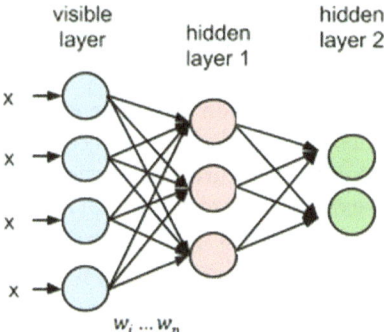

locus which processes input and starts by stochastically determining whether or not to transmit this data. Because inputs are passed from all visible nodes onto all hidden nodes, an RBM can be represented as a bipartite graph which has symmetry. This symmetry signifies that each visible node has a connection to each hidden node. Bipartite is a mathematical formalism for a network of nodes and has two parts or layers.

Each hidden node receives a stimulus (from an input node) equal to the input $\mathbf{x}$ multiplied by the respective weight $\mathbf{w}$ connecting them. This means, a single input $\mathbf{x}$ would, in the above example (Fig. 21.2), have 3 weights, resulting into 12 weights in total (4 input nodes × 3 hidden nodes). The weights between the two layers can be denoted by a matrix in which the numbers of rows are the same as the number of input nodes and the number of columns being the same as that of the output nodes.

The input to any hidden node is all the four inputs multiplied by the corresponding weights. Again, the sum of these products is added with a bias (which forces certain activations to occur at least), resulting in the activation function generating one output for each hidden node. If these two layers were portions of a deeper neural network (Fig. 21.3), the outputs of hidden layer number 1 would be passed as inputs to hidden layer number 2, and from there would be passed through various hidden layers until they get to a final classifying layer. The RBM nodes serve as an autoencoder and nothing more for plain feedforward actions. We train to regenerate data directly in an unsupervised manner and transfer between visible and hidden layers with several forward and reverse passes without deeper connectivity. The activations of the hidden layer are converted into the input in a backward pass during the reconstruction phase. They are reproduced using the same weights used in the forward pass, one per internode connection. Approximation of the initial input is the sum of those products to a visibility layer bias at each visible node, and the output of these operations is a reconstruction.

Consider an RBM which was fed only elephant and dog photos and only had two output nodes, one per animal. On its way forward, the query that the RBM is asking itself is: Do I give the elephant node or the dog node a stronger signal considering these pixels? And on the backward pass, the RBM asks: Provided the elephant which pixels would I anticipate?

This is joint probability, the simultaneous probability of a given x and of x given a, represented as the shared weights between the two layers of the RBM.

In a sense, the process of learning reconstructions is to learn that for a given set of images, which groups of pixels tend to co-occur. The activations that the nodes of hidden layers in the network produces represent significant co-occurrences; e.g., "nonlinear gray tube + big, floppy ears + wrinkles" might be one for elephant class.

21.3.2 Autoencoders

An autoencoder [15] is a neural network that tries to reconstruct its input. Autoencoders are composed of an input, a hidden, and an output layer. The output layer is a reconstruction of the input through the activations of the much fewer hidden nodes. It offers an elegant solution to dimensionality reduction and compression similar to PCA. We can omit the output layer after training and use the hidden layer as input features for downstream classification or another autoencoder.

Though the working of RBM and autoencoders may seem similar, they are slightly different. RBMs are made out of an input and hidden layer. We are trying to find a stochastic representation of the input. By sampling from the hidden layer, we can reproduce variants of samples encountered during training. The training of RBMs emerges through alternate sampling of both layers which is different from how one would train autoencoders, although back-propagation could still be used later to fine-tune the model.

An autoencoder works as a replicating MLP. So, if anyone feeds the autoencoder the vector (1,0,0,1,0), the autoencoder will attempt to output (1,0,0,1,0). The trick is the hidden layer. Let us consider that we have inputs in five dimensions as in our example. If two neurons inside the hidden layer are being utilized, then five features will be inputted into our autoencoder and "encode" them into two features in a way such that it can reconstruct the same five-dimensional input (Fig. 21.4).

So we go from (1,0,0,1,0) to (x, y) and from (x, y) to (1,0,0,1,0).

The autoencoder is trained with large quantities of data points, in the order of even thousands of millions, and it will scan through the search space for the

Fig. 21.4 An example of an
autoencoder

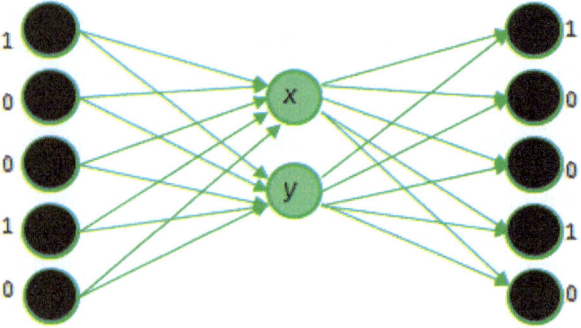

weights that will decrease the reconstruction error to the minimum possible value.
The weights are what we use to transform $(1,0,0,1,0)$ into $(\mathbf{x}, \mathbf{y})$ and $(\mathbf{x}, \mathbf{y})$ into
$(1,0,0,1,0)$. Autoencoders can be stacked on top of each other and trained in a
continuous way going from one autoencoder to the next. We train an autoencoder,
and then we take the middle layer produced by it and input it into the next
autoencoder and continue on in this way. This is the initial step for achieving deep
learning. The stacked autoencoders will be trained on how to represent the data
in a hierarchical way. The first autoencoder will output a basic representation, the
second level will combine the previous representation to produce a higher-level
representation, and the next autoencoders will work in the same way. For example,
let us consider images in general. A first-level autoencoder can be trained to notice
borders as features, the second level will merge those borders to learn detection of
traces and patterns, etc.

Nonlinear encoding functions f with nonlinear decoding functions g in an
autoencoder makes it capable of producing a generalized version of nonlinear PCA.
Sadly, the autoencoder may learn to do the copying task without capturing important
information about the data distribution if the encoder and the decoder were
permitted to have unnecessarily more capacity. Theoretically, it can be imagined
that an autoencoder with a one-dimensional code layer but having a very powerful
nonlinear encoder function f is able to learn to represent each training example
$\mathbf{x}_{(i)}$ with the code i in the one-dimensional code (hidden) layer. The decoder might
learn to map these integral indices back to the values of specific training examples.
This specific scenario practically might not happen, but it surely illustrates that an
autoencoder trained to perform the identity copying task will definitely fail to learn
any important characteristics about the dataset if the autoencoder is allowed too high
capacity.

The more outstanding features of the data distribution can be learned by under-
complete autoencoders with less code dimension than the input dimension. These
autoencoders do not learn any useful things when you give too much capacity to
the encoder and decoder. A similar problem exists when the hidden layer has same
dimensions as that of the input dimension. When the hidden code layer has larger
dimensions than the input, it is called overcomplete autoencoder. Even a linear

encoder and linear decoder can learn how simply the input can be duplicated to the output instead of learning any useful information about the data distribution in such cases. Ideally, one can successfully train any autoencoder design by choosing the encoder and decoder dimensions and capacity based on the distribution complexity to be constructed. Varied techniques seek to prevent autoencoders from developing the identity function and to improve their generalization capacity to gather and learn valuable information.

Regularized autoencoders use a loss function instead of constraints on model capacity, encouraging that the model needs different properties in addition to copying its input to its output. The other characteristics include the sparse presentation, the small values of the derivative, and the resilience to noise or to missing inputs. A regularized autoencoder can be nonlinear and overcomplete but can still continue to learn some interesting things about data distribution even when the model is good at learning a trivial identity function.

Through making hidden nodes sparse (when they have more hidden nodes than inputs) during the training process, an autoencoder may learn useful structures in the input data. This enables sparse input representations. Sparse autoencoders are generally used to learn features required for classification or a different task. A sparse-regularized autoencoder will respond to the particular statistical features of the dataset it has learnt, instead of simply working as an identity function. In this way, replicating with a sparsity penalty can produce a model that has managed to learn informative features as a by-product.

Denoising autoencoders receive partly corrupted input during training to recover the input. This technique was adopted with a particular approach to better representations. A better representation is one that can be robustly retrieved from corrupt data and is beneficial in obtaining the relevant clean information.

21.4   Types of Deep Neural Networks

With the emergence of new neural network models, it gets difficult to track them all. In the following few parts, we have mentioned certain basic deep networks among the different ones.

21.4.1   Deep Belief Networks

The terminology for the layered networks of mostly RBMs is the deep belief network (DBN). The networks were shown to be efficiently trained stack by stack, in which each RBM only needs to encode the previous layer's output. This method is also referred to as greedy training which implies local optimization alternatives for a decent but certainly not the perfect result. DBNs can be trained by contrastive divergence or back-propagation and can, like standard RBMs, learn to interpret data as a probabilistic model. The model can be used to create new data once it is trained or converged to (more) stability through unsupervised learning. It can even classify existing data if trained with contrastive divergence, because the neurons have been taught to search for different features.

The two most important properties of deep belief nets are the following:

(i) An efficient, layer-by-layer method is used for learning the top-down, generative weights that influence how the variables in one layer are a function of the variables in the layer above.
(ii) After learning, the values of the latent variables in every layer can be inferred by a single, bottom-up pass that starts with an observed data vector in the bottom layer and uses the generative weights in the reverse direction.

Deep belief nets learn the required weights layer by layer by using the values of the latent variables in one layer, when they are being inferred from data, as the input data for training the previous layer. This efficient, greedy learning can be followed by, or used in conjunction with, other learning methods that fine-tune all of the weights to enhance the generative or discriminative performance of the whole network.

21.4.2 Deep Convolutional Neural Networks

The convolutional neural networks (CNN), which were originally called neocognitron, are interestingly different types of networks. Their primary tasks are image processing but can also be modified to be used for various types of inputs like auditory signals, videos, etc. A traditional use case for CNNs is where the network is fed with some cat/dog images, and the data is classified by the network; e.g., it gives the output as "cat" if the picture of a "cat" is fed and gives output as "dog" when we feed it picture of a "dog." CNNs tend to begin with an input "scanner" which is not intended to parse all the training data at once. Like, while feeding a 200×200 image, the convention is not to make an input layer of 40,000 nodes. Rather, the sliding window is created which scans say 20×20 window of the image at a time which usually starts scanning from the upper left corner. It starts with the first pixel position, and as soon as the information is propagated down to the output node (and possibly use it for training), the next 20×20 pixels are fed to the network (i.e., it is done by shifting the scanner one pixel to the right). It should be noted that one would not shift the input by 20 pixels (or whatsoever is the scanner width), as we are not dividing the image into separate blocks of dimension 20×20, but rather sliding over the image. Next instead of normal layers, the input data is fed through convolutional layers, where all nodes are not connected to each other. Each node is only concerned about its close neighboring cells (the closeness depends on the implementation, but generally it is not more than a few). As they tend to become deeper, these convolutional layers also adapt a tendency to shrink , mostly the shrink happens by some easier divisible factors of the input (so 20 would probably shrink down to a layer of 10 followed by a layer of 5).

Generally, powers of two are very commonly used, because then the division would become clean and complete; according to definition, 32, 16, 8, 4, 2, and 1 are most commonly used as division factor. In addition to these convolutional layers, another layer that gets mostly used is the pooling layers. Pooling is a technique of details' filtration. The most commonly used pooling technique is termed as max pooling, where say like 2×2 pixels and pass on the pixel with the most amount of red. When CNNs are applied in audio, then basically the input audio waves are fed an inch over the length of the clip, segment by segment. Actual applied implementations of CNNs often add an FFNN as the final step to further operate on the data, which makes highly nonlinear abstractions possible. These networks are known as DCNNs, but the nomenclatures and abbreviations between these two often get used as interchangeably. Convolutional neural networks are biologically inspired category of multilayer perceptrons, constructed in a way to imitate the behavior of a visual cortex. These models diminish the challenges that are posed by the MLP architecture, by utilizing the strong spatially local correlation available in natural images. As opposed to MLPs, CNN have the following distinguishing features:

(i) **3D volumes of neurons**: The CNN layers have three dimensions of neurons: width, height, and depth. The neurons inside the layer are only connected to a limited area of the layer before it, termed as the receptive field. Different types of layers, both locally and fully connected, are piled to form the CNN architecture.

(ii) **Local connectivity**: Using the principle of responsive fields, CNNs leverage spatially local correlation by introducing a spatial pattern of a local connectivity between adjacent layer neurons. Therefore, the design ensures that the trained "filters" provide the best response to the spatially local input sequence. Stacking several such layers corresponds to nonlinear "filters" that become gradually "global" (i.e., responsive to a broader region of pixel space). It helps the network to first create good representations of smaller input sections and then compose larger field representations from those.

(iii) **Shared weights**: In CNNs, each filter is replicated across the entire visual field. These replicated units share the same parameterization (weight vector and bias) and form a feature map. This means that all the neurons in a given convolutional layer detect exactly the same feature. Replicating units in this way allows for features to be detected regardless of their position in the visual field, thus constituting the property of translation invariance.

Together, these characteristics enable convolution neural networks to produce a better generalization of vision problems. Weight sharing often significantly reduces the amount of free parameters learnt, thereby decreasing the space requirements for running the network. Decreasing the memory footprint makes it possible to train bigger, more efficient networks. However, this does not mean that the network will be trained faster as convolution operations are much more complex than those of fully connected layers. This would seem counterintuitive as for a fully connected network, each weight is updated only once during a single epoch of backpropagation. However, with convolution, the same kernel weights are updated for

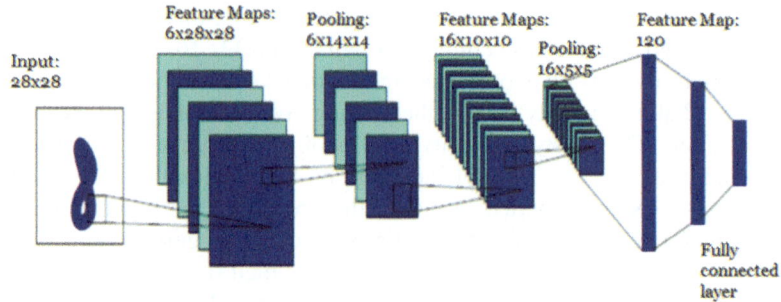

Fig. 21.5 Architecture of a convolutional neural network

each input pixel, and there are more than one kernel in a CNN. So, obviously the number of operations increases (Fig. 21.5).

Convolutional layers introduce a variety of filters to the input data. For example, the first convolution layer of the image might have many 6×6 filters. The effect of a filter that is applied across the whole image is called a feature map (FM), and the number of feature maps is equal to the total number of filters used. If the previous layer is also convolutional, the filters are applied across all of its FMs with different weights, so each input FM gets connected to each output FM. The basic idea behind the shared weights across the image is that the features will be detected regardless of their location, while the multiplicity of filters allows each of them to detect different sets of features. Subsampling layers decrease the input size. For instance, if a 28×28 image is used as input and the layer has a subsampling region set to 2×2, the output value would comprise a 14×14 image, which means a single pixel in the output image is formed from the condensation of 4 pixels (each 2×2 square) of the input image. There are several ways to subsample, but the most commonly used ones are max pooling, average pooling, and stochastic pooling. The last subsampling (or convolutional) layer is usually connected to one or more fully connected layers; the last of them represents the target data. Training is performed using modified back-propagation where the subsampling layers are taken into account and the convolutional filter weights are updated based on all values to which that filter gets applied (Fig. 21.6).

Nonetheless, there are still a few factors to consider while developing a CNN. There are many hyperparameters to choose from, such as learning rates and regularization, number of filters, filter shape, max-pooling structure, etc. From a memory and efficiency point of view, CNN is not much bigger than a standard two-layer network. During runtime, convolution operations are computationally expensive and require around 67% of the time because of the reasons stated above. CNNs are about three times slower than their fully connected counterparts (size-wise).

Fig. 21.6 Working of the convolution layers

Input Image Convolution Kernel Feature Map

$$\begin{bmatrix} -1 & -1 & -1 \\ -1 & 8 & -1 \\ -1 & -1 & -1 \end{bmatrix}$$

Image Convolved Feature

21.4.3 Deep Autoencoders

In the previous sections, the idea about the autoencoders has been introduced. A stacked autoencoder is a neural network which is made up of multiple layers of autoencoders in which each layer's output is connected to the inputs of the next layer. Usually, autoencoders are trained in an unsupervised, greedy layer-wise manner. Usually the autoencoders do not need any labels and learn to capture the latent feature spaces by reconstructing the input through a bottleneck layer. The weights can be learned using a variety of techniques ranging from "batch" gradient descent [12] to mini-batch stochastic gradient descent (SGD) [1] and to quasi-Newton methods like L-BFGS [10].

A deep autoencoder is nothing but an autoencoder which has more than two hidden layers. The hope is that the weights acquired in an unsupervised method to minimize reconstruction error for the representation learning process provide a good starting point for the network to be configured for a supervised discriminatory task such as classification or finding similarities, i.e., that the network learns something about the underlying distribution by looking at unlabeled samples, allowing it to discriminate across the intrinsic features. Nonetheless, for this new task, the weights still need to be "fine-tuned." So, we add a logistic regression layer or a softmax layer to the top of the network, and then we do supervised learning with a labeled dataset. The fine-tuning step will simultaneously perform gradient descent and adjust the weights for all layers in the network.

References

1. Bottou, L. (2010). Large-scale machine learning with stochastic gradient descent. *Proceedings of COMPSTAT' 2010* (pp. 177–186)
2. Cybenko, G. (1989). Approximation by superpositions of a sigmoidal function. *Mathematics of Control, Signals and Systems, 2*(4), 303–314.

3. Doering, A., Galicki, M., & Witte, H. (1997). Structure optimization of neural networks with the A*-Algorithm. *IEEE Transactions on Neural Networks, 8*(6), 1434–1445.
4. Håstad, J. (1987). *Computational limitations of small-depth circuits.* MIT Press, Cambridge.
5. Håstad, J., & Goldmann, M. (1991). On the power of small-depth threshold circuits. *Computational Complexity, 1*(2), 113–129.
6. Hochreiter, S., Bengio, Y., Frasconi, P., & Schmidhuber, J. (2001). *Gradient flow in recurrent nets: The difficulty of learning long-term dependencies.* IEEE Press.
7. Hornik, K., Stinchcombe, M., & White, H. (1989). Multilayer feedforward networks are universal approximators. *Neural Networks, 2*(5), 359–366.
8. Larochelle, H., Mandel, M., Pascanu, R., & Bengio, Y. (2012). Learning algorithms for the classification restricted boltzmann machine. *Journal of Machine Learning Research, 13*, 643–669.
9. Newman, M. (2006). Modularity and community structure in networks. *Proceedings of the National Academy of Sciences, 103*(23), 8577–8582.
10. Nocedal, J. (1980). Updating quasi-newton matrices with limited storage. *Mathematics of Computation, 35*(151), 773–782.
11. Pascanu, R., Mikolov, T., & Bengio, Y. (2012). Understanding the exploding gradient problem. ArXiv abs/1211.5063
12. Ruder, S. (2016). An overview of gradient descent optimization algorithms. arXiv preprint arXiv:1609.04747
13. Simon, H. A. (1962). The architecture of complexity. *Proceedings of the American Philosophical Society, 106*(6), 467–482.
14. Smolensky, P. (1986). Information processing in dynamical systems: Foundations of harmony theory. Technical report, DTIC Document.
15. Vincent, P., Larochelle, H., Bengio, Y., & Manzagol, P. (2008). Extracting and composing robust features with denoising autoencoders. *Proceedings of the 25th International Conference on Machine Learning* (pp. 1096–1103).

Chapter 22
Epilogue

With loads of data being generated every second all over the globe, the need of utilizing that data is increasing continuously. Data is also being sometimes referred to as a corporate currency in the current scenario. Most organizations and companies of the world accumulate huge amounts of raw data waiting to be processed. The underlying information within these heaps of data is unknown yet intriguing. With the infrastructure to hold and process this data, it has gained immense importance in the past few years. Thus, to deal with this ever-growing data, data science has to evolve constantly. With the advent of technology, humans are making new inventions and discoveries at every step. Thus, the knowledge associated with this field of data science is ever-growing. The aim of this book, primarily, has been to touch upon most of the traditional and well-known tools and methods as well as few of the newer members of the data science family. The book has been divided into broad categories, and each of them has been discussed fairly in detail. However, there are still innumerable ideas and topics that could not be included. Thus, we briefly mention few such developing areas which have been left out due to the limited scope of the book.

One field in particular is receiving attention from researchers, while creating innovative adaptations is the field of machine learning [3, 10]. Various subfields are now appearing within this large domain. One such emerging area is quantum machine learning [20], which lies at the intersection of quantum physics [11] and machine learning. As quantum physics refers to the study of energy and matter at a fundamental level, quantum machine learning aims to improve the capacities of machine learning intelligently at a much lower level. Similarly, another subfield, known as automated machine learning (AutoML) [4], aims to automate the end-to-end process of machine learning on real-world problems. It would be able to perform all the steps involved in a machine learning scenario automatically. Multimodal machine learning [1] is yet another direction in this field which can process and deal with information that arrives from various modalities. Probably approximately correct (PAC) learning [18] is a mathematical analysis model of machine learning, which aims with high probability, that a selected function will have low generalization error. The learner should be able to capture any concept given an arbitrary approximation ratio. Many such directions of machine learning are unfolding over time.

Diving deeper into learning, a specific tool that has evolved enormously over the past few years is artificial neural networks (ANN) [7]. Diverse array of ANN models are being developed vigorously and are being widely used. Capsule networks (CapsNet) [13, 14] attempt to model a hierarchical relationship in the data it learns, trying to mimic the learning in a human brain. It can recognize objects as a combination of individual smaller entities instead of a single one. Generative adversarial networks (GAN) [6], mainly used for generating new images, videos, text, etc., are ideally able to generate new content. It also learns to identify real and generated fake samples so as to improve the generation capability of the system. Recurrent neural networks (RNN) [9] are simple networks that are able to preserve historic information explicitly and use them to train itself. They are widely used to work on inputs of variable lengths. Residual networks (ResNet) [8] are another variation of ANNs where the connections between layers can jump from one layer to another, specifically termed as skip connections. Unlike some ANNs, ResNets reuse activations from previous layers and eventually avoid vanishing gradient problems. Memory-augmented neural network (MANN) [16] is based on neural Turing machine and memory network. They have an external memory associated with it which can learn especially long sequential data. Among the many ANN adaptations in literature, autoencoders (AEs) [19] are quite famously used for deep networks. One of its recent adaptation known as variational autoencoders [5] can handle variations in attributes or samples which was a bottleneck for the original AEs. Many such adaptations exist in the literature.

Moving over from a specific direction, we now touch upon more general areas. A phrase that we often come across is cognitive computing [12]. It aims to simulate human thought process using classical techniques from machine learning, data mining, etc. Human cognition involves real-time analysis of the environment, context and intent, among many other variables that inform a person's ability to solve problems. In general, cognitive computing is currently used to assist humans in their decision-making process. Mimicking of human cognition closely has particularly increased the significance of real-time data analysis [2]. This has unfolded a new direction of processing data that is nonstationary in nature. It appears as a stream of data that needs to be handled in real time. Multiple supervised and unsupervised learning technologies are being developed around this subfield of research. Explainable AI (XAI) [15] is a fairly new concept which attempts to exceed the black box outlook of learning algorithms. It provides an explanation of the concepts that are being learned by the system thus making human intervention within the black box possible. With the evolution of technology, humans and machines are getting access into depths of the data used for data science. This makes data security a major concern. Keeping that in mind, a new branch of technology referred as block chains [17] has developed in recent times. It aims that at any point of time, the digital information being processed is distributed as blocks over a cluster of systems, and a combination of these blocks can only construct the exact information. This ensures that hacking any single system will never divulge the entire data thus saving intricate information from falling into wrong hands. Every technology developed in today's world has a certain amount of risk associated with it. Humans need to assess and control the extent of good or harm technology can do.

Evolution of data science is an ongoing process which is opening doors to many new ideas and directions which were unimaginable a few decades back. At any given point of time, it is quite a far-fetched attempt to enclose this vast area of knowledge into a finite space, thus, leaving a scope for the readers to venture out and explore the unknown.

References

1. Baltrušaitis, T., Ahuja, C., & Morency, L.-P. (2018). Multimodal machine learning: A survey and taxonomy. *IEEE Transactions on Pattern Analysis and Machine Intelligence, 41*(2), 423–443.
2. Bifet, A., Gavaldá, R., Holmes, G., & Pfahringer, B. (2018). *Machine learning for data streams: With practical examples in MOA*. MIT Press.
3. Bishop, C. M. (2006). *Pattern recognition and machine learning* (2nd ed.). Springer, New York.
4. Das, S., & Cakmak, U. M. (2018). *Hands-on automated machine learning: A beginner's guide to building automated machine learning systems using autoML and python*. Packt Publishing Ltd.
5. Doersch, C. (2016). Tutorial on variational autoencoders. arXiv preprint arXiv:1606.05908
6. Ganguly, K. (2017). *Learning generative adversarial networks: Next-generation deep learning simplified*. Packt Publishing.
7. Haykin, S. (1999). *Neural networks: A comprehensive foundation* (1st ed.). Prentice Hall, Inc., New Jersey.
8. He, K., Zhang, X., Ren, S., & Sun, J. (2015). Deep residual learning for image recognition. *The Computing Research Repository*. abs/1512.03385
9. Medsker, L., & Jain, L. C. (1999). *Recurrent neural networks: Design and applications*. CRC Press.
10. Mitchell, T. M. (1997). *Machine learning* (1sd ed.). McGraw-Hill, Inc., New York.
11. Rae, A. (2005). *Quantum physics: A beginner's guide*. Oneworld Publications.
12. Raghavan, V. V., Gudivada, V. N., Govindaraju, V., & Rao, C. R. (2016). *Cognitive computing: Theory and applications* (vol. 35). Elsevier.
13. Sabour, S., Frosst, N., & Hinton, G. (2018). Matrix capsules with EM routing. In *6th International Conference on Learning Representations, ICLR*.
14. Sabour, S., Frosst, N., & Hinton, G. E. (2017). Dynamic routing between capsules. *The Computing Research Repository*. abs/1710.09829.
15. Samek, W. (2019). *Explainable AI: Interpreting, explaining and visualizing deep learning*. Springer.
16. Santoro, A., Bartunov, S., Botvinick, M., Wierstra, D., & Lillicrap, T. (2016). Meta-learning with memory-augmented neural networks. *International Conference on Machine Learning* (pp. 1842–1850).
17. Swan, M. (2015). *Blockchain: Blueprint for a new economy*. OReilly Media, Inc.
18. Valiant, L. (2013). *Probably approximately correct: Nature's algorithms for learning and prospering in a complex world*. Basic Books (AZ).

19. Vincent, P., Larochelle, H., Lajoie, I., Bengio, Y., & Manzagol, P. A. (2010). Stacked denoising autoencoders: Learning useful representations in a deep network with a local denoising criterion. *The Journal of Machine Learning Research, 11*, 3371–3408.
20. Wittek, P. (2014) *Quantum machine learning: What quantum computing means to data mining.* Academic.

Appendix A
Matrix Algebra

Since modeling of machine learning algorithms require an extensive use of matrix and vector operations, let us brush up them a little bit.

A.1 Definitions

A matrix of dimensions $(m \times n)$, m and n positive integers, is an array of elements a_{ij} arranged into m rows and n columns:

$$
\mathbf{A} = \begin{bmatrix} a_{11} & a_{12} & \dots & a_{1n} \\ a_{21} & a_{22} & \dots & a_{2n} \\ \vdots & \vdots & \dots & \vdots \\ a_{m1} & a_{m2} & \dots & a_{mn} \end{bmatrix}. \tag{A.1}
$$

If $m = n$, the matrix is said to be square; if $m < n$, the matrix has more columns than rows; if $m > n$, the matrix has more rows than columns. Further, if $n = 1$, the notation is used to represent a (column) vector $\vec{\mathbf{a}}$ of dimensions $(m \times 1)$; the elements a_i are said to be vector components. A square matrix $\mathbf{A}$ of dimensions $(n \times n)$ is said to be upper triangular if $a_{ij} = 0$ for i > j:

$$
\mathbf{A} = \begin{bmatrix} a_{11} & a_{12} & \dots & a_{1n} \\ 0 & a_{22} & \dots & a_{2n} \\ \vdots & \vdots & \dots & \vdots \\ 0 & 0 & \dots & a_{mn} \end{bmatrix}. \tag{A.2}
$$

Similarly, a matrix is said to be lower triangular if $a_{ij} = 0$ for $i < j$.

© The Author(s), under exclusive license to Springer Nature Singapore Pte Ltd. 2025
A. Ghosh, *Data Science and Cases in Sustainability*, Mathematics for Sustainable
Developments, https://doi.org/10.1007/978-981-96-8362-8

An $(n \times n)$ square matrix $\mathbf{A}$ is said to be diagonal if $a_{ij} = 0$ for $i \neq j$, i.e.,

$$\mathbf{A} = \begin{bmatrix} a_{11} & 0 & \dots & 0 \\ 0 & a_{22} & \dots & 0 \\ \vdots & \vdots & \dots & \vdots \\ 0 & 0 & \dots & a_{mn} \end{bmatrix}. \tag{A.3}$$

The product of a scalar α by an $(m \times n)$ matrix $\mathbf{A}$ is the matrix $\alpha\mathbf{A}$ whose elements are given by αa_{ij}. If an $(n \times n)$ diagonal matrix has all unit elements on the diagonal $(a_{ii} = 1)$ and all other elements are zero $(a_{ij} = 0, i \neq j)$, the matrix is said to be identity and is denoted by I. Thus for a 3×3 matrix

$$\mathbf{I} = \begin{bmatrix} 1 & 0 & 0 \\ 0 & 1 & 0 \\ 0 & 0 & 1 \end{bmatrix}. \tag{A.4}$$

If A is an $(n \times n)$ diagonal matrix with all equal elements on the diagonal $(a_{ii} = a)$, it follows that $\mathbf{A} = a\mathbf{I}$. A matrix is said to be null if all its elements are null and is denoted by O.

The transpose $\mathbf{A}^T$ of a matrix $\mathbf{A}$ of dimensions $(m \times n)$ is the matrix of dimensions $(n \times m)$ obtained from the original matrix by interchanging its rows and columns:

$$\mathbf{A}^T = \begin{bmatrix} a_{11} & a_{21} & \dots & a_{m1} \\ a_{12} & a_{22} & \dots & a_{m2} \\ \vdots & \vdots & \dots & \vdots \\ a_{1n} & a_{2n} & \dots & a_{mn} \end{bmatrix}. \tag{A.5}$$

The transpose of a column vector $\vec{\mathbf{a}}$ is the row vector $\mathbf{a}^T$.

An $(n \times n)$ square matrix $\mathbf{A}$ is said to be symmetric if $\mathbf{A}^T = \mathbf{A}$, and thus $a_{ij} = a_{ji}$. An $(n \times n)$ square matrix $\mathbf{A}$ is said to be skew-symmetric if $\mathbf{A}^T = -\mathbf{A}$.

A.2   Matrix Operations

Two matrices $\mathbf{A}$ and $\mathbf{B}$ of the same dimensions $(m \times n)$ are equal if $a_{ij} = b_{ij}$. If $\mathbf{A}$ and $\mathbf{B}$ are two matrices of the same dimensions, their sum is the matrix

$$\mathbf{C} = \mathbf{A} + \mathbf{B}, \tag{A.6}$$

whose elements are given by $c_{ij} = a_{ij} + b_{ij}$.

The following properties hold:

$$\mathbf{A} + \mathbf{0} = \mathbf{A};$$

$$\mathbf{A} + \mathbf{B} = \mathbf{B} + \mathbf{A};$$

$$(\mathbf{A} + \mathbf{B}) + \mathbf{C} = \mathbf{A} + (\mathbf{B} + \mathbf{C}).$$

If $\mathbf{A}$ is a square matrix, one may write

$$\mathbf{A} = \mathbf{A}_s + \mathbf{A}_a \tag{A.7}$$

where

$$\mathbf{A}_s = (\mathbf{A} + \mathbf{A}^T) \tag{A.8}$$

is a symmetric matrix representing the symmetric part of $\mathbf{A}$, and

$$\mathbf{A}_a = (\mathbf{A} - \mathbf{A}^T) \tag{A.9}$$

is a skew-symmetric matrix representing the skew-symmetric part of $\mathbf{A}$. Note that two matrices of the same dimensions and partitioned in the same way can be summed formally by operating on the blocks in the same position and treating them like elements.

The row-by-column product of a matrix $\mathbf{A}$ of dimensions $(m \times p)$ by a matrix $\mathbf{B}$ of dimensions $(p \times n)$ is the matrix of dimensions $(m \times n)$

$$\mathbf{C} = \mathbf{A}\mathbf{B} = [c_{ij}] \tag{A.10}$$

where

$$c_{ij} = \sum_{k=1}^{p} a_{ik} b_{kj}. \tag{A.11}$$

For a square matrix $\mathbf{A}$, the notation $\mathbf{A}^n$ for a positive integer n stands for the product $\mathbf{A}\mathbf{A}\ldots\mathbf{A}$ (n times) and $\mathbf{A}^0 = \mathbf{I}$. Matrix multiplication has the following properties:

(i) $\mathbf{A}\mathbf{B}\mathbf{C} = \mathbf{A}(\mathbf{B}\mathbf{C}) = (\mathbf{A}\mathbf{B})\mathbf{C}$.

(ii) $(\mathbf{A} + \mathbf{B})\mathbf{C} = \mathbf{A}\mathbf{C} + \mathbf{B}\mathbf{C}$.

(iii) $(\mathbf{A}\mathbf{B})^T = \mathbf{B}^T \mathbf{A}^T$ if $\mathbf{A}$ and $\mathbf{B}$ are real.

(iv) $(\mathbf{A}\mathbf{B})^\dagger = \mathbf{B}^\dagger \mathbf{A}^\dagger$ if $\mathbf{A}$ and $\mathbf{B}$ are complex, where $\mathbf{A}^\dagger$ is complex conjugate transpose of complex matrix $\mathbf{A}$.

(v) In general, matrix multiplication is not commutative, i.e., $\mathbf{A}\mathbf{B} \neq \mathbf{B}\mathbf{A}$.

A.3   Determinant of a Matrix

The determinant of a square matrix $\mathbf{A}$, denoted by $\det(\mathbf{A}) = |\mathbf{A}|$, is a scalar which provides some useful information about $\mathbf{A}$. The determinants of 2×2 and 3×3 matrices are defined, respectively, as:

$$de(\mathbf{A}) = \det \begin{bmatrix} a_{11} & a_{12} \\ a_{21} & a_{22} \end{bmatrix} = a_{11}a_{22} - a_{12}a_{21}. \tag{A.12}$$

$$de(\mathbf{A}) = \det \begin{bmatrix} a_{11} & a_{12} & a_{13} \\ a_{21} & a_{22} & a_{23} \\ a_{31} & a_{32} & a_{33} \end{bmatrix}$$

$$= a_{11} \begin{bmatrix} a_{22} & a_{23} \\ a_{32} & a_{33} \end{bmatrix} - a_{12} \begin{bmatrix} a_{21} & a_{23} \\ a_{31} & a_{33} \end{bmatrix} + a_{13} \begin{bmatrix} a_{21} & a_{22} \\ a_{31} & a_{32} \end{bmatrix}$$

$$= a_{11}a_{22}a_{33} - a_{11}a_{22}a_{32} - a_{12}a_{21}a_{33} + a_{12}a_{23}a_{31} + a_{13}a_{21}a_{32} - a_{13}a_{22}a_{31}. \tag{A.13}$$

For a general $n \times n$ matrix $\mathbf{A} = [a_{ij}]$, the determinant is defined as:

$$\det(\mathbf{A}) = \sum_{k=1}^{n} (-1)^{i+k} a_{ik} \det(\mathbf{A}_{ik}) \tag{A.14}$$

A.4   Matrix of Cofactors

Let A be a square matrix of order n; then by replacing every element a_{ij} of A by determinant of the matrix obtained by removing the ith row and jth column of matrix A matrix of cofactors C is calculated. Adjugate of A, $Adj A$ is defined as

$$Adj A = C^T, \tag{A.15}$$

where C is matrix of cofactors of matrix A.

A.5   Matrix Inversion

Let A be a square matrix of order n; then inverse of A A^{-1} is defined as

$$A^{-1} = \frac{Adj A}{|A|} \tag{A.16}$$

given $|A| \neq 0$. Inversion of a non-square matrix is dependent on rank of the matrix, and left inverse or right inverse of the matrix is calculated depending on the rank of the matrix. If $|A| = 0$, then matrix A is not invertible and called singular matrix. So, an invertible matrix is called non-singular matrix.

A.6   Matrix Orthogonality

Let $\mathbf{A}$ and $\mathbf{B}$ be two product-conforming real matrices. For example, $\mathbf{A}$ is $k \times m$, whereas $\mathbf{B}$ is $m \times n$. If their product $\mathbf{C}$ is the null matrix

$$\mathbf{C} = \mathbf{AB} = \mathbf{0}, \tag{A.17}$$

then the matrices are said to be orthogonal.

The projection operator $\mathbf{P}$ yields the orthogonal matrix of $\mathbf{A}$. The orthogonal matrix of $\mathbf{A}$ is given by

$$\tilde{\mathbf{A}} = \mathbf{PAP} \tag{A.18}$$

Suppose $\mathbf{A}$ is $m \times n$ with $m \geq n$ and that $\mathbf{A}^T\mathbf{A}$ is non-singular. Then the orthogonal projector matrix $\mathbf{P}$ is

$$\mathbf{P} = \mathbf{I} - \mathbf{A}(\mathbf{A}^T\mathbf{A})^{-1}\mathbf{A}^T \tag{A.19}$$

where $\mathbf{P}^T = \mathbf{P}$.

A.7   Eigenvalues and Eigenvectors

Eigenvectors of a matrix give the directions in which the spread of the matrix is maximum to minimum associated with eigenvalues. Eigenvalues inform how much spread is there in that associated direction. Let $\mathbf{A}$ be an $n \times n$ matrix. The number λ is an eigenvalue of $\mathbf{A}$ if there exists a nonzero vector $\mathbf{v}$ such that

$$\mathbf{A}.\mathbf{v} = \lambda\mathbf{v}. \tag{A.20}$$

In this case, vector $\mathbf{v}$ is called an eigenvector of $\mathbf{A}$ corresponding to λ.

We can rewrite the condition $\mathbf{A}.\mathbf{v} = \lambda\mathbf{v}$ as

$$(\mathbf{A} - \lambda\mathbf{I}).\mathbf{v} = 0; \tag{A.21}$$

where $\mathbf{I}$ is the $n \times n$ identity matrix. Now, in order for a nonzero vector $\mathbf{v}$ to satisfy this equation, $(\mathbf{A} - \lambda\mathbf{I})$ must not be invertible. That is, the determinant of $(\mathbf{A} - \lambda\mathbf{I})$

must equal 0. We call $p(\lambda) = \det(\mathbf{A} - \lambda\mathbf{I})$, the characteristic polynomial of $\mathbf{A}$. The eigenvalues of $\mathbf{A}$ are the roots of the characteristic polynomial of $\mathbf{A}$.

A.8   Singular Value Decomposition

Singular value decomposition of a matrix A informs in which direction the matrix is deviating from zero. To get the singular value decomposition, we can take advantage of the fact that for any matrix $\mathbf{A}$, $\mathbf{A}^T\mathbf{A}$ is symmetric. Symmetric matrices have the nice property that their eigenvectors form an orthonormal basis.

Since symmetric matrices have an orthonormal basis of eigenvectors, consider the eigenvectors $\mathbf{v}_i$ and corresponding eigenvalues λ_i. Let $\sigma_i = \sqrt{\lambda_i}$, and let $r_i = \dfrac{\mathbf{A}\mathbf{v}_i}{\sigma_i}$. Let us construct three matrices from these values: the diagonal matrix Σ, which has σ_i values on the diagonal (padded with zeros if we run out of σs); the matrix $\mathbf{U}$ with r_i values as columns; and the matrix $\mathbf{V}$ with $\mathbf{v}_i$ values as the columns.

Any matrix $\mathbf{A}$ can be rewritten as the product

$$\mathbf{A} = \mathbf{U}\Sigma\mathbf{V}^T \tag{A.22}$$

of three matrices, $\mathbf{U}$, Σ, and $\mathbf{V}^T$, where $\mathbf{U}$ and $\mathbf{V}$ have orthonormal columns and Σ is a diagonal matrix (the entries of which are known as singular values). This is known as singular value decomposition.

A.9   Diagonalization

A matrix $\mathbf{A}$ is diagonalizable if we can rewrite it (decompose it) as a product

$$\mathbf{A} = \mathbf{P}\mathbf{D}\mathbf{P}^{-1}, \tag{A.23}$$

where $\mathbf{P}$ is an invertible matrix (and thus $\mathbf{P}^{-1}$ exists) and $\mathbf{D}$ is a diagonal matrix (where all off-diagonal elements are zero) and the diagonal elements are eigenvalues of $\mathbf{A}$.

From this definition, we can deduce a few important things about diagonalization. First of all, since $\mathbf{P}$ is invertible, it must be square; therefore, this definition only makes sense for square matrices. A matrix that is not square is not diagonalizable, because the concept of diagonalization makes no sense for non-square matrices. The singular value decomposition (SVD) is essentially diagonalization in a more general sense. SVD plays a similar role to diagonalization, but it applies to matrices of any shape.

A.10   Trace of a Matrix

The sum of the elements of the principal diagonal of a square matrix is called the trace (or spur) of the matrix. Thus if A is a square matrix of order n, then

$$\text{trace}(\mathbf{A}) = a_{11} + a_{22} + a_{33} + \cdots + a_{nn} = \sum_{i=1}^{n} a_{ii}. \qquad (A.24)$$

A.11   Rank of a Matrix

The maximum order of the non-singular square sub-matrix of a matrix A is called the rank of A. The matrix A is said to be of rank r, if and only if it has at least one non-singular square sub-matrix of order r and all sub-matrices of $r + 1$ and higher orders are singular.

A.12   Positive Definite and Semi-definite Matrix

A square matrix $\mathbf{A}$ of order n is multiplied with $\vec{\mathbf{x}}$, $\vec{\mathbf{x}} \in \mathbb{R}^n$, and $\mathbf{A}$ is positive definite if $\vec{\mathbf{x}}^T \mathbf{A} \vec{\mathbf{x}} > 0$ and $\mathbf{A}$ is positive semi-definite if $\vec{\mathbf{x}}^T \mathbf{A} \vec{\mathbf{x}} \geq 0$.

Appendix B
Probability

The axiomatic development of probability involves the introduction of three concepts. These concepts taken together are called an experiment. The three concepts are:

(i) A set S consisting of all the possible outcomes of an experiment w_i, i.e., S={w_i}. A trial is the act of randomly drawing a single outcome. Hence, each trial produces one $w \in S$.

(ii) A, a set of certain subsets of S, is called an event. The event, which consists of a single outcome, is called an elementary event. The set S is called the certain event. Any subset from $U - S$, ς is called the impossible event where U is universal set. We say that an event with outcome w occurred if the outcome w of a trial is contained in A, i.e., if $w \in A$.

(iii) A real function $P(A)$ is defined on S. This function, called probability, satisfies the following axioms:

- $P(A) \geq 0$.
- $P(S) = 1$.
- If $A, B \in S$ and $A \cap B = \phi$; then $P(A \bigcup B) = P(A) + P(B)$.

Let S be a sample space and A be an event in S and P(A) is the probability satisfying the following axioms:

(i) The probability of any event ranges from zero to one, i.e., $0 \leq P(A) \leq 1$.
(ii) The probability of the entire space is 1, i.e., $P(S) = 1$.
(iii) If $A_1, A_2,\ldots$ is a sequence of mutually exclusive events in S, then $P(A_1 \cup A_2 \cup \ldots) = P(A_1) + P(A_2) + \ldots$
(iv) $S \Rightarrow$ Sample space
(v) $\bar{A} \Rightarrow$ A does not occur
(vi) $A \cup \bar{A} = S$
(vii) $A \cap B = \phi \Rightarrow$ A and B are mutually exclusive

© The Author(s), under exclusive license to Springer Nature Singapore Pte Ltd. 2025   359
A. Ghosh, *Data Science and Cases in Sustainability*, Mathematics for Sustainable
Developments, https://doi.org/10.1007/978-981-96-8362-8

(viii) $A \cup B \Rightarrow$ Event A occurs or B occurs or both A and B occur (at least one of the events A and B occurs).
(ix) $A \cap B \Rightarrow$ Both the events A and B occur
(x) $\bar{A} \cap \bar{B} \Rightarrow$ Neither A nor B occurs
(xi) $A \cap \bar{B} \Rightarrow$ Event A occurs and B does not occur
(xii) $\bar{A} \cap B \Rightarrow$ Event A does not occur and B occurs.

There are two types of probability. They are mathematical probability and statistical probability.

B.1 Mathematical Probability (or a Priori Probability)

If the probability of an event can be calculated even before the actual happening of the event, that is, even before conducting the experiment, it is called mathematical probability. If the random experiment results in "n" exhaustive, mutually exclusive and equally likely cases, out of which "m" are favorable to the occurrence of an event A, then the ratio m/n is called the probability of occurrence of event A, denoted by $P(A)$, and is given by

$$P(A) = \frac{m}{n} = \frac{\text{Number of cases favourable to event A}}{\text{Total number of exhaustive cases}}. \tag{B.1}$$

B.2 Statistical Probability (or a Posteriori Probability)

If the probability of an event can be determined only after the actual happening of the event, it is called statistical probability. If an event occurs m times out of n, its relative frequency is m/n. In the limiting case, when n becomes sufficiently large, it corresponds to a number which is called the probability of that event. In symbol,

$$P(A) = \lim_{n \to \infty} (m/n). \tag{B.2}$$

If a coin is tossed ten times, we may get six heads and four tails or four heads and six tails or any other result. In these cases, the probability of getting a head is not 0.5 as we consider in mathematical probability. However, if the experiment is carried out a large number of times, we should expect approximately equal number of heads and tails, and we can see that the probability of getting head approaches 0.5. The statistical probability calculated by conducting an actual experiment is also called a posteriori probability or empirical probability.

B.3   Conditional Probability

A fair die is about to be tossed. The probability that it lands with "5" showing up is 1/6; this is an unconditional probability. But the probability that it lands with "5" showing up, given that it always lands with an odd number showing up, is 1/3; this is a conditional probability.

Let A be any event with $P(A) > 0$. The probability that an event B occurs subject to the condition that A has already occurred is known as the conditional probability of occurrence of the event B on the assumption that the event A has already occurred and is denoted by the symbol $P(B/A)$ or $P(B|A)$ and is read as the probability of B given A. The same definition can be given as follows also: Two events A and B are said to be dependent when A can occur only when B is known to have occurred (or vice versa). The probability attached to such an event is called the conditional probability and is denoted by $P(B/A)$, or, in other words, probability of B given that A has occurred.

If two events A and B are dependent, then the conditional probability of B given A is

$$P(B/A) = P(B \cap A)/P(A). \tag{B.3}$$

Similarly, the conditional probability of A given B is obtained as

$$P(A/B) = P(A \cap B)/P(B). \tag{B.4}$$

For any two events A and B, $P(A \cap B) = P(B \cap A)$.
If the events A and B are independent, then

$$P(A/B) = P(A); \tag{B.5}$$

and

$$P(B/A) = P(B). \tag{B.6}$$

If two events A and B are independent, the probability that both of them occur is equal to the product of their individual probabilities, i.e.,

$$P(A \cap B) = P(A)P(B). \tag{B.7}$$

If two events A and B are not mutually exclusive, the probability of the event that either A or B or both occur is given as

$$P(A \cup B) = P(A) + P(B) - P(A \cap B). \tag{B.8}$$

B.4 Bayes' Theorem

Suppose, there are two events A_1 and A_2. Then Bayes' theorem states posterior probability of event B given event A_1 or A_2 are

$$P(A_1|B) = \frac{P(A_1)P(B|A_1)}{\sum_{i=1}^{2} P(A_i)P(B|A_i)}, \tag{B.9}$$

$$P(A_2|B) = \frac{P(A_2)P(B|A_2)}{\sum_{i=1}^{2} P(A_i)P(B|A_i)}, \tag{B.10}$$

There is an experiment of dice throwing, and there are two events, i.e., occurrence of even numbers denoted by A_1 and occurrence of odd numbers denoted by A_2. Another event of occurring number between 3 and 5 is denoted by B.

$$\sum_{i=1}^{2} P(A_i) = P(\sum_{i=1}^{2}(A_i)) = \frac{1}{2} + \frac{1}{2} = 1 \tag{B.11a}$$

$$B = B \bigcap \sum_{i=1}^{2}(A_i) = \sum_{i=1}^{2}(B \bigcap A_i) = (B \bigcap A_1) + (B \bigcap A_2) = \{4\} + \{3, 5\} = \{3, 4, 5\} \tag{B.11b}$$

$$P(B) = P(\sum_{i=1}^{2}(B \bigcap A_i)) = \sum_{i=1}^{2} P((B \bigcap A_i)) = \sum_{i=1}^{2} P(A_i)P(B|A_i)$$

$$= P(A_1)P(B|A_1) + P(A_2)P(B|A_2) = \frac{1}{2} \times \frac{1}{3} + \frac{1}{2} \times \frac{2}{3} = \frac{1}{6} + \frac{2}{6} = \frac{1}{2} \tag{B.11c}$$

$$P(A_1|B) = \frac{P(A_1)P(B|A_1)}{\sum_{i=1}^{2} P(A_i)P(B|A_i)} = \frac{\frac{1}{2} \times \frac{1}{3}}{\frac{1}{2}} = \frac{1}{6} \tag{B.11d}$$

$$P(A_2|B) = \frac{P(A_2)P(B|A_2)}{\sum_{i=1}^{2} P(A_i)P(B|A_i)} = \frac{\frac{1}{2} \times \frac{2}{3}}{\frac{1}{2}} = \frac{2}{3} \tag{B.11e}$$

For a sequence of mutually exclusive and exhaustive events $A_1, A_2, A_3, \cdots, A_n$ with $P(A_i) > 0)$ $\forall i = 1, 2, 3, \cdots, n$, the conditional probability of event B with A_j is

$$P(A_j|B) = \frac{P(A_j)P(B|A_j)}{\sum_{i=1}^{n} P(A_i)P(B|A_i)}, \tag{B.12}$$

where $P(B) > 0$.

Since $A_1, A_2, A_3, \cdots, A_n$ are mutually exclusive and exhaustive events, then

$$\sum_{i=1}^{n} P(A_i) = P(\sum_{i=1}^{n}(A_i)) = 1$$

(B.13a)

$$B = B \bigcap \sum_{i=1}^{n}(A_i) = \sum_{i=1}^{n}(B \bigcap A_i)$$

(B.13b)

$$P(B) = P(\sum_{i=1}^{n}(B \bigcap A_i)) = \sum_{i=1}^{n} P((B \bigcap A_i)) = \sum_{i=1}^{n} P(A_i)P(B|A_i)$$

(B.13c)

$$P(A_j|B) = \frac{P(A_j) \bigcap B}{P(B)} = \frac{P(A_j)P(B|A_j)}{\sum_{i=1}^{n} P(A_i)P(B|A_i)}$$

(B.13d)

Here $P(A_j)$ is the prior probability and $P(A_j|B)$ is posterior probability.

Appendix C
Statistics

A variable whose value is a number determined by the outcome of a random experiment is called a random variable. We can also say that a random variable is a function defined over the sample space of an experiment and generally assumes different values with a definite probability associated with each value. Generally, a random variable is denoted by capital letters like X, Y, and Z, ..., whereas the values the random variable take are denoted by the corresponding small letters like x, y, and z, ...

C.1 Discrete Random Variable

If a random variable takes only a finite or a countable number of values, it is called a discrete random variable. For example, when three coins are tossed, the number of heads obtained can be represented by a random variable X which can take values from 0, 1, 2, 3, a countable set. Such a variable is a discrete random variable.

C.2 Continuous Random Variable

A random variable X which can take any value between certain intervals is called a continuous random variable. Note that the probability of any single value at x, value of X is zero, i.e., $P(X = x) = 0$. Thus continuous random variable takes value only between two given limits. For example, the maximum life of electric bulbs is 2000 hours. For this, the continuous random variable will be $X = \{x | 0 \leq x \leq 2000\}$.

A. Ghosh, *Data Science and Cases in Sustainability*, Mathematics for Sustainable Developments, https://doi.org/10.1007/978-981-96-8362-8

C.3   Central Tendency Measures

A measure of central tendency is a single value that attempts to describe a set of data by identifying the central position within that set of data. As such, measures of central tendency are sometimes called measures of central location. They are also classed as summary statistics. The mean (often called the average) is most likely the measure of central tendency that we are most familiar with, but there are others, such as the median and the mode.

The mean, median, and mode are all valid measures of central tendency, but under different conditions, some measures of central tendency become more appropriate to use than others. In the following sections, we will look at the mean, mode, and median and learn how to calculate them and under what conditions they are most appropriate to be used.

C.3.1   Expectation or Mean

The expectation is the value, on average, of a random variable (or function of a random variable). The expectation in a probabilistic sense always averages over the possible values weighting by the probability of observing each value. The form of an expectation in the discrete case is particularly simple. For a discrete random variable X, the expectation is given as

$$E(X) = \sum_{x \in R(X)} xp(x);$$
(C.1)

with $p(x)$ as the probability of occurrence of the value x. $R(x)$ is the set of values that X can take. For a continuous random variable, the expectation is given as

$$E(X) = \int_{x=-\infty}^{\infty} xp(x)dx.$$
(C.2)

C.3.2   Median

The median is the middle most value for a set of data that has been arranged in order of magnitude. The median is less affected by outliers and skewed data. The median of {12,33,34,50,75,82,99} is 50.

C.3.3   Mode

The mode is the most frequent score in a data set. It represents the highest bar in a bar chart or histogram. You can, therefore, sometimes consider the mode as being the most popular option.

When we have a normally distributed sample, we can legitimately use both the mean and the median as the measure of central tendency. In fact, in any symmetrical distribution, the mean, median, and the mode are equal. However, in this situation, the mean is widely preferred as the best measure of central tendency, because it is the measure that includes all the values in the data set for its calculation, and any change in any of the scores will affect the value of the mean. This is not the case with the median or mode. If dealing with a distribution, and tests of normality show that the data is non-normal, it is customary to use the median instead of the mean. However, this is more a rule of thumb than a strict guideline. Sometimes, researchers wish to report the mean of a skewed distribution if the median and mean are not appreciably different (a subjective assessment), and if it allows easier comparisons to previous research to be made.

If we calculate mean, median, and mode of a random variable whose values are $\{12, 7, 2, 3, -5, 21, 37, 8, 3\}$, then the mean is 9.77, the median is 7, and the mode is 3.

C.3.4   Variance and Standard Deviation

The variance of a set of data signifies how much a particular data can deviate from the mean of the data set. It is given by

$$\text{Var}(X) = E(X^2) - [E(X)]^2. \tag{C.3}$$

The standard deviation of X denoted by σ_X is the square root of variance, expressed as

$$\sigma_X = \sqrt{\text{Var}(X)}. \tag{C.4}$$

C.4   Probability Distribution Function

Let X be a discrete random variable which assumes the values $x_1, x_2, \ldots, x_n$. With each of these values, we associate a number called the probability $p_i = p(X = x_i)$, $i = 1, 2, 3, \ldots, n$. This is called probability of x_i satisfying the conditions:

(i) $p_i \geq 0$ for all i, i.e., p_i' s are all nonnegative.
(ii) $\sum p_i = 1$, i.e., the total probability is one.

This function p_i or $p(x_i)$ is called the probability mass function of the discrete random variable X. The set of all possible ordered pairs $(x, p(x))$ is called the probability distribution of the random variable X.

Let C be a continuous random variable which assumes the values in [a, b], then to calculate the probability of C to be in [c, d], the area under the probability density function must be integrated.

$$P(c \leq C \leq d) = \int_c^d f(x)dx;$$ (C.5)

and

$$\int_{-\infty}^{\infty} f(x)dx = 1$$ (C.6)

C.4.1   Poisson Distribution

The act of counting the number of times that a certain random event takes place during a given time interval often involves a Poisson distribution. A discrete random variable X which obeys the Poisson distribution has the probability function:

$$P(X = n) = \frac{\lambda^n e^{-\lambda}}{n!};$$ (C.7)

λ is a parameter of the distribution. It is the expected number of events. Thus, $E(X) = \lambda$. A special feature of the Poisson distribution is that the variance and the expectation are equal, i.e., $\text{Var}(X) = E(X) = \lambda$. The Poisson random variable satisfies several conditions. The number of successes at two disjoint time intervals is independent. The probability of success during a small time interval is proportional to the entire length of the time interval. Apart from disjoint time intervals, the Poisson random variable also applies to disjoint regions of space (Fig. C.1).

C.4.2   Gaussian Distribution

A well-known example of a continuous random variable is the one with a Gaussian (or normal) probability density:

$$p(x) = \frac{1}{\sigma\sqrt{2\pi}} e^{\left(-\frac{(x-\mu)^2}{2\sigma^2}\right)}.$$ (C.8)

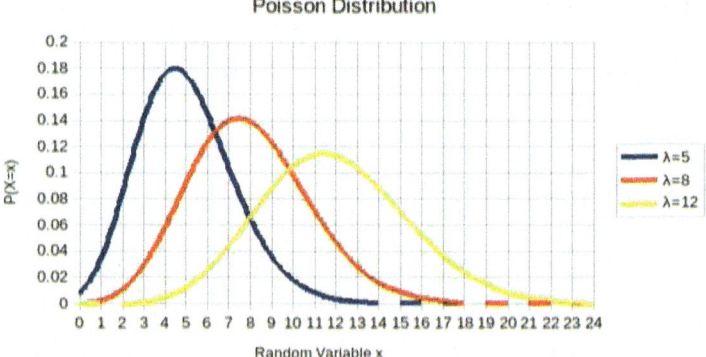

Fig. C.1 Poisson distribution

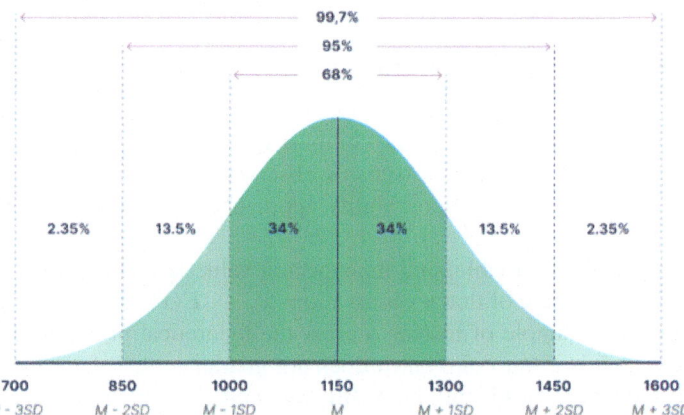

Fig. C.2 Gaussian distribution

The parameters μ and σ^2 are the expectation and the variance, respectively. Gaussian random variables occur whenever the underlying process is caused by the outcomes of many independent experiments and the associated random variables add up linearly (the central limit theorem) (Fig. C.2).

C.5 Binomial Distribution

Suppose, we toss a coin. Then the probability of getting head is p and probability of getting tail is q. If this random experiment is done N times, then the probability of getting n heads is

$$P(x = n) = \binom{N}{n} p^n q^{N-n}. \tag{C.9}$$

In case of binomial distribution result of every trial has two options, i.e., either true or false, pass or fail, 0 or 1. For that reason, this is called binomial. In (C.9) q is $(1 - p)$.

C.6 Bernoulli Distribution

If a random variable x has a probability p to take value 1 among two values $\{0, 1\}$ and the probability to take value 0 is $(1 - p)$, then the random variable x is said to be following a Bernoulli distribution. It is a special case of binomial distribution where $N = 1$ in (C.9). In pattern recognition, any unknown pattern can be classified using Bernoulli distribution in two class problem.

C.7 Laws of Large Numbers

The law of large numbers gives us the compass to navigate the randomness around us. It states that as the number of trials or observations increases and approaches infinity, the actual or observed probability approaches the theoretical or expected probability.

Let x be a random variable, and its expected value is $E(x)$; then if we take n observations, the average of that n observations $\bar{x}_n \rightarrow E(x)$ for $n \rightarrow \infty$.

If we take the example of tossing a coin, the theoretical probability of getting a head is 0.5 as it is a fair coin, but it does not guarantee that if we flip a coin 10 times, then number of heads will be 5. But we can be confident as we continue to flip the coin indefinitely the cumulative proportion of heads should get closer and closer to 50%. There could be a misconception that if the proportion of the number of heads is high initially, then the number of tails will be higher for future steps to make it up to the expected value. But this is not true, because each and every event is independent, i.e., its outcome is unaffected by all previous events. The high proportion of heads will be averaged by a huge number of events where heads and tails are flipped as the no of trials approaches infinity. The misconception is called *Gambler's Fallacy*. Though we cannot predict the outcome of a single coin flipped, we can say that over time the half of the outcome will be heads.

C.8   Central Limit Theorem

This basically states that as the sample size (N) becomes large, the following occur:

(i) The sampling distribution of the mean becomes approximately normal regardless of the distribution of the original variable.

(ii) The sampling distribution of the mean is centered at the population mean (μ) of the original variable. In addition, the standard deviation of the sampling distribution of the mean approaches $\sigma/\sqrt{N}$.

Appendix D
Correlation and Regression

Correlation is a measure of association between two variables. The variables are not designated as dependent or independent. The value of a correlation coefficient can vary from minus one to plus one. A minus one indicates a perfect negative correlation, while a plus one indicates a perfect positive correlation. A correlation of zero means there is no relationship between the two variables. When there is a negative correlation between two variables, as the value of one variable increases, the value of the other variable decreases, and vice versa. In other words, for a negative correlation, the variables work opposite to each other. When there is a positive correlation between two variables, as the value of one variable increases, the value of the other variable also increases. The variables move together.

Simple regression is used to examine the relationship between one dependent and one independent variable. After performing an analysis, the regression statistics can be used to predict the dependent variable when the independent variable is known. Regression goes beyond correlation by adding prediction capabilities.

People use regression on an intuitive level every day. In business, a well-dressed man is thought to be financially strong. A mother knows that more sugar in her children's diet results in higher energy levels. The ease of waking up in the morning often depends on how late you went to bed the night before. Quantitative regression adds precision by developing a mathematical formula that can be used for predictive purposes.

© The Author(s), under exclusive license to Springer Nature Singapore Pte Ltd. 2025   373
A. Ghosh, *Data Science and Cases in Sustainability*, Mathematics for Sustainable
Developments, https://doi.org/10.1007/978-981-96-8362-8

D.1   Covariance

In probability theory and statistics, covariance is a measure of how much two random variables change together. Covariance between variables x and y is given by the equation

$$s_{xy} = \frac{1}{n} \sum_{i=1}^{n} (x_i - \bar{x})(y_i - \bar{y});$$

(D.1)

where $\bar{x} = \frac{1}{n} \sum_{i=1}^{n} x_i$ is the mean of the data having n samples.

D.2   Correlation Coefficient

The correlation coefficient assesses how tightly the points hug to the best-fit line, and the sign corresponds to the direction of that line. It is calculated via:

$$r = \frac{1}{n-1} \sum_{i=1}^{n} \frac{(x_i - \bar{x})}{s_x} \frac{(y_i - \bar{y})}{s_y};$$

(D.2)

where s_x is the standard deviation of x given by

$$s_x = \sqrt{(\bar{x^2}) - (\bar{x})^2}.$$

(D.3)

Here, $\bar{x^2} = \frac{1}{n} \sum_{i=1}^{n} x_i^2$ is the mean square of the data. In other words, standard deviation is the root of variance (Fig. D.1).

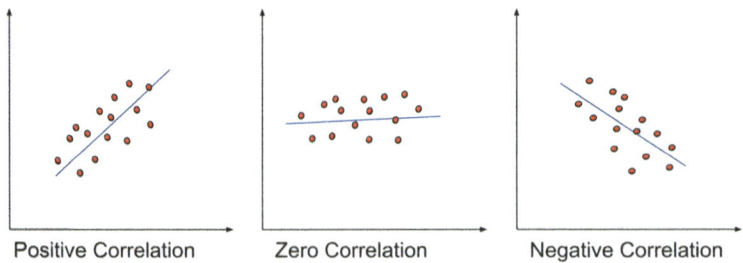

Positive Correlation   Zero Correlation   Negative Correlation

Fig. D.1 Types of correlation

D.3   Linear Regression

In statistics, linear regression is an approach for modeling the relationship between a scalar dependent variable y and one or more explanatory variables (or independent variable) denoted by x. The goal here is to find the equation of the best-fit line if there is one independent variable. In general, the best-fitted regression line is:

$$\hat{y} = a_0 + a_1 x. \tag{D.4}$$

This is referred as simple linear regression. The difference between the actual and the predicted values is residuals. Thus, the ith residual is

$$e_i = y_{i(\text{actual})} - y_{i(\text{predicted})}. \tag{D.5}$$

An important measure is the residual sum of squares:

$$\text{SSE} = \sum_{i=1}^{n} e_i^2. \tag{D.6}$$

The line for which SSE is minimum is the "best fit" line. This is the "least-squares criterion" which sense that the fitted regression line is the best. The parameters can be estimated by minimizing SSE as

$$a_1 = r \times \frac{s_y}{s_x} = \frac{\sum_{i=1}^{n} (x_i - \bar{x})(y_i - \bar{y})}{\sum_{i=1}^{n} [(x_i - \bar{x})^2]} \tag{D.7}$$

and

$$a_0 = \bar{y} - a_1 \bar{x}. \tag{D.8}$$

The regression line passes through the point of averages $(\bar{x}, \bar{y})$. In multilinear regression, the number of independent variable is more than one, and the correlation among the independent variables must be low (Fig. D.2).

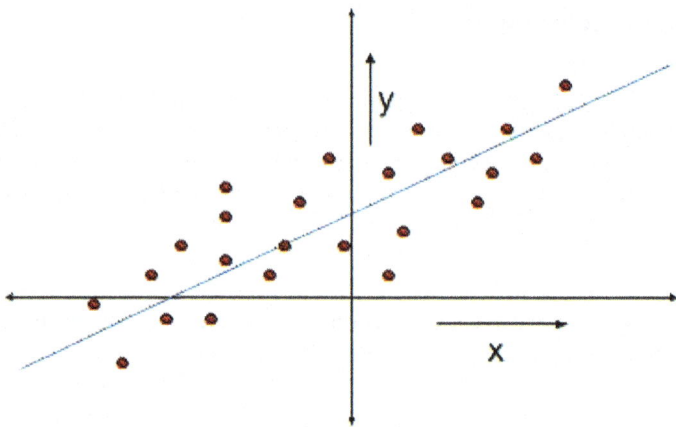

Fig. D.2 Linear regression

Appendix E
Vector Algebra

A field is a set F of numbers with the property that if $a, b \in F$, then $a \pm b$, ab, ba, and a/b are also in F (assuming, of course, that $b \neq 0$ in the expression a/b).

E.1 Definition of Vector Space

Roughly speaking, a vector space is a set of elements for which the operation of addition is defined and the operation of multiplication by a scalar is defined. The common example of directed line segments (arrows) in two or three dimensions fits this idea, because you can add such arrows by the parallelogram law and you can multiply them by numbers, changing their length.

In Fig. E.1, addition of T and D vectors with magnitude 15 and 10 with 38° angle between them results vector R with magnitude 23.69, and angle will be 15.1°. A vector space consists of a set V (elements of V are called vectors), a field F (elements of F are called scalars), and two operations.

- An operation called vector addition that takes two vectors $\mathbf{v}, \mathbf{w} \in V$ and produces a third vector, written $\mathbf{v} + \mathbf{w} \in V$.
- An operation called scalar multiplication that takes a scalar $c \in F$ and a vector $\mathbf{v} \in V$ and produces a new vector, written $c\mathbf{v} \in V$.

The operations must satisfy the following conditions.

(i) Commutativity of vector addition: $\mathbf{u} + \mathbf{v} = \mathbf{v} + \mathbf{u}$.
(ii) Associativity of vector addition: $(\mathbf{u} + \mathbf{v}) + \mathbf{w} = \mathbf{u} + (\mathbf{v} + \mathbf{w})$, $\forall \mathbf{u}, \mathbf{v}, \mathbf{w} \in V$.
(iii) Existence of a zero vector: There is a vector in V, written as $\mathbf{0}$ and called the zero vector, which has the property that $\mathbf{u} + \mathbf{0} = \mathbf{u}$, $\forall \mathbf{u} \in V$.
(iv) Existence of negatives: For every $\mathbf{u} \in V$, there is a vector in V, written $\sim \mathbf{u}$ and called the negative of $\mathbf{u}$, which has the property that $\mathbf{u} + (\sim \mathbf{u}) = \mathbf{0}$.
(v) Associativity of multiplication: $(ab)\mathbf{u} = a(b\mathbf{u})$ for any $a, b \in F$ and $\mathbf{u} \in V$.

© The Author(s), under exclusive license to Springer Nature Singapore Pte Ltd. 2025
A. Ghosh, *Data Science and Cases in Sustainability*, Mathematics for Sustainable Developments, https://doi.org/10.1007/978-981-96-8362-8

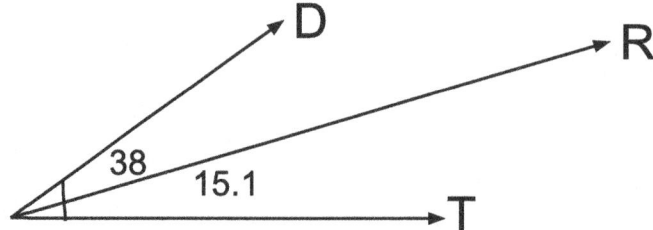

Fig. E.1 Vector addition using parallelogram law

(vi) Distributivity: $(a + b)\mathbf{u} = a\mathbf{u} + b\mathbf{u}$ (for scalar sums) and $a(\mathbf{u} + \mathbf{v}) = a\mathbf{u} + a\mathbf{v}$ (for vector sums).

E.1.1   Subspaces of Vector Spaces

A nonempty subset W of a vector space V is called a subspace of V, if W is a vector space under the operations addition and scalar multiplication defined in V.

E.1.2   Linear Combination of Vectors

A vector $\mathbf{v}$ in a vector space V is called a linear combination of vectors $\mathbf{u_1}, \mathbf{u_2}, \cdots, \mathbf{u_k}$ in V if $\mathbf{v}$ can be written in the form $\mathbf{v} = c_1\mathbf{u_1} + c_2\mathbf{u_2} + \cdots + c_k\mathbf{u_k}$, where $c_1, c_2, \cdots, c_k$ are scalars.

E.1.3   Spanning Sets

Let V be a vector space over R and $S = \{\mathbf{v_1}, \mathbf{v_2}, \cdots, \mathbf{v_k}\}$ be a subset of V. We say that S is a spanning set of V if every vector $\mathbf{v}$ of V can be written as a liner combination of vectors in S. In such cases, we say that S spans V.

E.1.4   Linear Dependence and Independence

A set of nonzero vectors is linearly dependent if one element of the set can be written as a linear combination of the others. The set is linearly independent if this cannot be done. Mathematically, let V be a vector space. A set of vectors $S = \{\mathbf{v_1}, \mathbf{v_2}, \cdots \mathbf{v_k}\}$ is said to be linearly independent if the equation

$$c_1\mathbf{v_1} + c_2\mathbf{v_2} + \cdots + c_k\mathbf{v_k} = \mathbf{0}$$

has only trivial solution $c_1 = 0, c_2 = 0, \ldots, c_k = 0$. We say S is linearly dependent, if S is not linearly independent.

E.2   Basis and Dimension

A line is thought of as one-dimensional, a plane two-dimensional, and surrounding space as three-dimensional. We will attempt to make this intuitive notion of dimension precise and extend it to general vector spaces.

E.2.1   Coordinate Systems of General Vector Spaces

A line is thought of as one-dimensional, because every point on that line can be specified by one coordinate. In the same way, a plane is thought of as two-dimensional, because every point on that plane can be specified by two coordinates and so on.

What defines this coordinate system? The most common form of defining a coordinate system is the use of coordinate axes. In case of the plane, the x and y axes are used most frequently. But there is also a way of specifying the coordinate system with vectors. This can be done by replacing each axis with a vector of length one that points in the positive direction of the axis. In case of the $x - y$ plane, the x and y-axes are replaced by the well-known unit vectors i and j, respectively. Let O be the origin of the system and P be any point in the plane. The point P can be specified by the vector $\overrightarrow{OP}$. Every vector, $\overrightarrow{OP}$ can be written as a linear combination of i and j as

$$\overrightarrow{OP} = a_i + b_j.$$

Informally stated, vectors such as $\vec{i}$ and $\vec{j}$ that specify a coordinate system are called "basis vectors" for that system. Although in the previous discussion our basis vectors were chosen to be of unit length and mutually perpendicular, this is not essential as long as linear combinations of the vectors chosen are capable of specifying all points in the plane. In our example, this only requires that the two vectors are not colinear. Different basis vectors however do change the coordinates of a point, as the following example demonstrates.

For example, let $S = \{i, j\}$, $U = \{i, 2j\}$ and $V = \{i + j, j\}$. Let the sets S, U, and V be three sets of basis vectors. Let P be the point $i + 2j$. Then, the coordinates of P relative to each set of basis vectors are: $S \rightarrow (1, 2)$; $U \rightarrow (1, 1)$; $T \rightarrow (1, 1)$.

E.2.2   Definition

If V is any vector space and $S = \{v_1, v_2, \cdots, v_n\}$ is a set of vectors in V, then S is called a basis for V if the following two conditions hold:

 (i) S is linearly independent,
 (ii) S spans V.

A basis is the vector space generalization of a coordinate system in two-space and three-space.

E.2.3   Dimension of General Vector Spaces

A nonzero vector space V is called finite-dimensional if it contains a finite set of vectors $v_1, v_2, \cdots, v_n$ that form a basis. If no such set exists, V is called infinite-dimensional.

E.2.4   Linear Transformation

Let V, W be vector spaces over a field F. A function that maps V into W, $T: V \rightarrow W$, is called a linear transformation from V to W if for all vectors $\mathbf{u}$ and $\mathbf{v}$ in V and all scalars $c \in F$

 (i) $T(\mathbf{u} + \mathbf{v}) = T(\mathbf{u}) + T(\mathbf{v})$;
 (ii) $T(c\mathbf{u}) = cT(\mathbf{u})$.

Appendix F
Optimization

Optimization is a technique of making best or most effective use of a situation or resources. In machine learning optimization, approaches have enjoyed prominence because of their wide applicability and attractive theoretical properties. Optimization formulations and methods are proving to be vital in designing algorithms to extract essential knowledge from huge volumes of data. We categorize optimization problems in two forms.

F.1 Unconstrained Optimization

In the unconstrained optimization, we deal with the problems of minimizing or maximizing of objective function that depends on real variables with no restrictions on their values. Mathematically, let $x \in \mathbb{R}^n$ be a real vector with $n \geq 1$ components, and let $f : \mathbb{R}^n \longrightarrow \mathbb{R}$ be a smooth function[1]. Then, the unconstrained optimization problem is (Fig. F.1)

$$(minimize \ or \ maximize) f(x), \qquad x \in \mathbb{R}^n.$$

Example: maximize $f : \mathbb{R} \longrightarrow \mathbb{R}$ *such that*

$$f(x) = \frac{x}{x^2 + 1}.$$

---

[1] A smooth function is a function which has continuous derivative up to some desired order in the domain.

© The Author(s), under exclusive license to Springer Nature Singapore Pte Ltd. 2025
A. Ghosh, *Data Science and Cases in Sustainability*, Mathematics for Sustainable Developments, https://doi.org/10.1007/978-981-96-8362-8

Fig. F.1 $f(x) = \frac{x}{x^2+1}$ which
is supposed to be maximized

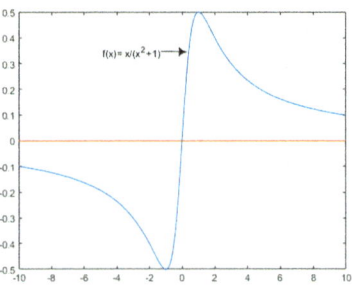

F.2   Constrained Optimization

Constrained optimization problems are the problems for which objective function is
to be minimized or maximized with subject to some constrains. Mathematically, let
$x \in R^n$ be a real vector with $n \geq 1$ components, and let $f : \mathbb{R}^n \longrightarrow \mathbb{R}$ be a smooth
function. Then the constrained problem is (Fig. F.2)

$$(minimize \ \ or \ \ maximize)_x f(x),$$

$$subject \ \ to \ \ x \in \Omega, \ \ \Omega \subseteq \mathbb{R}^n.$$

Example: minimize $f : \mathbb{R} \longrightarrow \mathbb{R}$ *such that*

$$f(x) = x^2,$$

$$Subject \ \ to \ \ x \geq f(x).$$

There are few necessary and sufficient condition in order to find minima and
maxima of a function.

Fig. F.2 The region in which
f(x) is to be minimized is
ABD

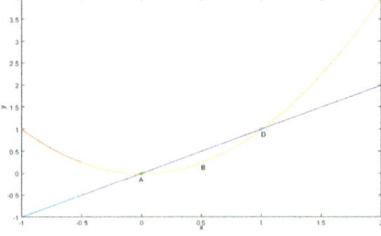

F.2.1   First-Order Necessary Condition (FONC)

Let $\Omega \subseteq \mathbb{R}^n$ and $f \in C^1$ be a real-valued function on Ω. if $x = x^*$ is a local minimizer[2] of f on Ω, then for any feasible direction d at x^*, we have

$$d^t \nabla f(x^*) \geqslant 0. \tag{F.1}$$

In the other words, if x^* is a local minimizer, then the rate of increase of f at x^* is nonnegative for any feasible direction d in Ω.

But this condition is not sufficient to guarantee a minimum, because it could also be a maximum or a saddle point[3]. To ensure a minimum, a second-order condition is necessary.

F.2.2   Second-Order Necessary Condition (SONC)

Let $\Omega \subset \mathbb{R}^n$, $f \in C^2$ be a function on Ω, x^* a local minimizer of f over Ω, and d a feasible direction at x^*. If $d^T \nabla f(x^*) = 0$, than

$$d^T F(x^*)d \geqslant 0; \tag{F.2}$$

where F is *Hessian* of f. This is not also a sufficient condition for the minima as this may also be attained at the saddle point.

Hence, there is a **second-order sufficient condition**.

F.2.3   Second-Order Sufficient Condition (SOSC) Interior Case

Let $f \in C^2$ be defined on a region in which x^* is an interior point. Suppose that

1. $\nabla f(x^*) = 0$,
2. $F(x^*) > 0$[4].

Then, x^* is a strict local minimizer of f.

---

[2] Suppose that $f : \mathbb{R}_n \longrightarrow \mathbb{R}$ is a real-valued function defined on some set $\Omega \subset \mathbb{R}_n$. A point $x^* \in \Omega$ is a local minimizer of over Ω if $\exists \epsilon > 0$ such that $f(x) \geq f(x^*)$, $\forall x \in \Omega \setminus \{x^*\}$ and $\|x - x^*\| < \epsilon$.

[3] Saddle pint is a point of a function or surface which is a stationary point but not an extremum.

[4] $F(x) > 0$ means $F(x)$ is a positive definite matrix which means $\forall x \in \mathbb{R}^n$ $x^T F(x)x > 0$.

Proof Because $f \in C^2$, we have $F(x^*) = F^T(x^*)$. Using assumption 2 and Rayleigh's inequality[5], it follows that if $d \neq 0$, then $0 < \lambda_{\min} F(x^*) \|d\|^2 \leq d^T F(x^*) d$. By the Taylor's theorem and assumption 1,

$$f(x^* + d) - f(x^*) = \frac{1}{2} d^T F(x^*) d + o(\|d\|^2) \geq \frac{\lambda_{\min} F(x^*)}{2} \|d\|^2 + o(\|d\|^2).$$
(F.3)

Hence for all d such that $\|d\|$ is sufficiently small,

$$f(x^* + d) > f(x^*),$$
(F.4)

which is a complete proof.

There are many numerical methods which are helpful in the processes of optimization.

F.3 Algorithms for Unconstrained Optimization

F.3.1 Gradient Method

Let a function be $f : \mathbb{R}^n \longrightarrow \mathbb{R}$. The gradient of f at $x = x_0 \in \mathbb{R}^n$ be denoted by ∇f. If it is not zero vector, then it is orthogonal to the tangent vector to an arbitrary smooth curve passing through x_0 on the level set[6] $f(x) = c$. Thus, the direction of maximum rate of increase of a real-valued differentiable function at a point is orthogonal to the level set of the function through that point. In the other words, the gradient act in such a direction that for a given small displacement, the function f increases more in the direction of the gradient than in any other direction. Thus, the ∇f points in the direction of maximum rate of increase of f at x. The direction in which $- \nabla f(x)$ points is the direction of maximum rate of decrease of f at x. Hence, negative gradient is a good direction to search if we want to find out minimizer[7].

F.3.1.1 Algorithm Generation

We proceed as follows. Let $x^{(0)}$ be a starting point, and consider the point $x^{(0)} - \alpha \nabla f(x^{(0)})$. Then by the Taylor's series expansion, we obtain.

$$f(x^{(0)}) - \alpha \nabla f(x^{(0)})) = f(x^{(0)}) - \alpha \|\nabla f(x^{(0)})\|^2 + o(\alpha).$$
(F.5)

---

[5] 1

[6] Level set of a function $f : \mathbb{R}^n \longrightarrow \mathbb{R}$ is the set of points corresponding to the level $f(x_0) = c$ for some constant c.

[7] The minimizing of f is equivalent to maximizing $- f$.

Thus, if $\nabla f(x^{(0)}) \neq 0$, then for sufficiently small $\alpha > 0$, we have

$$f(x^{(0)}) - \alpha \nabla f(x^{(0)})) < f(x^{(0)}) \tag{F.6}$$

This means that the point $x^{(0)} - \alpha \nabla f(x^{(0)})$ is an improvement over the point $x^{(0)}$ if we search for minimizer.

To formulate the algorithm that implement this idea, suppose that we are given a point $x^{(k)}$. To find the next point $x^{(k+1)}$, we start at $x^{(k)}$ and move by amount $- \alpha_k \nabla f(x^{(k)})$, where α_k is a positive scalar called step size. This leads to the following iterative algorithm.

$$x^{(k+1)} = x^{(k)} - \alpha_k \nabla f(x^{(k)}). \tag{F.7}$$

F.3.1.2   The Method of Steepest Descent

Suppose we would like to solve the optimization problem minimize $f : \mathbb{R}^n \longrightarrow \mathbb{R}$. The general idea behind most of the optimization problems is to find the search direction $d^{(k)}$ for the $(k + 1)$th iteration. For example,

$$x^{(k+1)} = x^{(k)} + \alpha_k d^{(k)}; \tag{F.8}$$

where the α_k, the step length, is chosen so that

$$\alpha_k = \mathrm{argmin}_{\alpha \geq 0} f(x^{(k)} - \alpha \nabla f(x^{(k)})). \tag{F.9}$$

The method of steepest descent provides the search direction as $d^{(k)} = -\nabla f(x^{(k)})$. Hence, the algorithm for the steepest decent method

$$x^{(k+1)} = x^{(k)} + \alpha_k \nabla f(x^{(K)}). \tag{F.10}$$

F.3.1.3   Steepest Descent Method with a Quadratic function

Let a function $f : \mathbb{R}^n \longrightarrow \mathbb{R}$ be of the form

$$f(x) = \frac{1}{2} x^T Q x - b^T x; \tag{F.11}$$

where $Q \in \mathbb{R}^{n \times n}$ is a symmetric positive definite matrix, $b, x \in \mathbb{R}^n$. The unique minimizer of f can be found by setting gradient of f to zero, where

$$\nabla f(x) = Qx - b. \tag{F.12}$$

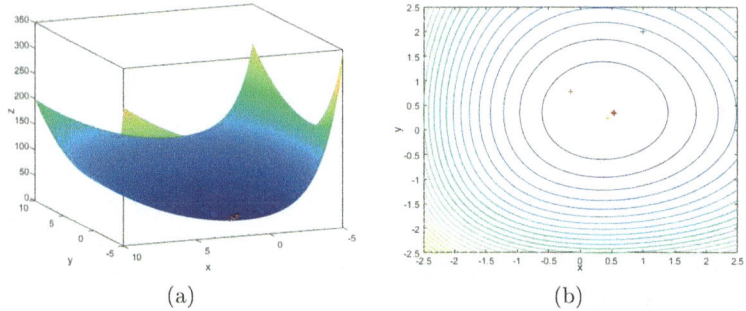

Fig. F.3 The 3D figure and contour of the function $f(x)$ and the sequence of solutions approaching to the minimum value of the function using MATLAB gradient method

The Hessian of f if $F(x) = Q = Q^T > 0$. Let $g^{(k)} = \nabla f(x^{(k)})$. Then the steepest descent algorithm for the quadratic function can be represented as

$$x^{(k+1)} = x^{(k)} - \alpha_k g(x^{(k)}); \tag{F.13}$$

where

$$\alpha_k = \operatorname{argmin}_{\alpha \geq 0}(f(x^{(k)} - \alpha g^{(k)})). \tag{F.14}$$

On applying FONC to the function $f(x^{(k)} - \alpha g^{(k)})$, we get

$$\alpha_k = \frac{g^{(k)T} g^k}{g^{(k)T} Q g^k}. \tag{F.15}$$

$$g^k = \nabla f(x^{(k)}) = Qx^{(k)} - b. \tag{F.16}$$

Example: Minimize the function $f(x) = x_1^2 + x_2^2 + e^{(-x_1)} + e^{(-x_2)}$
Initialize: $x = [1; 2]$
The optimal result we (got using MATLAB) as $x = (0.3517337, 0.3517337)$ and $f(x) = 1.654368$ (Fig. F.3).

F.3.1.4 Some Results Related to Gradient Method

(i) If $\{x^{(k)}\}_{k=0}^{k=\infty}$ is the steepest descent sequence for a given function $f : \mathbb{R}^n \longrightarrow \mathbb{R}$, then each vector $x^{(k)} - x^{(k+1)}$ is orthogonal to the vector $x^{(k+2)} - x^{(k+1)}$.

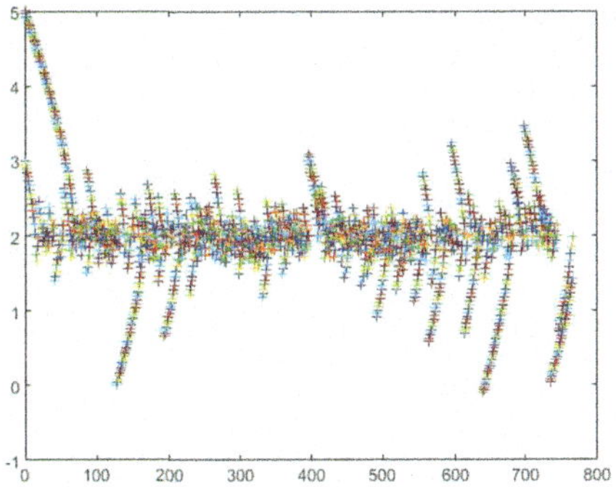

Fig. F.4 Visualization of non-convergence of the solution Using MATLAB

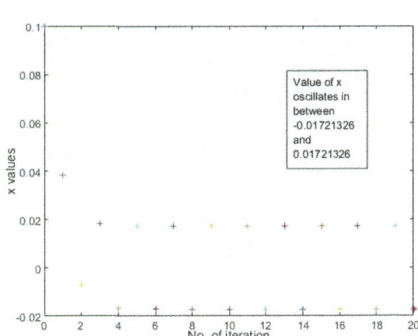

Fig. F.5 Functional oscillation visualization Using MATLAB

(ii) If $\{x^{(k)}\}_{k=0}^{k=\infty}$ is the steepest descent sequence for a given function $f : \mathbb{R}^n \longrightarrow \mathbb{R}$, and if $\nabla f(x^{(k)}) \neq 0$, then $f(x^{(k+1)}) < f(x^{(k)})$.[8]

(iii) In the steepest descent algorithm, we have $x^k \longrightarrow x^*$ for any x^0.

F.3.1.5   Limitations of Gradient Method

(i) For non-differentiable functions, gradient methods fail.
Example:
Minimize $f(x) = \sqrt{|x_1 - 2|} + \sqrt{|x_2 - 2|}$. It gives an oscillatory solution which does not converge onto the solution (Fig. F.4).

(ii) It may give oscillatory solution.
Example:
Minimize $f(x) = x^{4/3}$. Since the gradient of the function $\frac{4}{3}x^{1/3}$ has an inflation point at the $x^* = 0$, hence the function shows oscillation for all initial point except $[0, 0]$ (Fig. F.5).

---

[8] If $\nabla f(x^{(k)}) = 0$ then the point $x^{(k)} = 0$ satisfy the FONC, in that case $x^{(k)} = x^{(k+1)}$, which is used as the stopping criteria of the steepest gradient algorithm.

Fig. F.6 Sequential
minimization visualization
Using MATLAB

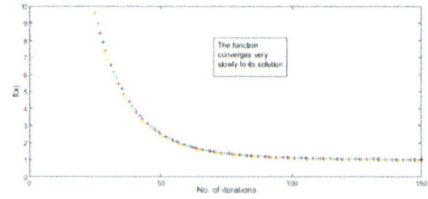

(iii) The convergence rate of gradient descent is $O(1/k)$ and is given by:

$$f(x^{(k)}) - f(x^*) \le \frac{\|x^{(0)} - x^*\|}{22 \times t \times k};$$
(F.17)

where k is the iteration number. Thus to get $f(x^{(k)}) - f(x^*) < \epsilon$, we need $O(\frac{1}{\epsilon})$ iterations. We can see that higher the degree of accuracy we want, the greater the number of iterations that would be required. Usually we want an accuracy of the order of e^5. This implies 10^5 number of iterations. That is why we usually use adaptive step sizes and stopping conditions.

Example:
Minimize $f(x) = (x_1 + 10x_2)^2 + 5(x_3 - x_4)^2 + (x_2 - 2x_3)^4 + 10(x_1 - x_4)^4$. For minimizing this problem of tolerance value, 10^{-3} takes more than 500 iterations (Fig. F.6).

F.3.2   Newton Method

In this method, we construct the quadratic approximation to the objective function on a given starting point that matches first and second derivative at that point. We then minimize the approximate quadratic function instead of the original objective function. We use the minimizer of the approximate function as the starting point in the next step and repeat the procedure iteratively.

We can obtain a quadratic approximation to the twice continuously differentiable objective function. Let $f : \mathbb{R}^n \longrightarrow \mathbb{R}$. Using Taylor's series expansion of f about the current point $x^{(k)}$, neglecting terms of order three and higher, we obtain

$$f(x) \approx f(x^{(k)}) + (x - x^{(k)})^T g^{(k)} + \frac{1}{2}(x - x^{(k)})^T F(x^{(k)})(x - x^{(k)}) = q(x);$$
(F.18)

where $g^{(k)} = \nabla f(x^{(k)})$. Applying the FONC to q yields

$$0 = \nabla q(x) = g^{(k)} + F(x^{(k)})(x - x^{(k)}).$$
(F.19)

Fig. F.7 3D representation of represents Newton's method Using MATLAB

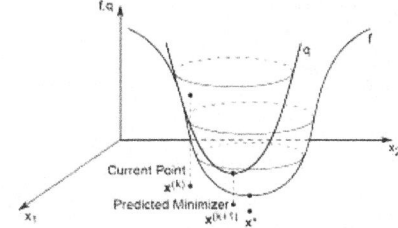

If $F(x^{(k)}) > 0$, then q achieves a minima at

$$x^{(k+1)} = x^{(k)} - F(x^{(k)})^{-1}g^{(k)}. \tag{F.20}$$

This recursive formula represents Newton's method (Fig. F.7).

F.3.2.1   Search Direction

Let $x^{(k)}$ be the sequence generated by Newton's method for minimizing a given function $f(x)$. If the Hessian $F(x^{(k)}) > 0$ and $g^{(k)} = \nabla f(x^{(k)}) \neq 0$, then the search direction

$$d^{(k)} = -F(x^{(k)})^{-1}g^{(k)} = x^{(k+1)} - x^{(k)}, \tag{F.21}$$

form $x^{(k)}$ to $x^{(k+1)}$ is a descent direction of f in the sense that there exists an $\alpha_0 > 0$ such that for all $\alpha \in (0, \alpha_0)$,

$$f(x^{(k)} + \alpha d^{(k)}) < f(x^{(k)}). \tag{F.22}$$

Hence, from the Newton's method, we get

$$x^{(k+1)} = x^{(k)} - \alpha_k F(x^{(k)})^{-1}g^{(k)}, \tag{F.23}$$

$$\alpha_k = \text{argmin}_{\alpha \geq 0} f(x^{(k)} - \alpha F(x^{(k)})^{-1}g^{(k)}). \tag{F.24}$$

Example 1:
Minimize $f(x) = (x_1 + 10x_2)^2 + 5(x_3 - x_4)^2 + (x_2 - 2x_3)^4 + 10(x_1 - x_4)^4$. The optimum value is obtained at the point $(-0.0004693586, 0.00004693586,$ $0.00003520188, 0.00003520188)$ in only 29 iterations with the tolerance value 10^{-7} due the fact that this function is twice differentiable which results the high convergence rate (of order more than 2).

Example 2:
Minimize $f(x) = e^{x_1} + e^{-x_1} + e^{x_2} + e^{-x_2}$.
Solution using MATLAB is $(1.387779e^{-17}, 6.120489e^{-8}) \simeq (0, 0)$.
Figure F.8 shows the function and sequence of x values approaching to optimal solution.

Fig. F.8 Function and
sequence of solution values
approaching to the optimal
solution Using MATLAB

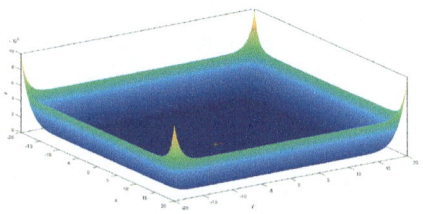

F.3.2.2   Some Result of Newton's Method

(i) Suppose that $f \in C^3$ and $x^* \in \mathbb{R}^n$ is a point such that $\nabla f(x^*) = 0$ and $F(x^*)$
is invertible. Then, for all $x^{(0)}$ sufficiently close to x^*, Newton's method is well
defined for all k and converges to x^* with an order of convergence at least 2.

(ii) For the quadratic case, the order of the convergence of Newton's method is ∞,
i.e., for every initial point $x^{(0)}$, we get the minima in only one step.

F.3.2.3   Levenberg-Marquardt (LM) Modification

If the Hessian matrix $F(x^{(k)})$ is not positive definite, then the Newton method may
fail to converge. Hence, a simple technique to ensure that the search direction is a
descent direction is to introduce the *Levenberg-Marquardt modification* of Newton's
algorithm.

$$x^{(k+1)} = x^{(k)} - (F(x^{(k)}) + \mu_k I)^{-1} g^{(k)}; \qquad (F.25)$$

where $\mu_k \geq 0$. If μ_k tends to zero the LM modification will behave as Newton's
method. But if μ_k attains a very large value, then the LM modification behaves as
steepest descent method.

F.3.2.4   Limitations of Newton's Method

(i) If the Hessian matrix is not positive definite, then we may not use Newton's
method.
Example:
Minimize $f(x) = x_1^{2/3} + x_2^{2/3}$.
   The Hessian of the function is $[\frac{-2}{9x_1^{4/3}}, 0; 0, \frac{-2}{9x_1^{4/3}}]$ which is not a positive
definite value at $x = [x_1; x_2]$. Hence, the method fails to converge (Fig. F.9).

(ii) If the starting point is not close to the solution, it may diverge.
Example:
Minimize $f(x) = (x_1 - 2)^2 + 5\log((x_2 - 2)^2 + 1))$. With the initial point
$[-1; -1]$, the function diverges and with the initial point $[1, 2.5]$, the function
converges (Fig. F.10).

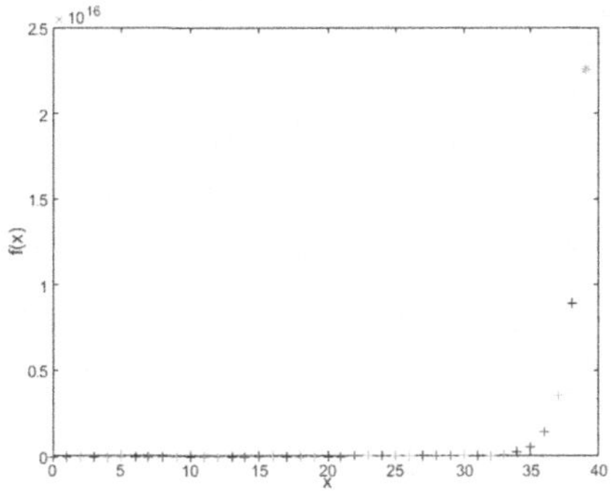

Fig. F.9 Visualization of non-convergence of the solution Using MATLAB

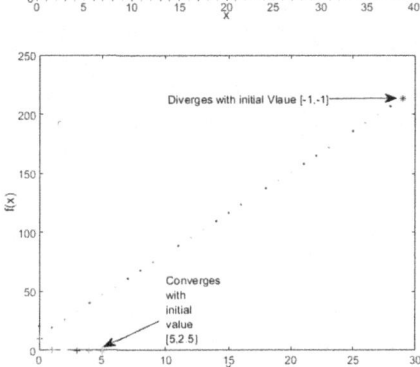

Fig. F.10 Visualization of convergence of the solution Using MATLAB

(iii) The main disadvantage is that each iteration of Newton's method requires solving a large system of linear equations, which for large-scale problems can be prohibitively expensive.

F.3.3 Quasi-Newton Method

A computational drawback of the Newton's method is the need to evaluate the $F(x^{(k)})$ and to solve $d^{(k)} = -(F^{(k)})^{-1}g^{(k)}$. To avoid the computation of $(F^{(k)})^{-1}$, quasi-Newton method uses an approximation to $(F^{(k)})^{-1}$ in place of true inverse. This approximation is updated at every step, so that it exhibits at least some properties of $(F^{(k)})^{-1}$. To get an idea about the properties that an approximation to $(F^{(k)})^{-1}$ should satisfy, consider the formula

$$x^{(k+1)} = x^{(k)} - \alpha H_k g^{(k)}; \tag{F.26}$$

where H_k is a real matrix of dimension $n \times n$ and α is a positive search parameter. Expanding f about $x^{(k)}$ yields

$$f(x^{(k)}) = f(x^{(k)}) + g^{(k)T}(x^{(k+1)} - x^{(k)}) + o(\|x^{(k+1)} - x^{(k)}\|) \quad \text{(F.27)}$$

$$= f(x^{(k)}) - \alpha g^{(k)T} H g^{(k)} + o(\|H_k g^{(k)}\|). \quad \text{(F.28)}$$

As α tends to zero, the second term on the right-hand side of this equation dominates the third. Thus, to guarantee a decrease in f for small α, we have to $g^{(k)T} H g^{(k)} > 0$.

F.3.3.1 Approximating the Inverse Hessian

Let the $H_0, H_1, H_2 \ldots$ be successive approximation of the inverse $(F^{(k)})^{-1}$ of the Hessian. Taking f (objective function) to be quadratic, with Hessian $F(x) = Q$ for all x, where $Q^T = Q$, and H_0 to be real symmetric positive definite matrix. After considering the condition that approximation must satisfy, we get

$$H_{k+1} \Delta g^{(i)} = \Delta x^{(i)}, \quad 0 \le i \le k. \quad \text{(F.29)}$$

Above consideration illustrates the basic idea behind the quasi-Newton method. Specifically, quasi-Newton algorithms have the form

$$d^{(k)} = -H_k g^{(k)}, \quad \text{(F.30)}$$

$$\alpha_k = \text{argmin}_{\alpha \ge 0} f(x^{(k)} + \alpha d^{(k)}), \quad \text{(F.31)}$$

$$x^{(k+1)} = x^{(k)} + \alpha_k d^{(k)}; \quad \text{(F.32)}$$

where $H_0, H_1, H_2\ldots$ are symmetric.

F.3.3.2 The Rank One Correction Formula

In the rank one correction formula, the correction term is symmetric and has the form $a_k z^{(k)} z^{(k)T}$, where $a_k \in \mathbb{R}$ and $z^{(k)} \in \mathbb{R}^n$. Therefore, the update equation is

$$H_{k+1} = H_k + a_k z^{(k)} z^{(k)T}. \quad \text{(F.33)}$$

Note that

$$\text{rank}(z^{(k)} z^{(k)T}) = 1, \quad \text{(F.34)}$$

and hence the name, "rank one correction." After doing certain calculations[9] with $z^{(k)}$ and $\triangle g^{(k)}$, we conclude the formula for H_{k+1} as:

$$H_{k+1} = H_k + \frac{(\triangle x^{(k)} - H_k \triangle g^{(k)})(\triangle x^{(k)} - H_k \triangle g^{(k)})^T}{\triangle g^{(k)T}(\triangle x^{(k)} - H_k \triangle g^{(k)T})}. \qquad (F.35)$$

F.3.3.3   DFL (Davidson, Fletcher and Powell) Method to Calculate Approximation of Inverse

The DFL formula to calculate approximation for the $(F(x^k))^{-1}$ is

$$H_{k+1} = H_k + \frac{\triangle x^{(k)} \triangle x^{(k)T}}{\triangle x^{(k)T} \triangle x^{(k)}} - \frac{[H_k \triangle g^{(k)}][H_k \triangle g^{(k)}]^T}{\triangle g^{(k)T} H_k \triangle g^{(k)}}. \qquad (F.36)$$

Example 1:
Minimize $f(x) = x_1^2 + \frac{x_2^2}{2} + 3$. The minimum value obtained using the MATLAB is $(0, 0)$.
Example 2:
Minimize $f(x) = \sqrt{|x_1 - 2|} + \sqrt{|x_2 - 2|}$. The optimum value obtained using the MATLAB (tolerance = 10^{-7}) is $(2, 2)$. The gradient method fails to get the optimal value of this function due to differentiability. Newton method fails to solve this problem as Hessian of this function is not a positive definite matrix (Fig. F.11).
Example 3:
Minimize $f(x) = x_1^{2/3} + x_2^{2/3}$. The optimum value obtained using MATLAB is $(5.382914e^{-9}, 5.382914e^{-9})$. Newton's method fails to solve this problem as Hessian of this function is not a positive definite matrix (Fig. F.12).

Fig. F.11 Visualization of drawback of Newton method Using MATLAB

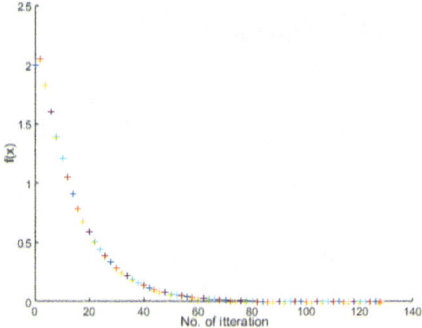

---

[9] Refer to book, "An Introduction to Optimization," by Edwin K. P. Chong and Stanislaw H. Zak.

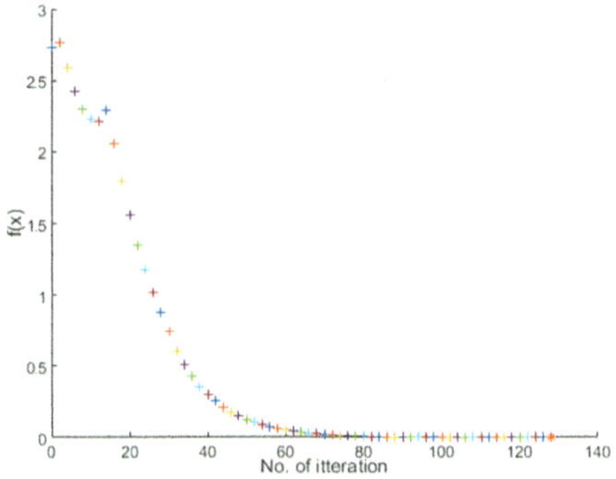

Fig. F.12 Visualization of drawback of Newton method Using MATLAB

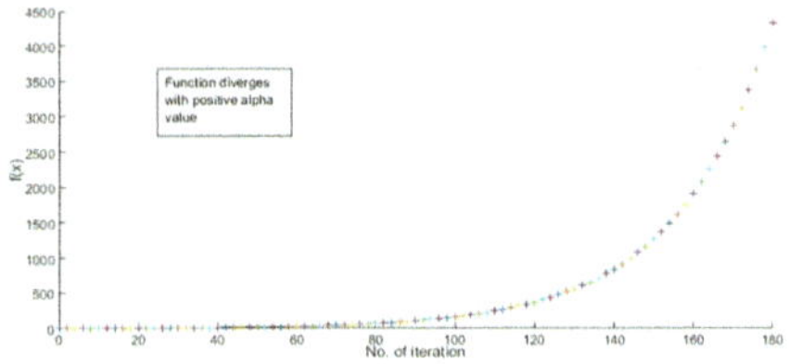

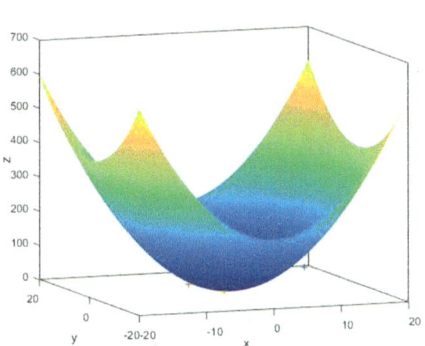

Fig. F.13 Sequential solve of the Hessian problem Using MATLAB

Fig. F.14 Function and sequence of solution values approaching to the optimal solution using MATLAB

F.3.3.4   Limitations

(i) It depends on the proper initialization value.

(ii) While calculating the direction, if the gradient of the function has a point of inflation or saddle point, the α value fails to get true value; hence, this method also fails. In order to avoid this, fix the α value with a small number like 0.1, 0.01, etc., and check the direction of convergence. If it increases, assign a negative small value to α.

   In the Example 3, $\alpha = -0.1$ was used (Figs. F.13, F.14, and F.15).

Fig. F.15 Contour plot of
function and sequence of
solution values approaching
to the optimal solution using
MATLAB

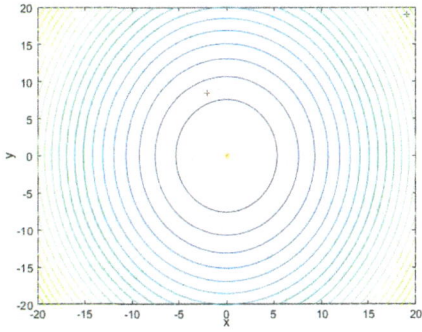

F.3.4   Conjugate Direction Method

The class of *conjugate direction method* can be viewed as being intermediate
between the method of steepest descent and Newton's method. In the *conjugate
direction method*, we specify the direction for search. Let the function $f : \mathbb{R}^n \longrightarrow \mathbb{R}$
be defined as

$$f(x) = \frac{1}{2}x^T Q x - x^T b;$$
(F.37)

where $Q = Q^T$ is a symmetric positive definite matrix, $x \in \mathbb{R}^n$. For any $x^{(k)}$, we
move in the conjugate direction[10] of Q to find $x^{(k+1)}$.

F.3.4.1   Basic Conjugate Direction Algorithm

Given a starting point $x^{(0)}$, and $Q - conjugate$ directions $d^{(0)}, d^{(1)}, d^{(2)} \ldots \ldots$
$d^{(n-1)}$. For $k \geq 0$,

$$g^{(k)} = \nabla f(x^{(k)}) = Q x^{(k)} - b,$$
(F.38)

$$\alpha_k = -\frac{g^{(k)T} d^{(k)}}{d^{(k)T} Q d^{(k)}},$$
(F.39)

$$x^{(k+1)} = x^{(k)} + \alpha_k d^{(k)}.$$
(F.40)

In order to find out conjugate directions of Q where our first direction $d^{(0)}$ search
is in the direction of the steepest direction of search. That is:

$$d^{(0)} = -g^{(0)}.$$
(F.41)

---

[10] The two direction d_1 and d_2 in $\mathbb{R}^n$ is said to be conjugate direction of Q if $d_1^T Q d_2 = 0$.

Thus

$$x^{(1)} = x^{(0)} + \alpha_0 d^{(0)};$$ (F.42)

where

$$\alpha_0 = \operatorname{argmin}_{\alpha \geq 0} f(x^{(0)} + \alpha d^{(0)}) = \frac{x^{(0)T} d^{(0)}}{d^{(0)T} Q d^{(0)}}.$$

In the next step, we search in the direction $d^{(1)}$ that is Q-conjugate to $d^{(0)}$. We choose $d^{(1)}$ as a linear combination of $g^{(1)}$ and $d^{(0)}$. In general, at the $(k+1)th$ step, we choose $d^{(k+1)}$ to be a linear combination of $g^{(k+1)}$ and $d^{(k)}$. Specifically, we choose

$$d^{(k+1)} = -g^{(k+1)} + \beta_k d^{(k)};$$ (F.43)

where

$$\beta_k = \frac{g^{(k+1)T} Q d^{(k)}}{d^{(k)T} Q d^{(k)}}.$$

F.3.4.2   Conjugate Gradient Algorithm for Non-quadratic Problems

Observe that Q appears only in the computation of the scalars α_k and β_k. Because

$$\alpha_k = \operatorname{argmin}_{\alpha \geq 0} f(x^{(k)} + \alpha d^{(k)}).$$ (F.44)

The closed-form formula for α_k in the algorithm can be replaced by the numerical line search procedure. Therefore, we need to bother ourselves with the formula f or β_k.

F.3.4.3   Hestenes-Stiefel Formula

The Hestenes-Stiefel's formula gives an alternative way to the term $Q d^{(k)}$ by replacing it with $(g^{(k+1)} - g^{(k)})/\alpha_k$ and make easier to the complex problem which are difficult to write in the form of $f(x) = x^T Q x + x^T b$. Hence,

$$\beta_k = \frac{g^{(k+1)T} [g^{(k+1)}) - g^{(k)}]}{d^{(k)T} [g^{(k+1)} - g^{(k)}]}.$$ (F.45)

Example: Minimize $f(x) = x^4 + 5x^3 + x^2 + 50x + 4$. It is not possible to write this function in the form of $f(x) = x^T Q x + x^T b$; hence, Hestenes-Stiefel formula is being used to calculate the optimum value of the function. The value of x for which the function is minimum is -4.307587.

F.3.4.4   Polak-Ribiere Formula

The Hestenes-Stiefel's formula may diverge for sufficiently small α_k. Hence the Polak-Ribiere's formula for finding β is

$$\beta_k = \frac{g^{(k+1)T}[g^{(k+1)} - g^{(k)}]}{g^{(k)T}g^{(k)}}. \tag{F.46}$$

F.3.4.5   Fltcher-Reeves Formula

We know that $g^{(k+1)T}g^{(k)} = 0$. Hence by putting this in Polak-Ribiere formula, we get

$$\beta_k = \frac{g^{(k+1)T}g^{(k+1)}}{g^{(k)T}g^{(k)}}. \tag{F.47}$$

This requires less computation.

Example 1:

Minimize $f(x) = x_1^2 + 5x_2$. The optimum value obtained using MATLAB is $(2.842171e^{-14}, -3.197442e^{-14}) \simeq (0, 0)$ with the initial value $(20, 20)$ (Figs. F.16 and F.17).

Some Results of Conjugate Direction Method

  (i) Solve quadratics of n variable in n steps.
 (ii) The usual implementation, the *conjugate direction algorithm*, requires no Hessian matrix evaluation.
(iii) No matrix and no storage of an $n \times n$ matrix is required.

Fig. F.16 Function and sequence of solutions approaching to the optimal solution using MATLAB

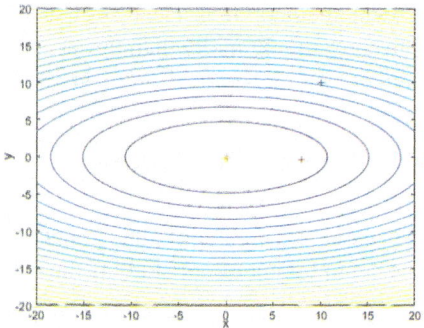

Fig. F.17 Contour plot of
function and sequence of
solutions approaching to the
optimal solution using
MATLAB

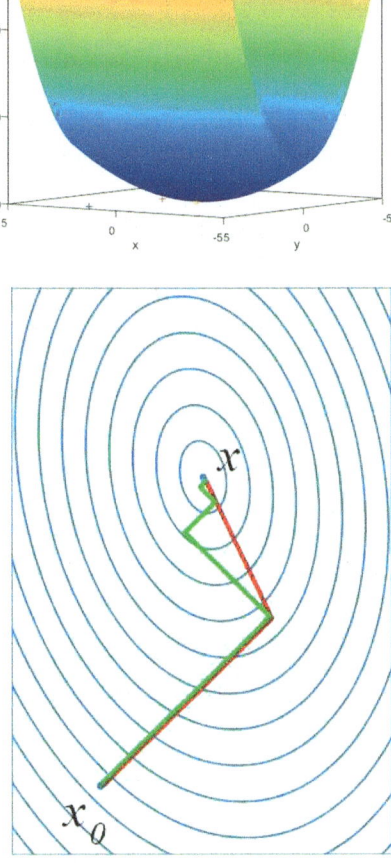

Fig. F.18 Conjugate gradient
(red) and steepest descent
method (green)

Comparison between the Steepest Descent Method and Conjugate Direction Method

Figure F.18 shows the path of convergence of conjugate gradient and steepest descent method. It is clear from the figure that the conjugate gradient method converges more rapidly than the steepest descent method.

F.4   Convex Optimization

In general, any figure is called convex if all its interior angle is not more than $180°$. The central point of our study is the convex function and convex sets.

Fig. F.19 Convex set

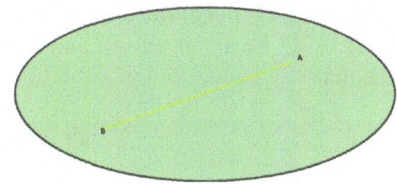

Fig. F.20 Convex function
$f(x) = x^2$

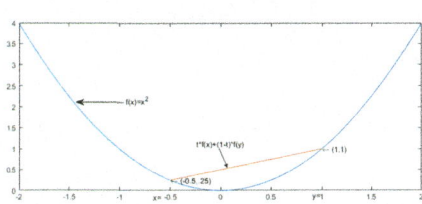

F.4.1   Definitions

(i) **Convex set**

A set $\Omega \subset \mathbb{R}^n$ is said to be convex if it contains all of its segments, that is,

$$\forall (x, y, t) \in \Omega \times \Omega \times [0, 1], tx + (1 - t)y \in \Omega; \tag{F.48}$$

where x and y are two data points in Ω and t is a real number in $[0, 1]$.

Example: The set of all the points in Fig. F.19 forms a convex set.

(ii) **Convex function**

A function $f : \mathbb{R}^n \longrightarrow \mathbb{R}$ is said to be convex if it lies below its chord, that is,

$$\forall (x, y, t) \in \Omega \times \Omega \times [0, 1], \ f(tx + (1 - t)y) \leq tf(x) + (1 - t)f(y). \tag{F.49}$$

Example: The function $f : \mathbb{R} \longrightarrow \mathbb{R}$ such that, $f(x) = x^2$ is a convex function (Fig. F.20).

F.4.1.1   Some Properties of Convexity

Separation Theorem

Let Ω be a closed convex set and $x_0 \in \mathbb{R}^n \setminus \Omega$. Then, there exists $w \in \mathbb{R}^n$ and $t \in \mathbb{R}$ such that

$$w^T x_0 \leq t \ \text{ and } \ \forall x \in \Omega, w^T x \geq t. \tag{F.50}$$

In other words, if $\Omega \subset \mathbb{R}^n$ is a closed convex set and there is a point $x \in \mathbb{R}^n$ which does not lie in the set Ω, then the set and point are separable.

Note that if Ω is not closed, then only we can guarantee that $w^T x_0 \leq w^T x, \forall x \in \Omega$.

Supporting Hyperplane Theorem

Let $\Omega \subset \mathbb{R}^n$ be a convex set, and $x_0 \in \det \mathbb{R}^n$. Then, there exists $w \in \mathbb{R}^n$, $w \neq 0$, such that

$$\forall x \in \Omega, w^T x \geq w^T x_0. \tag{F.51}$$

Now we will introduce the concept of sub-gradient.

Sub-gradient

Let $\Omega \subset \mathbb{R}^n$ and $f : \mathbb{R}^n \longrightarrow \mathbb{R}$. Then, $g \in \mathbb{R}^n$ is a sub-gradient of f at $x \in \Omega$ if for any $y \in \Omega$ one has

$$f(x) - f(y) \leq g^T(x - y). \tag{F.52}$$

Geometrically, sub-gradient at point x denotes the set of lines for which the function lie above them. The set of sub-gradient of f at x is denoted by $\delta f(x)$.

Existence of Sub-gradient

If $\Omega \subset \mathbb{R}^n$ be convex set, and $f : \mathbb{R}^n \longrightarrow \mathbb{R}$, $\delta f(x) \neq 0$, then f is convex. Conversely if f is convex, then for any $x \in \text{int}(\Omega)$, $\delta f(x) \neq 0$. Furthermore, if f is convex and differentiable at x, then $\nabla f(x) \in \delta f(x)$.

F.4.1.2 Why Convexity?

We are well aware of the fact that an optimization problem is solved only when we find the global minima. The key to the algorithmic success in minimizing convex functions is that these functions exhibit a local to global phenomenon. We have already seen one instance that $\nabla f(x) \in \delta f(x)$: the gradient $\nabla f(x)$ contains a priori local information about the function f around x, while the sub-differential $\delta f(x)$ gives a global information in the form of a linear lower bound on the entire function. Another instance of this local to global phenomenon is that local minima of convex functions are in fact global minima.

Local Minima Are Global Minima

If f is a convex function and x is a local minima of f, then x is the global minima of f. Furthermore, this happens iff $0 \in \delta f(x)$.

Proof One point is obvious: If the point x is the global minima, then it is also a local minima.

Conversely, we prove this by contraposition. Suppose that x is local minimizer of f but not the global minimizer. Then, for some $y \in \Omega$, we have $f(y) < f(x)$. By the assumption, the function f is convex, and hence $\forall t \in (0, 1)$

$$f(ty + (1 - t)x) \leq tf(y) + (1 - t)f(x). \tag{F.53}$$

Because $f(y) < f(x)$, we have

$$tf(y) + (1 - t)f(x) = t(f(y) - f(x)) + f(x) < f(x). \tag{F.54}$$

Thus, for all $t \in (0, 1)$

$$f(ty + (1 - t)x) < f(x). \tag{F.55}$$

Hence, there exist points that are arbitrarily close to x and have lower functional value. For example, the sequence $\{y_n\}$ of the points given by

$$y_n = \frac{1}{n}y + (1 - \frac{1}{n})x \tag{F.56}$$

converges to x, and $f(y_n) < f(x)$. Hence, x is not a local minimizer which is a contradiction. Hence, every local minima is global minima.

KKT's (Karush-Kuhn-Tucker) Sufficient Condition for Minima
Let $f : \mathbb{R}^n \longrightarrow \mathbb{R}$, $f \in C^1$, be a convex function on the set of feasible points.

$$\Omega = \{x \in \mathbb{R}^n : h(x) = 0, g(x) \leq 0\}, \tag{F.57}$$

where $h : \mathbb{R}^n \longrightarrow \mathbb{R}^m$, $g : \mathbb{R}^n \longrightarrow \mathbb{R}^p$, $h, g \in C^1$, and Ω is convex. Suppose that there exist $x^* \in \Omega$, $\lambda^* \in \mathbb{R}^m$ and $\mu^* \in \mathbb{R}^p$, such that

(i) $Df(x^*) + \lambda^{*t}Dh(x^*) + \mu^{*T}Dg(x^*) = 0^T$ (optimality condition),
(ii) $\mu^*g(x^*) = 0$ (complementary slackness property),
(iii) $\mu^*, \lambda^* \geq 0$ (non-negativity property),
(iv) $h(x^*) = 0$ and $g(x^*) \leq 0$ (feasibility condition).

Then, x^* is a global minimizer of f over Ω.

F.5   Algorithm for Constrained Optimization

We have already discussed algorithms for solving the *unconstrained*[11] optimization problem. Here, we will discuss algorithms for solving *constrained* optimization problem.

---

[11] Steepest descent method, Newton's method, conjugate gradient method, quasi-Newton's method

F.5.1   Lagrangian Algorithm

We consider an optimization method based on the Lagrangian function. The basic idea is to use gradient algorithm to update simultaneously the decision variable and Lagrangian multiplier vector.

F.5.1.1   Lagrangian Algorithm for the Equality Constraints

Consider the following optimization problem with equality constraints:

$$\text{minimize } f(x) \tag{F.58}$$

$$\text{subject to } h(x) = 0; \tag{F.59}$$

where $h : \mathbb{R}^n \longrightarrow \mathbb{R}^m$. Recall that for this problem, the Lagrangian function is given by

$$l(x, \lambda) = f(x) + \lambda^T h(x). \tag{F.60}$$

Assume that $f, h \in C^2$; denote the Hessian of Lagrangian by $L(x, \lambda)$.
    The Lagrangian algorithm for this problem is given by

$$x^{(k+1)} = x^{(k)} - \alpha_k (\nabla f(x^{(k)})) + Dh(x^{(k)})^T \lambda^{(k)}, \tag{F.61}$$

$$\lambda^{(k+1)} = \lambda^{(k)} + \beta_k h(x^{(k)}). \tag{F.62}$$

Notice that the update equation for $x^{(k)}$ is a gradient algorithm for minimizing the Lagrangian with respect to its argument, and the update equation for $\lambda^{(k)}$ is a gradient algorithm for maximizing the Lagrangian with respect to the λ argument. Because only the gradient is used, this method is called *first-order Lagrangian algorithm*.

F.5.1.2   Lagrangian Algorithm for Inequality Constraints

Consider the following optimization problem with inequality constraints:

$$\text{minimize } f(x)$$

$$\text{subject to } g(x) \leq 0,$$

where $g : \mathbb{R}^n \longrightarrow \mathbb{R}^p$ for this problem Lagrangian function is given by

$$l(x, \mu) = f(x) + \mu^T g(x). \tag{F.63}$$

Fig. F.21 Figure shows how
the points are converging to
the solution (feasible area is
ABCA)

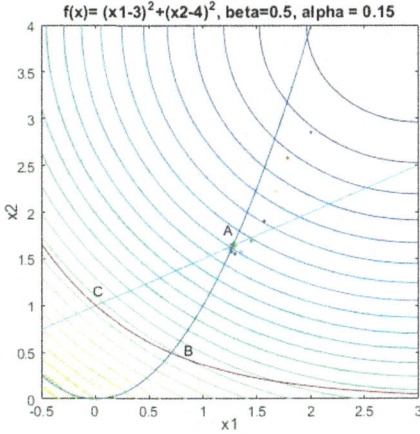

Fig. F.22 Divergence

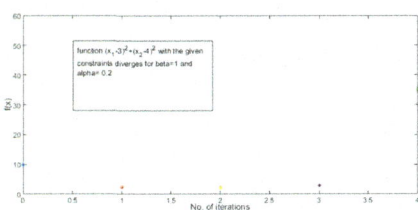

Assume that $f, g \in C^2$ denote the Hessian of the Lagrangian by $L(x, \mu)$. The
Lagrangian algorithm for this problem is given by

$$x^{(k+1)} = x^{(k)} - \alpha_k (\nabla f(x^{(k)})) + Dg(x^{(k)})^T \mu^{(k)}, \qquad (F.64)$$

$$\mu^{(k+1)} = [\mu^{(k)} + \beta_k g(x^{(k)})]_+. \qquad (F.65)$$

where $[.]_+ = max\{., 0\}$ (component wise). Notice that the update equation for $x^{(k)}$
is a gradient algorithm for minimizing the Lagrangian with respect to its argument,
and the update equation for $\mu^{(k)}$ is a projected gradient algorithm for maximizing
the Lagrangian with respect to the μ argument. The reason for the projection is that
KKT multiplier vector is required to be nonnegative to satisfy the KKT condition
(Figs. F.21, F.22, and F.23).
Example 1: Consider the following optimization problem

$$\text{Minimize } f(x) = (x_1 - 3)^2 + (x_2 - 4)^2 \qquad (F.66)$$

$$\text{Subject to } x_1^2 - x_2 \le 0, \qquad (F.67)$$

$$e^{-x_1} - x_2 \le 0, \qquad (F.68)$$

$$- x_1 + 2 \times x_2 - 2 \le 0. \qquad (F.69)$$

Table F.1 shows the value of $f(x)$ for different values of β, α, and x.

Fig. F.23 Convergence with slow rate

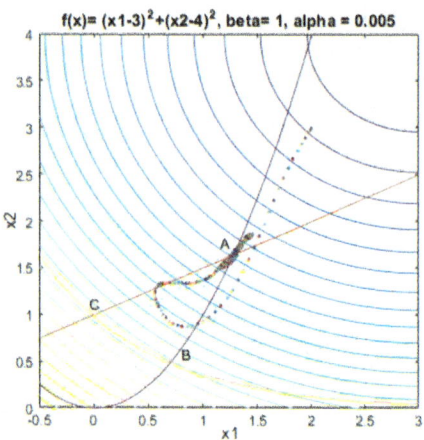

F.5.1.3   Limitations of Lagrangian Method

- Lagrange multipliers only are defined when gradients (multidimensional derivatives) exist. So we cannot use it for integer programming, i.e., when the answer must be integer, or if constraints or objective functions have no derivation, such as absolute values.
- It is initial value dependent (Fig. F.24).
- KKT conditions must be satisfied.

F.5.2   Penalty Method

Consider a general optimization problem

$$\text{minimize } f(x)$$

$$\text{subject to } x \in \Omega.$$

In this method, we approximate the constrained optimization problem above by the unconstrained optimization problem

$$\text{minimize } f(x) + \eta P(x);$$

where $\eta \in \mathbb{R}^+$ is a constant and $P : \mathbb{R}^n \longrightarrow \mathbb{R}$ is a given function. We then solve the associated unconstrained problem and use the solution as an approximation to minimizer of the original problem. The constant η is called *penalty parameter*, and function P is called the penalty function.

Table F.1 Values of β, α, x, and $f(x)$ in different iterations

β	α	x	No. of Iteration	$f(x)$
1.5	0.2	–	–	diverges
	0.15	–	–	diverges
	0.1	(1.281, 1.64)	37	8.523
	0.05	(1.282, 1.647)	55	8.51
	0.01	(1.287, 1.649)	211	8.462
	0.005	(1.287, 1.649)	301	8.447
1.0	0.2	–	–	diverges
	0.15	(1.281, 1.64)	33	8.524
	0.1	(1.281, 1.641)	31	8.522
	0.05	(1.28, 1.649)	57	8.534
	0.01	(1.283, 1.641)	256	8.495
	0.005	(1.299, 1.664)	301	8.447
0.5	0.2	(1.282, 1.64)	301	8.528
	0.15	(1.281, 1.64)	25	8.524
	0.1	(1.28, 1.64)	30	8.528
	0.05	(1.281, 1.64)	62	8.525
	0.01	(1.289, 1.649)	218	8.455
	0.005	(1.284, 1.645)	511	8.495
0.2	0.2	(1.282, 1.642)	76	8.513
	0.15	(1.282, 1.642)	66	8.51
	0.1	(1.282, 1.642)	59	8.512
	0.05	(1.28, 1.639)	64	8.532
	0.01	(1.285, 1.647)	235	8.48
	0.005	(1.291, 1.646)	301	8.466
0.1	0.2	(1.284, 1.644)	127	8.496
	0.15	(1.282, 1.644)	121	8.494
	0.1	(1.284, 1.644)	114	8.493
	0.05	(1.284, 1.644)	88	8.497
	0.01	(1.279, 1.637)	262	8.544
	0.005	(1.284, 1.646)	465	8.484
0.05	0.2	(1.288, 1.648)	212	8.464
	0.15	(1.288, 1.649)	205	8.461
	0.1	(1.288, 1.649)	200	8.460
	0.05	(1.287, 1.648)	186	8.463
	0.01	(1.28, 1.643)	269	8.512
	0.005	(1.282, 1.64)	488	8.521
0.01	0.2	(1.316, 1.683)	593	8.207
	0.15	(1.316, 1.683)	288	8.294
	0.1	(1.316, 1.683)	587	8.207
	0.05	(1.216, 1.682)	581	8.207
	0.01	(1.313, 1.676)	542	8.247
	0.005	(1.297, 1.664)	433	8.376

(continued)

Table F.1 (continued)

β	α	x	No. of Iteration	$f(x)$
0.005	0.2	(1.351, 1.725)	816	7.896
	0.15	(1.351, 1.725)	814	7.895
	0.1	(1.351, 1.725)	810	7.895
	0.05	(1.351, 1.724)	809	7.899
	0.01	(1.348, 1.718)	791	7.936
	0.005	(1.344, 1.707)	752	8.001

Fig. F.24 Divergence due to initial value

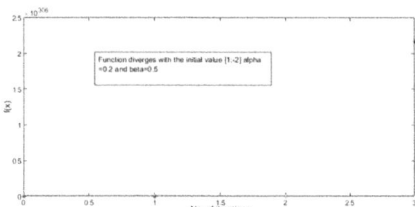

Penalty Function A function $P : \mathbb{R}^n \longrightarrow \mathbb{R}$ is called penalty function for the constrained optimization problem above if it satisfies the following three conditions:

(i) P is continuous.
(ii) $P(x) \geq 0, \forall x \in \mathbb{R}^n$.
(iii) $P(x) = 0$ if and only if x is feasible (i.e., $x \in \Omega$).

There are two alternative approaches to solve the constrained problem using penalty method.

1. **Interior Penalty Function Method:**
   The interior penalty method consists of a barrier term, which prevents the points leaving the feasible region of the objective function. The method generates the sequence of feasible points whose limit is the optimal solution to the original function. This method has a limitation that we need to find a feasible point to initiate the program due to which this method becomes a very complex problem to solve.
2. **Exterior Penalty Function Method:**
   In this method, penalty term is added to objective function for any violation of constraints. This method generates a sequence of infeasible points, hence its name, whose limit is an optimal solution of the original problem. This method is also known as penalty function parameter.

There are two fundamental issues associated with this method. The first issue is how well the unconstrained problem approximates the constrained one. The other issue, most important from the practical point of view is, how to solve the unconstrained problem when its objective function contains the penalty term.

F.5.2.1   The Concept of Penalty Function

Consider the following problem of constrained optimization

$$\text{minimize } f(x)$$

$$\text{subject to } h(x) = 0, \text{ and } g(x) \leq 0;$$

where $(f, g, h) : \mathbb{R}^n \longrightarrow \mathbb{R}$, are the continuous function. Suppose this problem is replaced with the following unconstrained problem,

$$\text{Minimize } f(x) + \mu(h^2(x) + g^2(x))$$

$$\text{subject to } x \in \mathbb{R}.$$

We can intuitively see that an optimal solution to the above problem must have $h^2(x)$ close to zero; otherwise a large penalty $\mu h^2(x)$ will be incurred. But the penalty term will be incurred where $g(x) \leq 0$ and $g(x) \geq 0$, that is, the penalty term is added to objective function if the point is inside or outside the feasible region. But a penalty is desired only if the point is not feasible, that is, if $g(x) \geq 0$. A suitable unconstrained problem is therefore given by:

$$\text{Minimize } f(x) + \mu(h^2(x) + \text{maximum}\{0, g(x)\}^2$$

$$\text{subject to } x \in \mathbb{R}^n.$$

Note that if $g(x) \leq 0$, then the maximum$\{0, g(x)\} = 0$ and no penalty is incurred on the other hand if $g(x) \geq 0$ then maximum$\{0, g(x)\} > 0$, and the penalty term $\mu g^2(x)$ is realized.

We can generalize penalty term as follows. Let the inequality constraints $g(x) = [g_1(x), g_2(x) \ldots g_m(x)]^T \leq 0$ and equality constraints are $h(x) = [h_1(x), h_2(x) \ldots h_n(x)] = 0$, such that $\forall \ 1 \leq i \leq m$ and $1 \leq j \leq n \ g_i(x) \leq 0$ and $h_j(x) = 0$ where $x \in \mathbb{R}^n$. Then the penalty term is

$$p(x) = \sum_{i=1}^{m} \max\{0, g_i(x)\}^q + \sum_{j=1}^{n} |h(x)|^q. \tag{F.70}$$

If $q = 1$, $p(x)$ is called "linear penalty function." This function may not be differentiable at the point $g_i(x) = 0$ and $h_i(x) = 0$ for some i.

Setting $q = 2$ is the most common form that is used in practice and is called "quadratic penalty function."

F.5.2.2   Subsequent Values of the Penalty Parameter

It is almost impossible to tell that how large value of μ must be to provide a solution to the problem without creating numerical difficulties in the computations. The most frequently used initial penalty parameters in the literature are $0.01, 0.1, 2, 5$, and 10. Once the initial value of μ_k is chosen, the subsequent values of μ_k have to be chosen such that $\mu_{k+1} > \mu_k$. For our convenience, the value of μ_k is chosen according to the relation

$$\mu_{k+1} = \beta\mu_k; \tag{F.71}$$

where $\beta_k > 1$. The values of β usually can be taken as $2, 5, 10, 100$, etc. (Figs. F.25, F.26, and F.27)

Example 1: Consider the optimization problem

$$\text{Minimize} f(x) = (x_1 - 5)^2 + (x_2 - 6)^2 \tag{F.72}$$

$$\text{subject to} x_1^2 - 4 \le 0, \tag{F.73}$$

$$e^{-x_1} - x_2 \le 0, \tag{F.74}$$

$$x_1 + 2x_2 - 4 \le 0. \tag{F.75}$$

Initialization:
$x = [3; 2], \ \mu = 1, \beta = 10, n = 6.$

Fig. F.25 The sequence of infeasible results from outside the feasible region (ABCDA)

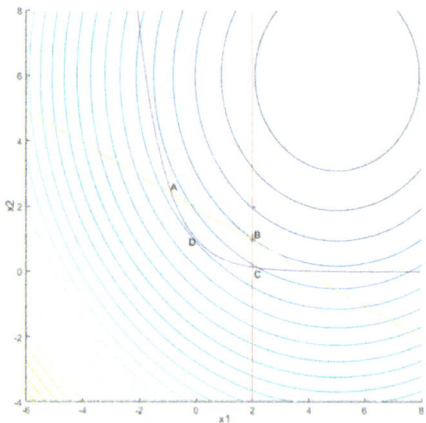

μ	x	$f(x)$
1	(2.0559275, 1.9776242)	24.84700400
10	(2.0034700, 1.1202584)	32.79107012
100	(2.0031607, 1.0123111)	33.85808635
1000	(2.0000312, 1.0012340)	33.98746892
10000	(2.0000031, 1.0001234)	33.99874740
100000	(2.0000003, 1.0000123)	33.9998590
1000000	(2.00000000, 1.0000012)	33.9999800

Example 2: Consider the following optimization problem:

$$\text{Minimize } f(x) = (x_1 - 3)^2 + (x_2 - 4)^2 \tag{F.76}$$

$$\text{Subject to } x_1^2 - x_2 \leq 0, \tag{F.77}$$

$$e^{-x_1} - x_2 \leq 0, \tag{F.78}$$

$$-x_1 + 2 \times x_2 - 2 \leq 0. \tag{F.79}$$

Initialization:
$x = [3; 2]$, $\mu = 1$, $\beta = 10$, $n = 7$.

Fig. F.26 The sequence of infeasible results from outside the feasible region (ABCDA)

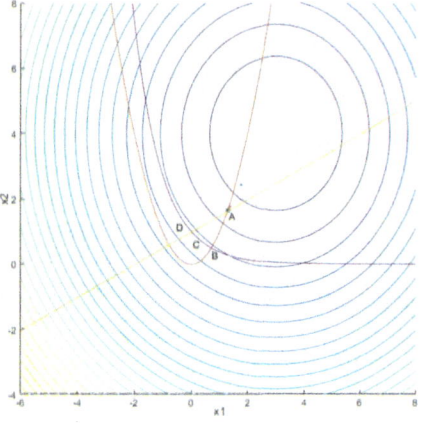

μ	x	$f(x)$
1	(1.762313, 2.438396)	3.970477
10	(1.3693190, 1.763400)	7.661502
100	(1.290853, 1.653829)	8.425699
1000	(1.281800, 1.641746)	8.513573
10000	(1.280879, 1.640524)	8.522504
100000	(1.280787, 1.640402)	8.523398
1000000	(1.280777, 1.6403900)	8.523488
10000000	(1.280777, 1.6403880)	8.5234970

Example 3: Consider the optimization problem

$$\text{Minimize } f(x) = (x_1 - 5)^2 + (x_2 - 6)^2 \tag{F.80}$$

$$\text{subject to } x_1^2 - 4 \le 0, \tag{F.81}$$

$$e^{-x_1} - x_2 \le 0, \tag{F.82}$$

$$x_1 + 2x_2 - 4 \le 0. \tag{F.83}$$

Initialization:
$x = [3; 2]$, $\mu = 1$, $\beta = 10$, $n = 7$.

Fig. F.27 The sequence of infeasible results from outside the feasible region (ABCDA)

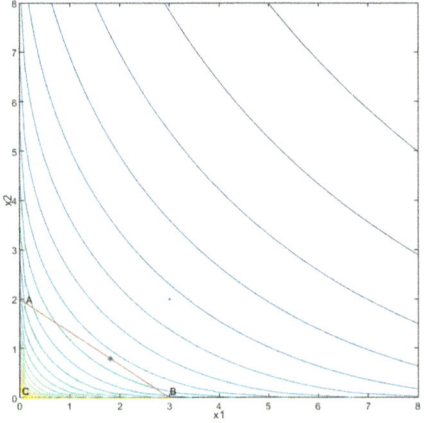

μ	x	$f(x)$
1	(1.812414, .8055175)	-0.8081556
10	(1.801136, 0.8005047)	-0.8050343
100	(1.800113, 0.8000500)	-0.8047502
1000	(1.800011, 0.8000050)	-0.8047221
10000	(1.800001, 0.8000005)	-0.8047193
100000	(1.800000, 0.8000001)	-0.804719
1000000	(1.800000, 0.8000001)	-0.804719

Limitations of Penalty Method

- It always gives approximate values, not exact solution.
- We do not know for which value of μ the required approximation we shall get due to which it complicates calculation.
- In the penalty method, we use the unconstrained optimization solving method[12]; hence, it includes limitations of the method used.

---

[12] Newton method, gradient method, etc.

Index

© The Author(s), under exclusive license to Springer Nature Singapore Pte Ltd. 2025   413
A. Ghosh, *Data Science and Cases in Sustainability*, Mathematics for Sustainable
Developments, https://doi.org/10.1007/978-981-96-8362-8

The manufacturer's authorised representative in the EU is Springer
Nature Customer Service Centre GmbH, Europaplatz 3, 69115 Heidelberg,
Germany. If you have any concerns regarding our products, please
contact ProductSafety@springernature.com

Printed and bound by CPI Group (UK) Ltd, Croydon, CR0 4YY
23/04/2026
02095585-0008